Die Grundlehren der mathematischen Wissenschaften

in Einzeldarstellungen
mit besonderer Berücksichtigung
der Anwendungsgebiete

Band 178

Siegfried Flügge

Practical Quantum Mechanics II

With 18 Figures

Springer-Verlag Berlin Heidelberg New York 1971

Prof. Dr. Siegfried Flügge
Physikalisches Institut der Universität Freiburg i. Br.

AMS Subject Classifications (1970)
Primary 81-01, 81-02, 81 A 06, 81 A 10, 81 A 63, 81 A 69, 81 A 75, 81 A 81
Secondary 34 E 20, 35 J 10

ISBN 3-540-05277-1 Springer-Verlag Berlin · Heidelberg · New York
ISBN 0-387-05277-1 Springer-Verlag New York · Heidelberg · Berlin

 Library of Congress Catalog Card Number 70-140508. Printed in Germany. Monophoto typesetting and offset printing: Zechnersche Buchdruckerei, Speyer. Bookbinding: Konrad Triltsch, Graphischer Betrieb, Würzburg

Contents Volume II

III. Particles with Spin

A. One-Body Problems 1

129. Construction of Pauli matrices 1
130. Eigenstates of Pauli matrices 3
131. Spin algebra 6
132. Spinor transformation properties 7
133. Spin electron in a central field 9
134. Quadrupole moment of a spin state 12
135. Expectation values of magnetic moments 14
136. Fine structure 16
137. Plane wave of spin $\frac{1}{2}$ particles 18
138. Free electron spin resonance 20

B. Two- and Three-Body Problems 22

139. Spin functions for two particles 22
140. Spin-dependent central force between nucleons 25
141. Powers of spin operators 26
142. Angular momentum eigenfunctions of two spin particles . 27
143. Tensor force operator 29
144. Deuteron with tensor interaction 31
145. Electrical quadrupole and magnetic dipole moments of deuteron 34
146. Spin functions of three particles 37
147. Neutron scattering by molecular hydrogen 40

IV. Many-Body Problems

A. Few Particles 43

148. Two repulsive particles on a circle 43
149. Three-atomic linear molecule 47
150. Centre-of-mass motion 51

151. Virial theorem 54
152. Slater determinant 55
153. Exchange in interaction terms with Slater determinant . . 57
154. Two electrons in the atomic ground state 58
155. Excited states of the helium atom 61
156. Excited S states of the helium atom 65
157. Lithium ground state 69
118. Exchange correction to lithium ground state 71
159. Dielectric susceptibility 74
160. Diamagnetic susceptibility of neon 76
161. Van der Waals attraction 78
162. Excitation degeneracy 80
163. Neutral hydrogen molecule 83
164. Scattering of equal particles 88
165. Anomalous proton-proton scattering 92
166. Inelastic scattering 95

B. Very Many Particles: Quantum Statistics 100

167. Electron gas in a metal 100
168. Paramagnetic susceptibility of a metal 102
169. Field emission, uncorrected for image force 105
170. Field emission, corrected for image force 108
171. White dwarf 113
172. Thomas-Fermi approximation 116
173. Amaldi correction for a neutral atom 121
174. Energy of a Thomas-Fermi atom 123
175. Virial theorem for the Thomas-Fermi atom 127
176. Tietz approximation of a Thomas-Fermi field 128
177. Variational approximation of Thomas-Fermi field 130
178. Screening of K electrons 131

V. Non-Stationary Problems

179. Two-level system with time-independent perturbation . . 135
180. Periodic perturbation of two-level system 137
181. Dirac perturbation method 140
182. Periodic perturbation: Resonance 142
183. Golden Rule for scattering 144
184. Born scattering in momentum space 147
185. Coulomb excitation of an atom 150
186. Photoeffect 153

187. Dispersion of light. Oscillator strengths 157
188. Spin flip in a magnetic resonance device 160

VI. The Relativistic Dirac Equation

189. Iteration of the Dirac equation 165
190. Plane Dirac waves of positive energy 167
191. Transformation properties of a spinor 171
192. Lorentz covariants 172
193. Parity transformation 175
194. Charge conjugation 177
195. Mixed helicity states 179
196. Spin expectation value 180
197. Algebraic properties of a Dirac wave spinor 181
198. Current in algebraic formulation 184
199. Conduction current and polarization current 186
200. Splitting up of Dirac equations into two pairs 188
201. Central forces in Dirac theory 191
202. Kepler problem in Dirac theory 195
203. Hydrogen atom fine structure 198
204. Radial Kepler solutions at positive kinetic energies . . . 203
205. Angular momentum expansion of plane Dirac wave . . . 206
206. Scattering by a central force potential 209
207. Continuous potential step 213
208. Plane wave at a potential jump 219
209. Reflected intensity at a potential jump 222

VII. Radiation Theory

210. Quantization of Schrödinger field 227
211. Scattering in Born approximation 229
212. Quantization of classical radiation field 231
213. Emission probability of a photon 234
214. Angular distribution of radiation 236
215. Transition probability 239
216. Selection rules for dipole radiation 240
217. Intensities of Lyman lines 243
218. Compton effect 245
219. Bremsstrahlung 249

Mathematical Appendix

Coordinate systems 257
Γ function 258
Bessel functions 260
Legendre functions 264
Spherical harmonics 267
The hypergeometric series 271
The confluent series 274
Some functions defined by integrals 276

Index for Volumes I and II 279

Contents Volume I

I. General Concepts

1. Law of probability conservation 1
2. Variational principle of Schrödinger 2
3. Classical mechanics for space averages 5
4. Classical laws for angular motion 6
5. Energy conservation law 8
6. Hermitian conjugate 9
7. Construction of an hermitian operator 10
8. Derivatives of an operator 12
9. Time rate of an expectation value 13
10. Schrödinger and Heisenberg representations 14
11. Time dependent hamiltonian 17
12. Repeated measurement 18
13. Curvilinear coordinates 19
14. Momentum space wave functions 20
15. Momentum space: Periodic and aperiodic wave functions 22

II. One-Body Problems without Spin

A. One-Dimensional Problems 25

16. Force-free case: Basic solutions 26
17. Force-free case: Wave packet 28
18. Standing wave 32
19. Opaque division wall 35
20. Opaque wall described by Dirac δ function 39
21. Scattering at a Dirac δ function wall 40
22. Scattering at a symmetric potential barrier 42
23. Reflection at a rectangular barrier 44
24. Inversion of reflection 47
25. Rectangular potential hole 48
26. Rectangular potential hole between two walls 52
27. Virtual levels 57

28. Periodic potential 62
29. Dirac comb 64
30. Harmonic oscillator 68
31. Oscillator in Hilbert space 72
32. Oscillator eigenfunctions constructed by Hilbert space operators 74
33. Harmonic oscillator in matrix notation 76
34. Momentum space wave functions of oscillator 79
35. Anharmonic oscillator 80
36. Approximate wave functions 85
37. Potential step 86
38. Pöschl-Teller potential hole 89
39. Potential hole of modified Pöschl-Teller type 94
40. Free fall of a body over earth's surface 101
41. Accelerating electrical field 105

B. Problems of Two or Three Degrees of Freedom without Spherical Symmetry 107

42. Circular oscillator 107
43. Stark effect of a two-dimensional rotator 110
44. Ionized hydrogen molecule 113
45. Oblique incidence of a plane wave 118
46. Symmetrical top 121

C. The Angular Momentum 125

47. Infinitesimal rotation 125
48. Components in polar coordinates 126
49. Angular momentum and Laplacian 129
50. Hilbert space transformations 130
51. Commutators in Schrödinger representation 132
52. Particles of spin 1 133
53. Commutation with a tensor 135
54. Quadrupole tensor. Spherical harmonics 136
55. Transformation of spherical harmonics 139
56. Construction of Hilbert space for an angular momentum component 141
57. Orthogonality of spherical harmonics 143

D. Potentials of Spherical Symmetry 144

a) Bound States 144

58. Angular momentum expectation values 147

59. Construction of radial momentum operator 149
60. Solutions neighbouring eigenfunctions 152
61. Quadrupole moment 154
62. Particle enclosed in a sphere 155
63. Square well of finite depth 159
64. Wood-Saxon potential 162
65. Spherical oscillator 166
66. Degeneracy of the spherical oscillator 168
67. Kepler problem 171
68. Hulthén potential 175
69. Kratzer's molecular potential 178
70. Morse potential 182
71. Rotation correction of Morse formula 186
72. Yukawa potential hole 189
73. Isotope shift in x-rays 191
74. Muonic atom ground state 193
75. Central-force model of deuteron 196
76. Momentum space wave functions for central force potentials 200
77. Momentum space integral equation for central force potentials 202
78. Momentum space wave functions for hydrogen 204
79. Stark effect of a three-dimensional rotator 206

b) Problems of Elastic Scattering 208

80. Interference of incident and scattered waves 208
81. Partial wave expansion of plane wave 210
82. Partial wave expansion of scattering amplitude 213
83. Scattering at low energies 216
84. Scattering by a constant repulsive potential 218
85. Anomalous scattering 222
86. Scattering resonances 225
87. Contribution of higher angular momenta 229
88. Shape-independent approximation 231
89. Rectangular hole: Low-energy scattering 235
90. Low-energy scattering and bound state 238
91. Deuteron potential with and without hard core 240
92. Low-energy cross section with and without hard core . . 243
93. Low-energy scattering by a modified Pöschl-Teller potential hole 244
94. Radial integral equation 248
95. Variational principle of Schwinger 252

96. Successive approximations to partial-wave phase shift . . 254
97. Calogero's equation 258
98. Linearization of Calogero's equation 259
99. Scattering length for a negative-power potential 260
100. Second approximation to Calogero equation 263
101. Square-well potential: Scattering length 266
102. Scattering length for a Yukawa potential 268
103. Improvement of convergence in a spherical harmonics series 271
104. Collision-parameter integral 272
105. Born scattering: Successive approximation steps 275
106. Scattering by a Yukawa potential 278
107. Scattering by an exponential potential 282
108. Born scattering by a charge distribution of spherical symmetry 285
109. Hard sphere: High energy scattering 288
110. Rutherford scattering formula 290
111. Partial wave expansion for the Coulomb field 293
112. Anomalous scattering 298
113. Sommerfeld-Watson transform 299
114. Regge pole 301

E. The Wentzel-Kramers-Brillouin (WKB) Approximation . . 303

115. Eikonal expansion 303
116. Radial WKB solutions 305
117. WKB boundary condition of Langer 306
118. Oscillator according to WKB approach 310
119. WKB eigenvalues in a homogeneous field 312
120. Kepler problem in WKB approach 314
121. WKB phases in the force-free case 316
122. Calculation of WKB phases 317
123. Coulomb phases by WKB method 318
124. Quasipotential 320

F. The Magnetic Field 322

125. Introduction of a magnetic field 322
126. Current in presence of a magnetic field 324
127. Normal Zeeman effect 326
128. Paramagnetic and diamagnetic susceptibilities without spin 328

Index for Volumes I and II 333

III. Particles with Spin

A. One-Body Problems

Problem 129. Construction of Pauli matrices

A particle of spin $\frac{1}{2}$ has three basic properties:

1. It bears an intrinsic vector property that does not depend upon space coordinates.
2. This vector is an angular momentum (= spin) to be added to the orbital momentum of the particle.
3. If one of the components of the spin is measured, the result can be only one of its two eigenvalues, $+\frac{1}{2}\hbar$ or $-\frac{1}{2}\hbar$.

These properties can be described by using a two-component wave function and correspondingly 2×2 matrices for the spin operators. These matrices shall be constructed.

Solution. Let $\boldsymbol{S}=\frac{\hbar}{2}\boldsymbol{\sigma}$ be the spin vector operator; then according to property 2 the three components have to obey the commutation relations of angular momentum operators,

$$S_x S_y - S_y S_x = \hbar i S_z \quad \text{(etc. cyclic)} \tag{129.1 a}$$

or, written in the dimensionless operators σ_i,

$$\sigma_x\sigma_y - \sigma_y\sigma_x = 2i\sigma_z \quad \text{(etc. cyclic)}. \tag{129.1 b}$$

Since, according to property 3, each σ_i has eigenvalues $+1$ and -1 only, it ought to be possible to represent these operators by 2×2 matrices in a two-dimensional Hilbert space. They cannot be all diagonal in the same Hilbert coordinate system since they do not commute. Let us choose the Hilbert coordinate system in such a way that

$$\sigma_z = \begin{pmatrix} 1 & 0 \\ 0 & -1 \end{pmatrix} \tag{129.2}$$

is diagonal; then the unit vectors of the coordinate directions are

$$\alpha=\begin{pmatrix}1\\0\end{pmatrix} \quad \text{and} \quad \beta=\begin{pmatrix}0\\1\end{pmatrix} \tag{129.3}$$

so that

$$\sigma_z \alpha=\alpha; \qquad \sigma_z \beta=-\beta. \tag{129.4}$$

If a particle is in the state with Hilbert vector $\alpha(\beta)$ its spin points in the positive (negative) z direction.

We now write σ_x and σ_y in the general form

$$\sigma_x=\begin{pmatrix}a_{11} & a_{12}\\ a_{21} & a_{22}\end{pmatrix}; \quad \sigma_y=\begin{pmatrix}b_{11} & b_{12}\\ b_{21} & b_{22}\end{pmatrix}. \tag{129.5}$$

To determine their matrix elements we first use the two commutators (129.1 b) linear in σ_x and σ_y, viz.

$$\sigma_x\sigma_z-\sigma_z\sigma_x=-2i\sigma_y, \quad \text{or:} \begin{pmatrix}0 & -2a_{12}\\ 2a_{21} & 0\end{pmatrix}=-2i\begin{pmatrix}b_{11} & b_{12}\\ b_{21} & b_{22}\end{pmatrix}$$

and

$$\sigma_y\sigma_z-\sigma_z\sigma_y=+2i\sigma_x, \quad \text{or:} \begin{pmatrix}0 & -2b_{12}\\ 2b_{21} & 0\end{pmatrix}=+2i\begin{pmatrix}a_{11} & a_{12}\\ a_{21} & a_{22}\end{pmatrix}.$$

This yields,

$$a_{11}=a_{22}=b_{11}=b_{22}=0; \quad b_{12}=-ia_{12}; \quad b_{21}=+ia_{21} \tag{129.6}$$

so that there yet remain only two matrix elements a_{12} and a_{21} still to be fixed. The third commutation relation,

$$\sigma_x\sigma_y-\sigma_y\sigma_x=2i\sigma_z, \quad \text{or:} \begin{pmatrix}2ia_{12}a_{21} & 0\\ 0 & -2ia_{12}a_{21}\end{pmatrix}=2i\begin{pmatrix}1 & 0\\ 0 & -1\end{pmatrix},$$

leads to one relation between them, viz.

$$a_{12}a_{21}=1. \tag{129.7}$$

Eqs. (129.6) and (129.7) yet leave one complex number, say a_{12}, undetermined. We arbitrarily fix that parameter:

$$a_{12}=1; \tag{129.8}$$

then we obtain the representation by the three Pauli matrices,

$$\sigma_x=\begin{pmatrix}0 & 1\\ 1 & 0\end{pmatrix}; \quad \sigma_y=\begin{pmatrix}0 & -i\\ i & 0\end{pmatrix}; \quad \sigma_z=\begin{pmatrix}1 & 0\\ 0 & -1\end{pmatrix}. \tag{129.9}$$

In terms of the eigenvectors of σ_z, Eq. (129.3), we may replace (129.9) by the equivalent relations

$$\begin{aligned} &\sigma_x \alpha = \beta; \quad \sigma_y \alpha = i\beta; \quad \sigma_z \alpha = \alpha; \\ &\sigma_x \beta = \alpha; \quad \sigma_y \beta = -i\alpha; \quad \sigma_z \beta = -\beta. \end{aligned} \tag{129.10}$$

Problem 130. Eigenstates of Pauli matrices

To determine the eigenvectors of the operators σ_x and σ_y and to show that $|a_{12}|^2 = 1$ is a necessary condition. What are the properties of the two "shift operators"

$$\sigma_+ = \sigma_x + i\sigma_y; \; \sigma_- = \sigma_x - i\sigma_y \tag{130.1}$$

and of the absolute square of the spin vector operator

$$\boldsymbol{\sigma}^2 = \sigma_x^2 + \sigma_y^2 + \sigma_z^2 \,? \tag{130.2}$$

Solution. Writing a instead of a_{12}, the last problem gave the results

$$\sigma_x = \begin{pmatrix} 0 & a \\ 1/a & 0 \end{pmatrix}; \quad \sigma_y = \begin{pmatrix} 0 & -ia \\ i/a & 0 \end{pmatrix} \tag{130.3}$$

and hence,

$$\sigma_+ = \begin{pmatrix} 0 & 2a \\ 0 & 0 \end{pmatrix}; \quad \sigma_- = \begin{pmatrix} 0 & 0 \\ 2/a & 0 \end{pmatrix}. \tag{130.4}$$

Let

$$\psi = \begin{pmatrix} u \\ v \end{pmatrix} = u \begin{pmatrix} 1 \\ 0 \end{pmatrix} + v \begin{pmatrix} 0 \\ 1 \end{pmatrix} = u\alpha + v\beta \tag{130.5}$$

be a two-component wave function, then

$$\begin{aligned} &\sigma_x \begin{pmatrix} u \\ v \end{pmatrix} = \begin{pmatrix} av \\ u/a \end{pmatrix}; \quad \sigma_y \begin{pmatrix} u \\ v \end{pmatrix} = \begin{pmatrix} -iav \\ iu/a \end{pmatrix}; \\ &\sigma_+ \begin{pmatrix} u \\ v \end{pmatrix} = \begin{pmatrix} 2v \\ 0 \end{pmatrix}; \quad \sigma_- \begin{pmatrix} u \\ v \end{pmatrix} = \begin{pmatrix} 0 \\ 2u \end{pmatrix}. \end{aligned} \tag{130.6}$$

The eigenvectors of σ_x satisfy the equation $\sigma_x \psi = \lambda \psi$ with λ an eigenvalue, i.e. in components

$$av = \lambda u \quad \text{and} \quad u/a = \lambda v.$$

These relations are compatible with one another only for $\lambda = \pm 1$. Thus we find the eigensolutions

$$\begin{aligned} &\text{for } \lambda_1 = +1: \; \psi_1 = u\begin{pmatrix} 1 \\ 1/a \end{pmatrix} = u\left(\alpha + \frac{1}{a}\beta\right); \\ &\text{for } \lambda_2 = -1: \; \psi_2 = u\begin{pmatrix} 1 \\ -1/a \end{pmatrix} = u\left(\alpha - \frac{1}{a}\beta\right). \end{aligned} \tag{130.7}$$

The probabilities that the spin may point upward (i.e. in the $+z$ direction) and downward are proportional to the absolute squares of the factors in front of the Hilbert vectors α and β, viz. $1:1/|a|^2$. Since there is no reason why the one should be greater than the other, it follows that

$$|a|^2 = 1. \tag{130.8}$$

The same consideration is possible for σ_y.

Let us, from now on, definitely choose $a=1$. Then we have eigenvalues $\lambda_1 = +1$ and $\lambda_2 = -1$ for each of the three σ_i's and the following eigenvectors.

$$\sigma_x = \begin{pmatrix} 0 & 1 \\ 1 & 0 \end{pmatrix}; \quad \psi_1 = 2^{-\frac{1}{2}}\begin{pmatrix} 1 \\ 1 \end{pmatrix}; \quad \psi_2 = 2^{-\frac{1}{2}}\begin{pmatrix} 1 \\ -1 \end{pmatrix}; \tag{130.9a}$$

$$\sigma_y = \begin{pmatrix} 0 & -i \\ i & 0 \end{pmatrix}; \quad \psi_1 = 2^{-\frac{1}{2}}\begin{pmatrix} 1 \\ i \end{pmatrix}; \quad \psi_2 = 2^{-\frac{1}{2}}\begin{pmatrix} 1 \\ -i \end{pmatrix}; \tag{130.9b}$$

$$\sigma_z = \begin{pmatrix} 1 & 0 \\ 0 & -1 \end{pmatrix}; \quad \psi_1 = \begin{pmatrix} 1 \\ 0 \end{pmatrix}; \quad \psi_2 = \begin{pmatrix} 0 \\ 1 \end{pmatrix}. \tag{130.9c}$$

The three Pauli matrices are hermitian operators, $\sigma_i^\dagger = \sigma_i$, with real eigenvalues. The two operators

$$\sigma_+ = \begin{pmatrix} 0 & 2 \\ 0 & 0 \end{pmatrix} \quad \text{and} \quad \sigma_- = \begin{pmatrix} 0 & 0 \\ 2 & 0 \end{pmatrix} \tag{130.10}$$

are not hermitian, $\sigma_+^\dagger = \sigma_-$ and $\sigma_-^\dagger = \sigma_+$. They have no eigenvalues or eigenvectors, because they cannot even be diagonalized. This may be seen as follows.

The most general unitary matrix in two dimensions may be written (apart from an irrelevant phase factor)

$$U = \begin{pmatrix} \cos\vartheta; & \sin\vartheta\, e^{i\xi} \\ -\sin\vartheta\, e^{i\eta}; & \cos\vartheta\, e^{i(\xi+\eta)} \end{pmatrix}$$

with real parameters ϑ, ξ, η. The unitary transformation of σ_+ then becomes

$$U^\dagger \sigma_+ U = 2\,e^{i\eta}\begin{pmatrix} -\sin\vartheta\cos\vartheta; & \cos^2\vartheta\, e^{i\xi} \\ -\sin^2\vartheta\, e^{-i\xi}; & \sin\vartheta\cos\vartheta \end{pmatrix},$$

and this cannot be made diagonal for any choice of the real parameters since $\sin\vartheta$ and $\cos\vartheta$ never vanish at the same argument ϑ.

If the operators σ_+ and σ_- are applied to the Hilbert vectors α and β we find according to (130.6):

$$\begin{array}{llll} \sigma_+\alpha=0; & \sigma_-\alpha=2\beta; & \sigma_+\sigma_-\alpha=4\alpha; & \sigma_-\sigma_+\alpha=0; \\ \sigma_+\beta=2\alpha; & \sigma_-\beta=0; & \sigma_+\sigma_-\beta=0; & \sigma_-\sigma_+\beta=4\beta. \end{array} \tag{130.11}$$

When commuted with σ_z, these operators are essentially reproduced,

$$\sigma_+\sigma_z - \sigma_z\sigma_+ = -2\sigma_+; \qquad \sigma_-\sigma_z - \sigma_z\sigma_- = +2\sigma_-. \tag{130.12}$$

The operators σ_+ and σ_-, in angular momentum normalization,

$$S_+ = \frac{\hbar}{2}\sigma_+; \qquad S_- = \frac{\hbar}{2}\sigma_- \tag{130.13}$$

have the same property as the analogous operators L_+ and L_- (Problem 56) to shift the z component by one (in units $\hbar$):

$$\begin{array}{ll} S_+\alpha=0; & S_-\alpha=\hbar\beta; \\ S_+\beta=\hbar\alpha; & S_-\beta=0. \end{array}$$

State β with spin component $-\frac{1}{2}\hbar$ is altered by S_+ into state α with $+\frac{1}{2}\hbar$, and inversely by S_-. Both, $S_+\alpha$ and $S_-\beta$ must vanish because, according to this shift rule, they would lead to states with $+\frac{3}{2}\hbar$ and $-\frac{3}{2}\hbar$ not existing in the Hilbert space used.

We finally investigate the absolute square of the spin operator,

$$\boldsymbol{\sigma}^2 = \sigma_x^2 + \sigma_y^2 + \sigma_z^2 = \tfrac{1}{2}(\sigma_+\sigma_- + \sigma_-\sigma_+) + \sigma_z^2. \tag{130.14}$$

It is easily seen that all three σ_i^2 are the unit matrix so that

$$\boldsymbol{\sigma}^2 = 3\begin{pmatrix} 1 & 0 \\ 0 & 1 \end{pmatrix}$$

is diagonal with its only value 3 for whatever Hilbert vector we choose, or—in less sophisticated wording—it is $=3$. This can also be seen from the second form in (130.14) using the relations for $\sigma_+\sigma_-$ and $\sigma_-\sigma_+$ given in (130.11).

Eq. (130.14) leads to $S^2=\frac{3}{4}\hbar^2$ or, with S being the quantum number describing the spin, to $\hbar^2 S(S+1)$ with $S=\frac{1}{2}$. This is the meaning of a state to have "spin $\frac{1}{2}$".

Problem 131. Spin algebra

To show that the three Pauli matrices together with unity form the complete basis of an algebra.

Solution. If 1, $\sigma_x, \sigma_y, \sigma_z$ form a complete basis, no number outside the algebra can be generated either by adding or by multiplying any pair of numbers of the form

$$N=a_0+a_1\sigma_x+a_2\sigma_y+a_3\sigma_z \tag{131.1}$$

with complex number coefficients a_i. Obviously the rule obtains for addition, but it still has to be proved for a product of two numbers. For this purpose we shall construct a multiplication table of the Pauli matrices.

Let i,k,l be an arbitrary cyclic permutation of the three subscripts x, y, z, then the σ_i's satisfy the commutation relations

$$\sigma_i\sigma_k-\sigma_k\sigma_i=2i\sigma_l \tag{131.2}$$

and the normalization relations

$$\sigma_i^2=1. \tag{131.3}$$

Besides, the Pauli matrices obey anticommutation rules,

$$\sigma_i\sigma_k+\sigma_k\sigma_i=0 \quad (i\neq k) \tag{131.4}$$

as may easily be verified. Then, by addition and subtraction of (131.2) and (131.4), we find the products

$$\sigma_i\sigma_k=i\sigma_l; \quad \sigma_k\sigma_i=-i\sigma_l. \tag{131.5}$$

Hence, any product of two basis elements leads back again, except for a complex number factor, to another basis element.

Multiplication table

first factor	second factor 1	σ_x	σ_y	σ_z
1	1	σ_x	σ_y	σ_z
σ_x	σ_x	1	$i\sigma_z$	$-i\sigma_y$
σ_y	σ_y	$-i\sigma_z$	1	$i\sigma_x$
σ_z	σ_z	$i\sigma_y$	$-i\sigma_x$	1

It should be noted that the product of all three Pauli matrices therefore simply becomes

$$\sigma_x \sigma_y \sigma_z = i. \tag{131.6}$$

Further, another argument may now be added for σ_+ and σ_- having no eigenvalues. From the multiplication table one gets

$$\sigma_\pm^2 = (\sigma_x \pm i\sigma_y)^2 = \sigma_x^2 - \sigma_y^2 \pm i(\sigma_x\sigma_y + \sigma_y\sigma_x) = 0.$$

This is an interesting result showing that in the Pauli algebra the square of a non-vanishing number may be zero. It is, by the way, corroborated by the fact that neither for σ_+ nor σ_- do any reciprocal numbers exist.

A number N for which the relation

$$N = N^2 \tag{131.7}$$

holds is called an idempotent number. Such numbers belong to our algebra, viz.

$$\tfrac{1}{2}(1+\sigma_i) \quad \text{and} \quad \tfrac{1}{2}(1-\sigma_i) \qquad (i = x, y, z). \tag{131.8}$$

In matrix representation we have e.g.

$$P_+ = \tfrac{1}{2}(1+\sigma_z) = \begin{pmatrix} 1 & 0 \\ 0 & 0 \end{pmatrix}; \qquad P_- = \tfrac{1}{2}(1-\sigma_z) = \begin{pmatrix} 0 & 0 \\ 0 & 1 \end{pmatrix};$$

applied to the basic Hilbert vectors α and β they give

$$P_+\alpha = \alpha; \qquad P_-\alpha = 0;$$
$$P_+\beta = 0; \qquad P_-\beta = \beta.$$

Thus they suppress either β or α in a state vector of mixed spin orientations:

$$P_+(u\alpha + v\beta) = u\alpha; \qquad P_-(u\alpha + v\beta) = v\beta$$

and leave us with the projection of the state vector upon one of the basic directions in Hilbert space. They are therefore called projection operators.

NB. The Pauli algebra is essentially the same as the algebra of quaternions which uses $i\sigma_k$ instead of σ_k as basis elements.

Problem 132. Spinor transformation properties

How can it be proved that the spin of a one-particle state,

$$\mathbf{s} = \int d^3x\, \psi^\dagger \boldsymbol{\sigma} \psi, \tag{132.1}$$

is a vector? It should be noted that the σ_i's shall not transform with the coordinates under space rotation, but the transformation properties of **s** shall rest entirely upon the wave function.

Solution. In consequence of the group property of space rotations, it suffices to investigate an infinitesimal rotation,

$$x_i' = x_i + \sum_k{}' \varepsilon_{ik} x_k; \qquad \varepsilon_{ki} = -\varepsilon_{ik} \tag{132.2}$$

with infinitesimal angles of rotation about the three axes,

$$\varepsilon_{12} = \alpha_3; \qquad \varepsilon_{23} = \alpha_1; \qquad \varepsilon_{31} = \alpha_2. \tag{132.3}$$

If **s** is a vector, it has to obey the same transformation rule,

$$s_i' = s_i + \sum_k{}' \varepsilon_{ik} s_k. \tag{132.4}$$

This has to be achieved by transforming only the wave function,

$$\psi' = (1+\xi)\psi; \qquad \psi'^{\dagger} = \psi^{\dagger}(1+\xi^{\dagger}) \tag{132.5}$$

with infinitesimal ξ. This transformation shall now be determined.

We begin by stating that $\psi^{\dagger}\psi$ is a scalar so that

$$\psi'^{\dagger}\psi' = \psi^{\dagger}(1+\xi^{\dagger})(1+\xi)\psi = \psi^{\dagger}\psi$$

or

$$\xi^{\dagger} = -\xi. \tag{132.6}$$

Putting (132.4) and (132.5) in (132.1), we get

$$s_i' = \int d^3x\, \psi^{\dagger}(1-\xi)\sigma_i(1+\xi)\psi = s_i + \int d^3x\, \psi^{\dagger}(\sigma_i\xi - \xi\sigma_i)\psi.$$

Comparison of this expression with (132.4) leads to the determining equations for the operator ξ:

$$\sigma_i\xi - \xi\sigma_i = \sum_k{}' \varepsilon_{ik}\sigma_k \qquad (i,k=1,2,3). \tag{132.7}$$

It can easily be shown that these equations are solved unambiguously by

$$\xi = \frac{i}{2}(\varepsilon_{12}\sigma_3 + \varepsilon_{23}\sigma_1 + \varepsilon_{31}\sigma_2) \tag{132.8}$$

because, using the commutation rules, we find for the left-hand side of Eq. (132.7), e. g. with $i=1$:

$$\frac{i}{2}[\sigma_1,\ \varepsilon_{12}\sigma_3 + \varepsilon_{23}\sigma_1 + \varepsilon_{31}\sigma_2] = \frac{i}{2}\{-\varepsilon_{12}\cdot 2i\sigma_2 + \varepsilon_{31}\cdot 2i\sigma_3\}$$
$$= \varepsilon_{12}\sigma_2 + \varepsilon_{13}\sigma_3$$

in agreement with the right-hand side. Analogous results are found for $i=2$ or 3.

The operator (132.8) can be written in a simpler form using the angles of rotation in the notation (132.3),

$$\xi = \frac{i}{2} \sum_k \alpha_k \sigma_k . \tag{132.9}$$

Applied to any two-component wave function,

$$\psi = \begin{pmatrix} u \\ v \end{pmatrix} = u\alpha + v\beta \tag{132.10}$$

this leads to the transformed wave function

$$\psi' = \begin{pmatrix} u' \\ v' \end{pmatrix} = u'\alpha + v'\beta \tag{132.10'}$$

with

$$\begin{aligned} u' &= \left(1 + \frac{i}{2}\alpha_3\right) u + \frac{i}{2}(\alpha_1 - i\alpha_2) v ; \\ v' &= \frac{i}{2}(\alpha_1 + i\alpha_2) u + \left(1 - \frac{i}{2}\alpha_3\right) v . \end{aligned} \tag{132.11}$$

A two-component function with these transformation properties is called a *spinor*.

Problem 133. Spin electron in a central field

To determine the wave functions of a spinning electron in a central spin-independent force field. The wave functions must be eigenfunctions of the two operators

$$\boldsymbol{J}^2 = (\boldsymbol{L} + \boldsymbol{S})^2 \quad \text{and} \quad J_z = L_z + S_z \tag{133.1}$$

with $\boldsymbol{L}$ the orbital momentum and $\boldsymbol{S}$ the spin of the electron.

Solution. We begin with the z component of the angular momentum. Since

$$L_z = \frac{\hbar}{i} \frac{\partial}{\partial \varphi} ; \quad S_z = \frac{\hbar}{2} \sigma_z ; \quad \sigma_z = \begin{pmatrix} 1 & 0 \\ 0 & -1 \end{pmatrix}$$

we can write

$$J_z = \hbar \begin{pmatrix} -i \dfrac{\partial}{\partial \varphi} + \dfrac{1}{2} ; & 0 \\ 0 ; & -i \dfrac{\partial}{\partial \varphi} - \dfrac{1}{2} \end{pmatrix} . \tag{133.2}$$

The eigenfunctions of this operator have the form

$$\psi = \begin{pmatrix} C_1 \, e^{i(m_j - \frac{1}{2})\varphi} \\ C_2 \, e^{i(m_j + \frac{1}{2})\varphi} \end{pmatrix} \tag{133.3}$$

with C_1 and C_2 still arbitrary functions of the variables r and ϑ. This can easily be seen by letting J_z operate on ψ; the result is

$$J_z \psi = \hbar m_j \psi \tag{133.4}$$

so that $\hbar m_j$ is an eigenvalue of the z component of the total angular momentum $\boldsymbol{J}$. A better understanding of Eq. (133.3) is achieved by using the notation

$$\psi = C_1 \, e^{i(m_j - \frac{1}{2})\varphi} \alpha + C_2 \, e^{i(m_j + \frac{1}{2})\varphi} \beta \tag{133.3'}$$

because it is then seen that the first term describes the dependence of ψ on the coordinates if the spin points upward, and the second term if it points downward. This coordinate dependence determines the component $\hbar m_l$ of the orbital momentum $\boldsymbol{L}$ which makes $m_l = 0, \pm 1, \pm 2, \ldots$ an integer. With spin upward we have $m_j = m_l + \frac{1}{2}$ and with spin downward $m_j = m_l - \frac{1}{2}$ so that m_j becomes a half-integer. These are the addition rules well-known for the vector model. The characteristic feature of the wave function (133.3) or (133.3′) is that m_j is a "good quantum number" but that m_l is not, because the state vector ψ is a mixture of two parts with different values of m_l.

We now investigate the operator $\boldsymbol{J}^2$. We have

$$\boldsymbol{J}^2 = \boldsymbol{L}^2 + \boldsymbol{S}^2 + 2(\boldsymbol{L} \cdot \boldsymbol{S}) = \boldsymbol{L}^2 + \tfrac{3}{4}\hbar^2 + \hbar \begin{pmatrix} L_z & L_+ \\ L_- & -L_z \end{pmatrix}$$

if the Pauli matrices are used for $\boldsymbol{S} = \dfrac{\hbar}{2}\boldsymbol{\sigma}$. We therefore have to solve the eigenvalue problem

$$\boldsymbol{J}^2 \psi = \begin{pmatrix} \boldsymbol{L}^2 + \frac{3}{4}\hbar^2 + \hbar L_z; & \hbar L_+ \\ \hbar L_-; & \boldsymbol{L}^2 + \frac{3}{4}\hbar^2 - \hbar L_z \end{pmatrix} \psi = \hbar^2 j(j+1)\psi \tag{133.5}$$

if, in analogy to $\boldsymbol{L}^2$, we arbitrarily call the eigenvalue $\hbar^2 j(j+1)$. In order to make ψ simultaneously an eigenfunction of J_z and $\boldsymbol{J}^2$ it must have the form (133.3) where now the dependence of C_1 and C_2 upon the variable ϑ has to be determined. This can be done by putting

$$\psi = \begin{pmatrix} f(r) \, Y_{l, m_j - \frac{1}{2}}(\vartheta, \varphi) \\ g(r) \, Y_{l, m_j + \frac{1}{2}}(\vartheta, \varphi) \end{pmatrix}. \tag{133.6}$$

The spherical harmonics in (133.6) depend upon φ just in the way of Eq. (133.3). When $\boldsymbol{J}^2\psi$ is formed from (133.5) and (133.6), and we apply the general formulae (cf. Problem 56)

$$L_+ Y_{l,m} = -\hbar\sqrt{(l+m+1)(l-m)}\, Y_{l,m+1}; \tag{133.7a}$$

$$L_- Y_{l,m} = -\hbar\sqrt{(l+m)(l-m+1)}\, Y_{l,m-1}; \tag{133.7b}$$

$$L_z Y_{l,m} = \hbar m\, Y_{l,m}; \tag{133.7c}$$

$$L^2 Y_{l,m} = \hbar^2 l(l+1)\, Y_{l,m}, \tag{133.7d}$$

we obtain

$$J^2\psi = \hbar^2 \begin{pmatrix} f(r)\left[l(l+1)+\frac{3}{4}+(m_j-\frac{1}{2})\right] Y_{l,m_j-\frac{1}{2}} \\ -g(r)\sqrt{(l+m_j+\frac{1}{2})(l-m_j+\frac{1}{2})}\, Y_{l,m_j-\frac{1}{2}} \\ \\ -f(r)\sqrt{(l+m_j+\frac{1}{2})(l-m_j-\frac{1}{2})}\, Y_{l,m_j+\frac{1}{2}} \\ +g(r)\left[l(l+1)+\frac{3}{4}-(m_j+\frac{1}{2})\right] Y_{l,m_j+\frac{1}{2}} \end{pmatrix}$$

so that the eigenvalue problem (133.5) leads to two linear algebraic equations for $f(r)$ and $g(r)$, viz.

$$\begin{aligned} f\left[l(l+1)+\tfrac{3}{4}+(m_j-\tfrac{1}{2})-j(j+1)\right] - g\sqrt{(l+\tfrac{1}{2})^2-m_j^2} &= 0; \\ -f\sqrt{(l+\tfrac{1}{2})^2-m_j^2} + g\left[l(l+1)+\tfrac{3}{4}-(m_j+\tfrac{1}{2})-j(j+1)\right] &= 0. \end{aligned} \tag{133.8}$$

The possibility of thus eliminating the spherical harmonics shows that the function (133.6) suffices to solve the problem. Thus l is still a "good" quantum number even in spin theory.

The equations (133.8) are compatible with one another only if $f(r)$ and $g(r)$ differ only by a constant factor. We write

$$f(r) = A\,F(r); \qquad g(r) = B\,F(r). \tag{133.9}$$

Then (133.8) permits to determine the ratio B/A. As the linear equations (133.8) are homogeneous, their determinant must vanish,

$$[j(j+1)-(l+\tfrac{1}{2})^2]^2 - m_j^2 - [(l+\tfrac{1}{2})^2 - m_j^2] = 0. \tag{133.10}$$

Obviously, this condition is independent of m_j. That is one of the simplest consequences of a very general theorem of Wigner and Eckart. There are two different values of $j(j+1)$ satisfying (133.10), viz.

Solution I. $j = l+\frac{1}{2}$; $\quad B = -A\sqrt{\dfrac{l+\frac{1}{2}-m_j}{l+\frac{1}{2}+m_j}}$;

$$\psi_{\mathrm{I}} = \frac{F_l(r)}{\sqrt{2l+1}} \begin{pmatrix} \sqrt{l+\frac{1}{2}+m_j}\, Y_{l,m_j-\frac{1}{2}} \\ -\sqrt{l+\frac{1}{2}-m_j}\, Y_{l,m_j+\frac{1}{2}} \end{pmatrix}. \tag{133.11}$$

Solution II. $j = l-\frac{1}{2}$; $\quad B = A\sqrt{\dfrac{l+\frac{1}{2}+m_j}{l+\frac{1}{2}-m_j}}$;

$$\psi_{\mathrm{II}} = \frac{F_l(r)}{\sqrt{2l+1}} \begin{pmatrix} \sqrt{l+\frac{1}{2}-m_j}\, Y_{l,m_j-\frac{1}{2}} \\ \sqrt{l+\frac{1}{2}+m_j}\, Y_{l,m_j+\frac{1}{2}} \end{pmatrix}. \tag{133.12}$$

Both solutions are normalized. Since in all component functions a spherical harmonic of the same order l is split off, and since the potential is supposed not to depend upon spin, the function F_l is to be determined from the radial Schrödinger equation,

$$-\frac{\hbar^2}{2m}\left(\chi_l'' - \frac{l(l+1)}{r^2}\chi_l\right) + V(r)\chi_l = E_l\chi_l \tag{133.13}$$

with

$$\chi_l = r F_l(r). \tag{133.14}$$

NB. The formulae (133.11) and (133.12) may be applied to states with $l=0$ if attention is paid to the fact that the spherical harmonics then vanish if $m_j \mp \frac{1}{2}$ does not equal zero. Thus we get

$$\psi_{\mathrm{I}} = \frac{F_0(r)}{\sqrt{4\pi}}\begin{pmatrix}1\\0\end{pmatrix} \quad \text{for } l=0, m_j=+\tfrac{1}{2}$$

and

$$\psi_{\mathrm{I}} = -\frac{F_0(r)}{\sqrt{4\pi}}\begin{pmatrix}0\\1\end{pmatrix} \quad \text{for } l=0, m_j=-\tfrac{1}{2}.$$

The other function, ψ_{II}, vanishes identically in both cases so that no solutions with negative j are originated. The same results are found by applying the operators $\boldsymbol{J}^2$ and J_z to a two-component function that depends on the radius only.

Problem 134. Quadrupole moment of a spin state

To determine the quadrupole moment of a one-electron state in a potential field of spherical symmetry, taking the electron spin into account.

Solution. The eigenfunctions (cf. Eqs. (133.11, 12)) have the absolute squares

$$|\psi_{\mathrm{I}}|^2 = \frac{|F_l(r)|^2}{2l+1}\left\{(l+\tfrac{1}{2}+m_j)|Y_{l,m_j-\frac{1}{2}}|^2 + (l+\tfrac{1}{2}-m_j)|Y_{l,m_j+\frac{1}{2}}|^2\right\} \tag{134.1a}$$

if $j=l+\frac{1}{2}$, and

$$|\psi_{\mathrm{II}}|^2 = \frac{|F_l(r)|^2}{2l+1}\left\{(l+\tfrac{1}{2}-m_j)|Y_{l,m_j-\frac{1}{2}}|^2 + (l+\tfrac{1}{2}+m_j)|Y_{l,m_j+\frac{1}{2}}|^2\right\} \tag{134.1b}$$

if $j=l-\frac{1}{2}$. It should be noted that $|m_j| \leqq j$, and that for $l=0$ the function ψ_{II} vanishes identically.

Since (134.1a, b) do not depend upon the angle φ, the argument of Problem 61 still holds, so that the non-diagonal elements of the quadrupole tensor have vanishing averages and that

$$\langle Q_{xx}\rangle = \langle Q_{yy}\rangle = -\tfrac{1}{2}\langle Q_{zz}\rangle. \tag{134.2}$$

Thus again, we need only calculate $\langle Q_{zz}\rangle$ according to the formula

$$\langle Q_{zz}\rangle = \int d^3x\, |\psi|^2 r^2 (3\cos^2\vartheta - 1). \tag{134.3}$$

With the relation

$$\oint d\Omega (3\cos^2\vartheta - 1)|Y_{l,m}|^2 = \frac{2l(l+1) - 6m^2}{(2l-1)(2l+3)} \tag{134.4}$$

proved in Problem 61, and the abbreviation

$$\int_0^\infty dr\, r^4 |F_l(r)|^2 = \langle r^2\rangle, \tag{134.5}$$

we get from (134.1a, b):

$$\langle Q_{zz}\rangle = \frac{\langle r^2\rangle}{2l+1}\left\{\left(l+\frac{1}{2}\pm m_j\right)\frac{2l(l+1)-6(m_j-\frac{1}{2})^2}{(2l-1)(2l+3)} + \left(l+\frac{1}{2}\mp m_j\right)\frac{2l(l+1)-6(m_j+\frac{1}{2})^2}{(2l-1)(2l+3)}\right\}$$

with the upper signs for $j=l+\frac{1}{2}$, the lower ones for $j=l-\frac{1}{2}$ and $l\geqq 1$. An elementary reordering in the curly bracket leads to

$$\langle Q_{zz}\rangle = \langle r^2\rangle \frac{2l(l+1) - 6\left(m_j^2+\frac{1}{4}\right) \pm \frac{12}{2l+1} m_j^2}{(2l-1)(2l+3)}, \tag{134.6}$$

a formula to be compared with (61.8).

It is possible, by using j instead of l, to write instead of (134.6) one comprehensive formula,

$$\langle Q_{zz}\rangle = \langle r^2\rangle \cdot \frac{1}{2}\left(1 - \frac{3m_j^2}{j(j+1)}\right) \tag{134.7}$$

wich obtains for both signs, $j=l\pm\frac{1}{2}$, equally. Numerical results for the lowest values of j are listed in the accompanying table.

States	j	$\langle Q_{zz}\rangle/\langle r^2\rangle$ for $m_j=\pm\frac{1}{2}$	$m_j=\pm\frac{3}{2}$	$m_j=\pm\frac{5}{2}$	$m_j=\pm\frac{7}{2}$
$S_{\frac{1}{2}}, P_{\frac{1}{2}}$	$\frac{1}{2}$	0	—	—	—
$P_{\frac{3}{2}}, D_{\frac{3}{2}}$	$\frac{3}{2}$	$+\frac{2}{5}$	$-\frac{2}{5}$	—	—
$D_{\frac{5}{2}}, F_{\frac{5}{2}}$	$\frac{5}{2}$	$+\frac{16}{35}$	$+\frac{4}{35}$	$-\frac{20}{35}$	—
$F_{\frac{7}{2}}, G_{\frac{7}{2}}$	$\frac{7}{2}$	$+\frac{10}{21}$	$+\frac{6}{21}$	$-\frac{2}{21}$	$-\frac{14}{21}$

It can be seen that for $j=\frac{1}{2}$ spherical symmetry obtains for S as well as for P states. Generally, it may be stated that with increasing values of $|m_j|$ the configuration passes from oblong to oblate figure of the electron distribution.

The sum of the moments in each horizontal line of the table vanishes, since it leads to a closed shell of spherical symmetry. This can quite generally be shown as

$$\sum_{m_j=\frac{1}{2}}^{j} 1 = j+\tfrac{1}{2} \quad \text{and} \quad \sum_{m_j=\frac{1}{2}}^{j} m_j^2 = \tfrac{1}{3} j(j+\tfrac{1}{2})(j+1)$$

which by combination lead to

$$\sum_{m_j=\frac{1}{2}}^{j} \langle Q_{zz}\rangle = \tfrac{1}{2}\langle r^2\rangle \left\{ \sum_{m_j=\frac{1}{2}}^{j} 1 - \frac{3}{j(j+1)} \sum_{m_j=\frac{1}{2}}^{j} m_j^2 \right\} = 0.$$

Problem 135. Expectation values of magnetic moments

For the spin electron in a central field there shall be derived the expectation values of all three components of the vectors $\boldsymbol{S}, \boldsymbol{L}, \boldsymbol{J}$ and of the magnetic moment.

Solution. If

$$u = \begin{pmatrix} u_1 \\ u_2 \end{pmatrix}$$

is the eigenspinor of a state, we have

$$u^\dagger S_x u = \frac{\hbar}{2}(u_1^* u_2 + u_2^* u_1); \qquad u^\dagger S_y u = \frac{\hbar}{2i}(u_1^* u_2 - u_2^* u_1);$$

$$u^\dagger S_z u = \frac{\hbar}{2}(u_1^* u_1 - u_2^* u_2). \tag{135.1}$$

The eigenspinors of $\boldsymbol{J}^2$ and J_z have been determined in Problem 133; they are

$$u_1 = \frac{F_l(r)}{\sqrt{2l+1}} A_{j,l} Y_{l,m_j-\frac{1}{2}}; \qquad u_2 = \frac{F_l(r)}{\sqrt{2l+1}} B_{j,l} Y_{l,m_j+\frac{1}{2}} \tag{135.2}$$

with

$$A_{l+\frac{1}{2},l} = \sqrt{l+\tfrac{1}{2}+m_j}; \qquad B_{l+\frac{1}{2},l} = \sqrt{l+\tfrac{1}{2}-m_j} \tag{135.3a}$$

and

$$A_{l-\frac{1}{2},l} = \sqrt{l+\tfrac{1}{2}-m_j}; \qquad B_{l-\frac{1}{2},l} = \sqrt{l+\tfrac{1}{2}+m_j}. \tag{135.3b}$$

In the expressions (135.1) underlying the expectation values of S_x and S_y this leads to products of different spherical harmonics so that $\langle S_x \rangle = 0$ and $\langle S_y \rangle = 0$. On the other hand, we find

$$\langle S_z \rangle = \frac{\hbar}{2} \int_0^\infty dr\, r^2 \frac{|F_l(r)|^2}{2l+1} (A_{j,l}^2 - B_{j,l}^2) \tag{135.4}$$

and for the normalization integral

$$\int_0^\infty dr\, r^2 \frac{|F_l(r)|^2}{2l+1} (A_{j,l}^2 + B_{j,l}^2) = \int_0^\infty dr\, r^2 |F_l(r)|^2 = 1. \tag{135.5}$$

Thus we get

$$\langle S_z \rangle = \frac{\hbar}{2} \frac{A_{j,l}^2 - B_{j,l}^2}{A_{j,l}^2 + B_{j,l}^2} \tag{135.6}$$

which, in a state with $j = l + \frac{1}{2}$, is

$$\langle S_z \rangle_+ = \hbar m_j \cdot \frac{1}{2l+1} = \hbar m_j \cdot \frac{1}{2j} \tag{135.7 a}$$

and in a state with $j = l - \frac{1}{2}$,

$$\langle S_z \rangle_- = -\hbar m_j \cdot \frac{1}{2l+1} = -\hbar m_j \cdot \frac{1}{2(j+1)}. \tag{135.7 b}$$

The expectation values of the three components of orbital momentum follow in a similar way from

$$u^\dagger \boldsymbol{L} u = u_1^* \boldsymbol{L} u_1 + u_2^* \boldsymbol{L} u_2 .$$

Since the operators $L_x \pm i L_y$ change the first subscript of the spherical harmonics from l into $l \pm 1$, these components again have vanishing expectation values (cf. Problem 58). For the component L_z we have

$$\langle L_z \rangle = \hbar \frac{\int d^3x \{u_1^* (m_j - \frac{1}{2}) u_1 + u_2^* (m_j + \frac{1}{2}) u_2\}}{\int d^3x \{u_1^* u_1 + u_2^* u_2\}}$$

or

$$\langle L_z \rangle = \hbar \cdot \frac{A_{j,l}^2 (m_j - \frac{1}{2}) + B_{j,l}^2 (m_j + \frac{1}{2})}{A_{j,l}^2 + B_{j,l}^2}.$$

Using (135.6), this can be written much more simply

$$\langle L_z \rangle = \hbar m_j - \langle S_z \rangle, \tag{135.8}$$

a formula which we might well have started with because of u being eigenspinor of $J_z = L_z + S_z$ with the eigenvalue $\hbar m_j$.

The expectation values of J_x and J_y of course vanish, as do these two components of $\boldsymbol{L}$ and $\boldsymbol{S}$.

The *magnetic moment* operator is

$$\boldsymbol{M} = -\frac{e}{2mc}(\boldsymbol{L}+2\boldsymbol{S}) \tag{135.9}$$

for an electron of charge $-e$. Its expectation values in x and y direction vanish again, but

$$\langle M_z\rangle = -\frac{e}{2mc}\{\langle L_z\rangle + 2\langle S_z\rangle\} \tag{135.10}$$

may, according to (135.8), be written

$$\langle M_z\rangle = -\frac{e\hbar}{2mc}(m_j + \langle S_z\rangle)$$

which, with (135.7a, b) leads, for the states $j=l+\frac{1}{2}$, to

$$\langle M_z\rangle_+ = -\frac{e\hbar}{2mc}m_j\left(1+\frac{1}{2j}\right) = -\frac{e\hbar}{2mc}m_j\frac{2j+1}{2j} \tag{135.11a}$$

and, for the states $j=-\frac{1}{2}$, to

$$\langle M_z\rangle_- = -\frac{e\hbar}{2mc}m_j\left(1-\frac{1}{2(j+1)}\right) = -\frac{e\hbar}{2mc}m_j\frac{2j+1}{2j+2}. \tag{135.11b}$$

NB. The last formulae show that a closed subshell (n, l) with either $j=l+\frac{1}{2}$ or $j=l-\frac{1}{2}$ has no resultant magnetic moment.

The factor of $-\frac{e\hbar}{2mc}m_j$ in (135.11a, b) is called the Landé g-factor of the state. It permits $\langle M_z\rangle$ to be written in the form

$$\langle M_z\rangle = -\frac{e\hbar}{2mc}m_j g(j) = -\frac{e}{2mc}\langle J_z\rangle g(j)$$

thus describing the deviation, originated by the spin, from the classical Maxwell relation between magnetic moment and mechanical moment, i.e. angular momentum.

Problem 136. Fine structure

The interaction of the intrinsic magnetic moment of an electron,

$$\boldsymbol{\mu} = -g\frac{e}{mc}\boldsymbol{S}, \tag{136.1}$$

and its orbital momentum $\boldsymbol{L}$ is described by a hamiltonian term,

$$H' = \frac{g}{2m^2c^2}\,\frac{1}{r}\,\frac{dV(r)}{dr}(\boldsymbol{S}\cdot\boldsymbol{L})\,. \tag{136.2}$$

The level splitting due to this interaction shall be determined.

NB. The so-called g factor of the electron is almost 1. Its exact value has been found to be $g=1.001145$. Since the complete theory of fine structure cannot be given in this unrelativistic treatment, this g factor should not here be taken too seriously. The same holds for the factor 2 in the denominator of (136.2), the so-called Thomas factor, not to be explained by unrelativistic considerations.

Solution. The electron wave function in a central force field is a simultaneous eigenfunction of the operators J_z and $\boldsymbol{J}^2$, the angular structure of which has been given in the preceding problem. The operator $(\boldsymbol{S}\cdot\boldsymbol{L})$ occurring in the hamiltonian (136.2) is then to be reduced to the quantum numbers j and l of the state $\psi=|j,l\rangle$ by

$$\boldsymbol{J}^2|j,l\rangle=\{\boldsymbol{L}^2+\boldsymbol{S}^2+2(\boldsymbol{L}\cdot\boldsymbol{S})\}|j,l\rangle$$

or

$$\hbar^2\{j(j+1)-[l(l+1)+\tfrac{3}{4}]\}|j,l\rangle=2(\boldsymbol{L}\cdot\boldsymbol{S})|j,l\rangle\,.$$

The term (136.2) of the hamiltonian therefore simply adds to $V(r)$ an energy perturbation

$$V'(r) = \frac{g\hbar^2}{4m^2c^2}\,\frac{1}{r}\,\frac{dV(r)}{dr}\{j(j+1)-l(l+1)-\tfrac{3}{4}\}\,; \tag{136.3}$$

since it depends upon the quantum numbers j and l, it will differ for different values $j=l\pm\frac{1}{2}$ with the same l.

First-order perturbation calculation gives a contribution

$$E'_{j,l}=\langle j,l|\,V'|j,l\rangle \tag{136.4}$$

to the energy of a level. In the notation of Eqs. (133.11) and (133.12), in the normalization

$$\int_0^\infty dr\,r^2\,|F_l(r)|^2=1 \tag{136.5}$$

this leads to

$$E'_{j,l} = \frac{g\hbar^2}{4m^2c^2}\{j(j+1)-l(l+1)-\tfrac{3}{4}\}\int_0^\infty dr\,r^2\,|F_l(r)|^2\,\frac{1}{r}\,\frac{dV}{dr}\,. \tag{136.6}$$

The energy splitting between two levels of the same l, but with different j values then becomes proportional to the difference

$$(l+\tfrac{1}{2})(l+\tfrac{3}{2})-(l-\tfrac{1}{2})(l+\tfrac{1}{2})=2l+1$$

so that

$$\Delta E = \frac{g\hbar^2}{4m^2c^2}(2l+1)\int_0^\infty dr\, r^2 |F_l(r)|^2 \frac{1}{r}\frac{dV}{dr}. \tag{136.7}$$

The level of the smaller j value is the lower one (normal doublet).

Some estimate of the integral (136.7) may be made by using the potential

$$V = -\frac{Ze^2}{r}; \quad \frac{1}{r}\frac{dV}{dr} = \frac{Ze^2}{r^3}. \tag{136.8}$$

Since this expression holds in the neighbourhood of the nucleus in all atoms, and since $F_l \propto r^l$ in this domain, the integrand in (136.7) becomes $\propto r^{2l-1}$ so that the integral converges for $l=1,2,3,\ldots$ but diverges logarithmically for S states ($l=0$). As there is no level splitting but only a shift in an S state, this result is not of primary importance for the evaluation of spectroscopic data. The difficulty does not occur in a rigorous relativistic treatment of the problem (cf. Problem 203).

Without evaluating in detail integrals of the type (136.7) it may safely be said that the result is of the order of Ze^2/a^3 with a a length of the order of atomic radii. Since term energies in atoms are of the order Ze^2/a, we then have roughly

$$\Delta E/E \propto \lambda^2/a^2$$

with $\lambda=\hbar/mc$ the Compton wavelength. This is a small quantity, hence the effect is a fine structure only and may be treated as a first-order perturbation as has been done above.

Problem 137. Plane wave of spin $\frac{1}{2}$ particles

To expand a plane wave of spin $\frac{1}{2}$ particles with either positive or negative helicity into a series of spherical harmonics. Let the wave run in z direction.

Solution. The two-component wave spinors

$$\psi_+ = \begin{pmatrix}1\\0\end{pmatrix} e^{ikz}; \quad \psi_- = \begin{pmatrix}0\\1\end{pmatrix} e^{ikz} \tag{137.1}$$

describe plane waves in z direction. In the state ψ_+ the particles have spin in the direction of propagation; we then speak of *helicity* $h=+1$. In ψ_- we have the opposite spin direction, $h=-1$. If we decompose these spinors into eigenspinors of angular momentum, we have $m_l=0$ in both cases and $m_j=+\frac{1}{2}$ for ψ_+, $m_j=-\frac{1}{2}$ for ψ_-.

It has been shown in problem 133 that, for a given orbital momentum l, there exist two eigenspinors of J_z and $\boldsymbol{J}^2$, viz.

$$u^{\mathrm{I}}_{l,m_j}=\frac{F_l(r)}{\sqrt{2l+1}}\begin{pmatrix}\sqrt{l+\frac{1}{2}+m_j}\,Y_{l,m_j-\frac{1}{2}}\\ -\sqrt{l+\frac{1}{2}-m_j}\,Y_{l,m_j+\frac{1}{2}}\end{pmatrix}\quad \text{if } j=l+\tfrac{1}{2} \tag{137.2a}$$

and

$$u^{\mathrm{II}}_{l,m_j}=\frac{F_l(r)}{\sqrt{2l+1}}\begin{pmatrix}\sqrt{l+\frac{1}{2}-m_j}\,Y_{l,m_j-\frac{1}{2}}\\ \sqrt{l+\frac{1}{2}+m_j}\,Y_{l,m_j+\frac{1}{2}}\end{pmatrix}\quad \text{if } j=l-\tfrac{1}{2} \tag{137.2b}$$

with $F_l(r)$ satisfying the radial Schrödinger equation. In the present force-free case the latter runs

$$F_l''+\frac{2}{r}F_l'+\left(k^2-\frac{l(l+1)}{r^2}\right)F_l=0 \tag{137.3}$$

and has solutions regular at the origin, in arbitrary normalization,

$$F_l=\frac{1}{kr}j_l(kr)\,. \tag{137.4}$$

The plane wave then shall be composed of solutions (137.2a) and (137.2b) in the form

$$\psi=\sum_{l=0}^{\infty}\left(A_l u^{\mathrm{I}}_{l,m_j}+B_l u^{\mathrm{II}}_{l,m_j}\right). \tag{137.5}$$

Let us begin with the case of helicity $h=+1$ where $m_j=+\frac{1}{2}$. Eq. (137.5) then may be written

$$\psi_+=\frac{1}{kr}\sum_{l=0}^{\infty}\frac{1}{\sqrt{2l+1}}j_l(kr)\begin{pmatrix}(A_l\sqrt{l+1}+B_l\sqrt{l})\,Y_{l,0}\\ (-A_l\sqrt{l}+B_l\sqrt{l+1})\,Y_{l,1}\end{pmatrix}. \tag{137.6}$$

In order to make the second line of the spinor vanish according to Eq. (137.1) we have to put

$$B_l=\sqrt{\frac{l}{l+1}}\,A_l \tag{137.7}$$

so that

$$\psi_+=\frac{1}{kr}\begin{pmatrix}1\\0\end{pmatrix}\sum_{l=0}^{\infty}A_l\sqrt{\frac{2l+1}{l+1}}\,j_l(kr)\,Y_{l,0}\,. \tag{137.8}$$

Comparing with the expansion of the plane wave (cf. (81.13)),

$$e^{ikz} = \frac{1}{kr}\sum_{l=0}^{\infty}\sqrt{4\pi(2l+1)}\,i^l j_l(kr)\,Y_{l,0} \tag{137.9}$$

there follows

$$A_l = \sqrt{4\pi(l+1)}\,i^l\,, \tag{137.10}$$

so that

$$\psi_+ = \sqrt{4\pi}\sum_{l=0}^{\infty} i^l\left(\sqrt{l+1}\,u^{\mathrm{I}}_{l,\frac{1}{2}} + \sqrt{l}\,u^{\mathrm{II}}_{l,\frac{1}{2}}\right) \tag{137.11}$$

is the correct expansion.

In the opposite case, $h=-1$ and $m_j=-\frac{1}{2}$, Eq. (137.5) yields

$$\psi_- = \frac{1}{kr}\sum_{l=0}^{\infty}\frac{1}{\sqrt{2l+1}}\,j_l(kr)\begin{pmatrix}(A_l\sqrt{l}+B_l\sqrt{l+1})\,Y_{l,-1}\\ (-A_l\sqrt{l+1}+B_l\sqrt{l})\,Y_{l,0}\end{pmatrix}. \tag{137.12}$$

Now, according to (137.1), the first line of the spinor should vanish, i.e.

$$B_l = -\sqrt{\frac{l}{l+1}}\,A_l\,. \tag{137.13}$$

Comparison with (137.9) now renders

$$A_l = -\sqrt{4\pi(l+1)}\,i^l \tag{137.14}$$

so that we arrive at the expansion

$$\psi_- = -\sqrt{4\pi}\sum_{l=0}^{\infty} i^l\left(\sqrt{l+1}\,u^{\mathrm{I}}_{l,-\frac{1}{2}} - \sqrt{l}\,u^{\mathrm{II}}_{l,-\frac{1}{2}}\right). \tag{137.15}$$

Problem 138. Free electron spin resonance

A free electron is put inside a cavity in which there exist two magnetic fields, viz. a constant homogeneous field $\mathcal{H}_0$ in z direction, and a field $\mathcal{H}'$ rotating in the x, y plane:

$$\left.\begin{aligned} &\mathcal{H}_x = 0\,; && \mathcal{H}_y = 0\,; && \mathcal{H}_z = \mathcal{H}_0\,;\\ &\mathcal{H}'_x = \mathcal{H}'\cos\omega t\,; && \mathcal{H}'_y = \mathcal{H}'\sin\omega t\,; && \mathcal{H}'_z = 0\,. \end{aligned}\right\} \tag{138.1}$$

At the time $t=0$, the electron has its spin in $+z$ direction, and the field $\mathcal{H}'$ is switched on. The probability P of the electron having its spin inverted into the $-z$ direction shall be determined as a function of time.

Solution. The hamiltonian of the problem runs

$$H=\mu(\sigma_z \mathscr{H}_0+\sigma_x \mathscr{H}'_x+\sigma_y \mathscr{H}'_y)$$

where $-\mu\boldsymbol{\sigma}$ is the intrinsic magnetic moment of the electron with $\mu=e\hbar/(2mc)$ except for quantum electrodynamical corrections. We write

$$\sigma_x \mathscr{H}'_x+\sigma_y \mathscr{H}'_y=\tfrac{1}{2}\mathscr{H}'(\sigma_+ \mathrm{e}^{-i\omega t}+\sigma_- \mathrm{e}^{i\omega t})$$

with $\sigma_\pm=\sigma_x\pm i\sigma_y$. Thus the Schrödinger equation becomes

$$-\frac{\hbar}{i}\frac{\partial\psi}{\partial t}=\mu\{\mathscr{H}_0\sigma_z+\tfrac{1}{2}\mathscr{H}'(\mathrm{e}^{-i\omega t}\sigma_+ +\mathrm{e}^{i\omega t}\sigma_-)\}\psi. \tag{138.2}$$

The solution can be expressed by the eigenfunctions of σ_z, i. e.

$$\psi(t)=u(t)\alpha+v(t)\beta. \tag{138.3}$$

Putting (138.3) into (138.2) and using the relations [cf. (129.10)]

$$\sigma_z\alpha=\alpha; \qquad \sigma_+\alpha=0; \qquad \sigma_-\alpha=2\beta;$$
$$\sigma_z\beta=-\beta; \qquad \sigma_+\beta=2\alpha; \qquad \sigma_-\beta=0$$

we obtain

$$-\frac{\hbar}{i}(\dot{u}\alpha+\dot{v}\beta)=\mu\mathscr{H}_0(u\alpha-v\beta)+\mu\mathscr{H}'(\mathrm{e}^{-i\omega t}v\alpha+\mathrm{e}^{i\omega t}u\beta).$$

Introducing the abbreviations

$$\frac{\mu\mathscr{H}_0}{\hbar}=\omega_0; \qquad \frac{\mu\mathscr{H}'}{\hbar}=\omega', \tag{138.4}$$

and separating into α and β parts, we find

$$\begin{aligned} i\dot{u}&=\omega_0 u+\omega'\mathrm{e}^{-i\omega t}v;\\ i\dot{v}&=-\omega_0 v+\omega'\mathrm{e}^{i\omega t}u. \end{aligned} \tag{138.5}$$

This system is solved by

$$u=A\,\mathrm{e}^{-i(\Omega+\frac{1}{2}\omega)t}; \qquad v=B\,\mathrm{e}^{-i(\Omega-\frac{1}{2}\omega)t}. \tag{138.6}$$

A straightforward calculation leads to two solutions $\Omega_1=+\Omega$ and $\Omega_2=-\Omega$ with

$$\Omega=\sqrt{(\omega_0-\tfrac{1}{2}\omega)^2+\omega'^2} \tag{138.7}$$

with amplitudes A_1, B_1, respectively A_2, B_2:

$$\psi(t)=(A_1\mathrm{e}^{-i\Omega t}+A_2\mathrm{e}^{i\Omega t})\mathrm{e}^{-\frac{i\omega}{2}t}\alpha+(B_1\mathrm{e}^{-i\Omega t}+B_2\mathrm{e}^{i\Omega t})\mathrm{e}^{\frac{i\omega}{2}t}\beta \tag{138.8}$$

and

$$B_{1,2}=A_{1,2}\frac{\pm\Omega-(\omega_0-\frac{1}{2}\omega)}{\omega'}. \tag{138.9}$$

In (138.8) we introduce the initial condition $\psi(0)=\alpha$ or

$$A_1+A_2=1; \qquad B_1+B_2=0, \tag{138.10}$$

which with (138.9) finally leads to

$$\psi(t)=\left\{\cos\Omega t-\frac{\omega_0-\frac{1}{2}\omega}{\Omega}\,i\sin\Omega t\right\}e^{-\frac{i\omega}{2}t}\alpha-\frac{\omega'}{\Omega}\,i\sin\Omega t\,e^{\frac{i\omega}{2}t}\beta. \tag{138.11}$$

The probability of spin flip at time t is therefore

$$P=\left(\frac{\omega'}{\Omega}\right)^2\sin^2\Omega t, \tag{138.12}$$

its time average being independent of time,

$$\overline{P}=\frac{1}{2}\,\frac{\omega'^2}{\Omega^2}=\frac{1}{2}\,\frac{\omega'^2}{(\omega_0-\frac{1}{2}\omega)^2+\omega'^2}. \tag{138.13}$$

If the homogeneous field $\mathcal{H}_0$ and thus, according to (138.4), the Larmor frequency ω_0 is continually varied, the average flip probability becomes a maximum if

$$\omega_0=\frac{1}{2}\,\omega, \quad \text{i.e. } \mathcal{H}_0=\frac{\hbar\omega}{2\mu}. \tag{138.14}$$

We call this the resonance field, $\mathcal{H}_{\mathrm{res}}$, and find

$$\overline{P}=\frac{1}{2}\,\frac{\mathcal{H}'^2}{(\mathcal{H}_0-\mathcal{H}_{\mathrm{res}})^2+\mathcal{H}'^2}. \tag{138.15}$$

At resonance, $\overline{P}=\frac{1}{2}$, independent of the strength of the rotating field $\mathcal{H}'$, the width of the resonance region, however, being determined by $\mathcal{H}'$.

NB. The method may either be used to determine μ from the resonance field strength or, if μ is sufficiently well known, to determine the difference between the field applied and the field acting on the electron inside a molecule. In a similar way, proton resonance may be used to detect molecular structures.

B. Two- and Three-Body Problems

Problem 139. Spin functions for two particles

To determine the spin eigenfunctions for a system of two particles of spin $\frac{1}{2}$ (say, a neutron and a proton) which for a total spin vector operator

$$\boldsymbol{S}=\frac{\hbar}{2}(\boldsymbol{\sigma}_n+\boldsymbol{\sigma}_p) \tag{139.1}$$

simultaneously diagonalize its component S_z and its absolute square, $\boldsymbol{S}^2$.

Solution. Let α_n, β_n be the Hilbert basis vectors for the neutron, and α_p, β_p for the proton. Then the spin function χ of the two-particle system is bound to be of the form

$$\chi = A\,\alpha_n\alpha_p + B\,\alpha_n\beta_p + C\,\beta_n\alpha_p + D\,\beta_n\beta_p. \tag{139.2}$$

From the definition of the spin operators (Problem 129) it follows that

$$\frac{2}{\hbar} S_z\chi = (\sigma_{nz} + \sigma_{pz})\chi = A\,\alpha_n\alpha_p + B\,\alpha_n\beta_p - C\,\beta_n\alpha_p - D\,\beta_n\beta_p$$
$$+ A\,\alpha_n\alpha_p - B\,\alpha_n\beta_p + C\,\beta_n\alpha_p - D\,\beta_n\beta_p.$$

Each of the four terms of χ, Eq. (139.2), therefore is an eigenfunction of S_z, viz.

$$\begin{aligned}
&\alpha_n\alpha_p \quad \text{for the eigenvalue } +2 \text{ of } \tfrac{2}{\hbar}S_z \text{ or } +\hbar \text{ of } S_z;\\
&\alpha_n\beta_p \quad \text{for the eigenvalue } \ \ 0 \text{ of } \tfrac{2}{\hbar}S_z \text{ or } \ \ 0 \text{ of } S_z;\\
&\beta_n\alpha_p \quad \text{for the eigenvalue } \ \ 0 \text{ of } \tfrac{2}{\hbar}S_z \text{ or } \ \ 0 \text{ of } S_z;\\
&\beta_n\beta_p \quad \text{for the eigenvalue } -2 \text{ of } \tfrac{2}{\hbar}S_z \text{ or } -\hbar \text{ of } S_z.
\end{aligned} \tag{139.3}$$

The spin components of $+1, 0, -1$ in units $\hbar$ are in agreement with the half-classical vector model. The two functions $\alpha_n\beta_p$ and $\beta_n\alpha_p$ are still degenerate so that any linear combination of them still belongs to the eigenvalue zero.

We now proceed to investigate the operator

$$\left(\frac{2}{\hbar}\right)^2 \mathbf{S}^2 = \boldsymbol{\sigma}_n^2 + \boldsymbol{\sigma}_p^2 + 2(\boldsymbol{\sigma}_n\cdot\boldsymbol{\sigma}_p) = 6 + 2(\sigma_{nx}\sigma_{px} + \sigma_{ny}\sigma_{py} + \sigma_{nz}\sigma_{pz}).$$

We find

$$\begin{aligned}
\sigma_{nx}\sigma_{px}\chi &= \ \ A\,\beta_n\beta_p + B\,\beta_n\alpha_p + C\,\alpha_n\beta_p + D\,\alpha_n\alpha_p;\\
\sigma_{ny}\sigma_{py}\chi &= -A\,\beta_n\beta_p + B\,\beta_n\alpha_p + C\,\alpha_n\beta_p - D\,\alpha_n\alpha_p;\\
\sigma_{nz}\sigma_{pz}\chi &= \ \ A\,\alpha_n\alpha_p - B\,\alpha_n\beta_p - C\,\beta_n\alpha_p + D\,\beta_n\beta_p
\end{aligned}$$

and thence,

$$\left(\frac{2}{\hbar}\right)^2 \mathbf{S}^2\,\alpha_n\alpha_p = 8\,\alpha_n\alpha_p; \tag{139.4a}$$

$$\left(\frac{2}{\hbar}\right)^2 \mathbf{S}^2(B\,\alpha_n\beta_p + C\,\beta_n\alpha_p) = 4(B+C)(\alpha_n\beta_p + \beta_n\alpha_p); \tag{139.4b}$$

$$\left(\frac{2}{\hbar}\right)^2 \mathbf{S}^2\,\beta_n\beta_p = 8\,\beta_n\beta_p. \tag{139.4c}$$

The functions $\alpha_n \alpha_p$ and $\beta_n \beta_p$ therefore are eigenfunctions to the eigenvalue $2\hbar^2$ of the operator $\boldsymbol{S}^2$. In the usual notation,

$$\boldsymbol{S}^2 \chi = \hbar^2 S(S+1)\chi, \tag{139.5}$$

they belong to the quantum number $S=1$ or, in the half-classical language of the vector model, to the total spin $S=1$ (in units $\hbar$) with its component S_z either $+1$ or -1.

From (139.4b) we construct two more eigenfunctions of $\boldsymbol{S}^2$ with $S_z=0$. Let λ be the still unknown eigenvalue of $\boldsymbol{S}^2/\hbar^2$, then we have

$$(B+C)(\alpha_n \beta_p + \beta_n \alpha_p) = \lambda(B\alpha_n \beta_p + C\beta_n \alpha_p).$$

This yields two linear equations for B and C,

$$(B+C) = \lambda B \quad \text{and} \quad (B+C) = \lambda C.$$

Their determinant must vanish,

$$\begin{vmatrix} 1-\lambda; & 1 \\ 1; & 1-\lambda \end{vmatrix} = 0 \quad \text{or} \quad 1-\lambda = \pm 1.$$

The two eigenfunctions of $\boldsymbol{S}^2$ with $S_z=0$ therefore become

a) for $\lambda=2$: $B=C$; $\quad \chi = \alpha_n \beta_p + \beta_n \alpha_p$; $\quad S=1$; (139.6a)

b) for $\lambda=0$: $B=-C$; $\quad \chi = \alpha_n \beta_p - \beta_n \alpha_p$; $\quad S=0$. (139.6b)

The results have been collected in the following table where the functions have been normalized according to the rules

$$\langle \alpha | \alpha \rangle = \langle \beta | \beta \rangle = 1; \quad \langle \alpha | \beta \rangle = 0.$$

Triplet, $S=1$ (symmetrical spin function)	$S_z=+1$	$\alpha_n \alpha_p$
	0	$\frac{1}{\sqrt{2}}(\alpha_n \beta_p + \beta_n \alpha_p)$
	-1	$\beta_n \beta_p$
Singlet, $S=0$ (antisymmetrical spin function)	$S_z=0$	$\frac{1}{\sqrt{2}}(\alpha_n \beta_p - \beta_n \alpha_p)$

NB. From $(\boldsymbol{\sigma}_n + \boldsymbol{\sigma}_p)^2 = 6 + 2(\boldsymbol{\sigma}_n \cdot \boldsymbol{\sigma}_p)$ it follows that the triplet and singlet spin functions, say χ_t and χ_s, given in the table, are eigenfunctions also of the operator $(\boldsymbol{\sigma}_n \cdot \boldsymbol{\sigma}_p)$ so that

$$(\boldsymbol{\sigma}_n \cdot \boldsymbol{\sigma}_p)\chi_t = \chi_t; \quad (\boldsymbol{\sigma}_n \cdot \boldsymbol{\sigma}_p)\chi_s = -3\chi_s.$$

Of these relations use will be made in the following problem.

Problem 140. Spin-dependent central force between nucleons

In a reasonable approximation, the interaction energy between a neutron and a proton in an S state may be described by a central force, different for symmetrical and antisymmetrical spin states. Such an interaction shall be expressed in terms of a spin-dependent potential

a) using the spin exchange operator Σ_{np},

b) using the spin vectors $\boldsymbol{\sigma}_n$ and $\boldsymbol{\sigma}_p$ of the two particles.

Solution. A central force means that the interaction energy must depend only upon the distance r between the two particles. This energy shall be different for different spin-state symmetry, say, $V_t(r)$ in the triplet case of parallel spins and $V_s(r)$ in the singlet case of antiparallel spins.

a) Let $\chi(s_n, s_p)$ be a two-particle function. Then the spin exchange operator is defined by

$$\Sigma_{np}\chi(s_n, s_p) = \chi(s_p, s_n). \tag{140.1}$$

For the symmetrical triplet state, $\chi_t(s_n, s_p) = \chi_t(s_p, s_n)$, we therefore have

$$\Sigma_{np}\chi_t = \chi_t \tag{140.2a}$$

and for the antisymmetrical singlet state, $\chi_s(s_n, s_p) = -\chi_s(s_p, s_n)$,

$$\Sigma_{np}\chi_s = -\chi_s. \tag{140.2b}$$

Hence both kinds of functions are eigenfunctions of the exchange operator, with its respective eigenvalues $+1$ and -1. As the three triplet and one singlet functions form a complete orthogonal set, Eqs. (140.2a, b) explain the exchange operator completely and uniquely.

An interaction energy

$$V = V_1(r) + V_2(r)\Sigma_{np}$$

yields, according to (140.2a, b),

$$V\chi_t = (V_1 + V_2)\chi_t; \qquad V\chi_s = (V_1 - V_2)\chi_s$$

so that

$$V_t = V_1 + V_2 \quad \text{and} \quad V_s = V_1 - V_2$$

are the interactions in the triplet and singlet states, respectively. Thence,

$$V = \tfrac{1}{2}(V_t + V_s) + \tfrac{1}{2}(V_t - V_s)\Sigma_{np}. \tag{140.3}$$

b) At the end of the preceding problem we have shown that the spin functions χ_t and χ_s are eigenfunctions also of the operator $(\boldsymbol{\sigma}_n \cdot \boldsymbol{\sigma}_p)$, viz.

$$(\boldsymbol{\sigma}_n \cdot \boldsymbol{\sigma}_p)\chi_t = \chi_t; \qquad (\boldsymbol{\sigma}_n \cdot \boldsymbol{\sigma}_p)\chi_s = -3\chi_s. \tag{140.4}$$

It follows that Σ_{np} may be expressed linearly by $(\boldsymbol{\sigma}_n \cdot \boldsymbol{\sigma}_p)$. Indeed,

$$\Sigma_{np} = \tfrac{1}{2}(1 + (\boldsymbol{\sigma}_n \cdot \boldsymbol{\sigma}_p)) \tag{140.5}$$

leads to the wanted eigenvalues (140.2a, b). Again, since there exist no other spin functions of the two-nucleon system, both operators are completely described by the eigenvalue problems (140.2a, b) and (140.4) so that (140.5) holds in full generality.

Replacing Σ_{np} in (140.3) according to (140.5), we arrive at the result

$$V = \tfrac{1}{4}(3V_t + V_s) + \tfrac{1}{4}(V_t - V_s)(\boldsymbol{\sigma}_n \cdot \boldsymbol{\sigma}_p). \tag{140.6}$$

Problem 141. Powers of spin operators

To show that the operator $(\boldsymbol{\sigma}_1 \cdot \boldsymbol{\sigma}_2)^n$ for two particles 1 and 2 can be linearly expressed by $(\boldsymbol{\sigma}_1 \cdot \boldsymbol{\sigma}_2)$.

Solution. The operator $(\boldsymbol{\sigma}_1 \cdot \boldsymbol{\sigma}_2)$ is completely described by the two eigenvalue relations

$$(\boldsymbol{\sigma}_1 \cdot \boldsymbol{\sigma}_2)\chi_t = \chi_t; \quad (\boldsymbol{\sigma}_1 \cdot \boldsymbol{\sigma}_2)\chi_s = -3\chi_s \tag{141.1}$$

for the three triplet and one singlet spin functions since these form a complete orthogonal set. It therefore suffices to investigate the application of $(\boldsymbol{\sigma}_1 \cdot \boldsymbol{\sigma}_2)^n$ to these four spin functions. By iteration of (141.1) we get at once,

$$(\boldsymbol{\sigma}_1 \cdot \boldsymbol{\sigma}_2)^n \chi_t = \chi_t; \quad (\boldsymbol{\sigma}_1 \cdot \boldsymbol{\sigma}_2)^n \chi_s = (-3)^n \chi_s. \tag{141.2}$$

It follows that

$$(\boldsymbol{\sigma}_1 \cdot \boldsymbol{\sigma}_2)^n = A + B(\boldsymbol{\sigma}_1 \cdot \boldsymbol{\sigma}_2) \tag{141.3}$$

can be linearly expressed by $(\boldsymbol{\sigma}_1 \cdot \boldsymbol{\sigma}_2)$. Putting (141.3) in (141.2) we find, according to (141.1),

$$(A+B)\chi_t = \chi_t; \quad (A-3B)\chi_s = (-3)^n \chi_s.$$

It follows that

$$A + B = 1; \quad A - 3B = (-3)^n$$

or that

$$A = \tfrac{1}{4}[3 + (-3)^n]; \quad B = \tfrac{1}{4}[1 - (-3)^n]. \tag{141.4}$$

Thus, e. g., we get

$$(\boldsymbol{\sigma}_1 \cdot \boldsymbol{\sigma}_2)^2 = 3 - 2(\boldsymbol{\sigma}_1 \cdot \boldsymbol{\sigma}_2); \quad (\boldsymbol{\sigma}_1 \cdot \boldsymbol{\sigma}_2)^3 = -6 + 7(\boldsymbol{\sigma}_1 \cdot \boldsymbol{\sigma}_2).$$

NB. The representation (140.6) of a spin-dependent force in the preceding problem is unique, because replacing it by a power series in $(\boldsymbol{\sigma}_n \cdot \boldsymbol{\sigma}_p)$ would change nothing in the result. – The solution of the problem becomes even simpler if powers of the exchange operator Σ_{12} are considered.

Problem 142. Angular momentum eigenfunctions for two spin particles

To construct the triplet state eigenfunctions of the operators J_z and $\boldsymbol{J}^2$ for a system of two spin $\frac{1}{2}$ particles. Use $\hbar=1$ as unit of angular momentum.

Solution. Any eigenfunction for a triplet state can certainly be written in the form

$$\psi = \sum_{l=0}^{\infty} \{f_l(r)\, Y_{l,m-1}\, \chi_{1,1} + g_l(r)\, Y_{l,m}\, \chi_{1,0} + h_l(r)\, Y_{l,m+1}\, \chi_{1,-1}\}, \tag{142.1}$$

each of the three possible spin functions multiplying by a space factor formally written as a spherical harmonics expansion, with only the one restriction to generality that the second subscript of each Y is so chosen as to make

$$J_z \psi = m \psi, \tag{142.2}$$

thus making ψ an eigenfunction of the operator J_z.

Let us now apply

$$\boldsymbol{J}^2 = \{\boldsymbol{L} + \tfrac{1}{2}(\boldsymbol{\sigma}_1 + \boldsymbol{\sigma}_2)\}^2 = \boldsymbol{L}^2 + \boldsymbol{L}\cdot(\boldsymbol{\sigma}_1 + \boldsymbol{\sigma}_2) + \tfrac{1}{4}[6 + 2(\boldsymbol{\sigma}_1 \cdot \boldsymbol{\sigma}_2)]$$

to the function (142.1). It is then suitable to define the operators

$$\sigma_+ = \sigma_x + i\sigma_y \quad \text{and} \quad \sigma_- = \sigma_x - i\sigma_y \tag{142.3}$$

in analogy to (cf. Problem 56)

$$L_+ = L_x + i L_y \quad \text{and} \quad L_- = L_x - i L_y. \tag{142.4}$$

Then $\boldsymbol{J}^2$ may equally well be written

$$\boldsymbol{J}^2 = \boldsymbol{L}^2 + \tfrac{1}{2}(L_+ \sigma_- + L_- \sigma_+) + L_z \sigma_z + \tfrac{3}{2} + \tfrac{1}{2}(\boldsymbol{\sigma}_1 \cdot \boldsymbol{\sigma}_2), \tag{142.5}$$

where $\sigma_\pm = \sigma_{1\pm} + \sigma_{2\pm}$ and $\sigma_z = \sigma_{1z} + \sigma_{2z}$. Application of these operators to the triplet spin functions yields

$$\sigma_+ \begin{pmatrix} \chi_{1,1} \\ \chi_{1,0} \\ \chi_{1,-1} \end{pmatrix} = 2\sqrt{2} \begin{pmatrix} 0 \\ \chi_{1,1} \\ \chi_{1,0} \end{pmatrix}; \qquad \sigma_- \begin{pmatrix} \chi_{1,1} \\ \chi_{1,0} \\ \chi_{1,-1} \end{pmatrix} = 2\sqrt{2} \begin{pmatrix} \chi_{1,0} \\ \chi_{1,-1} \\ 0 \end{pmatrix};$$

$$\sigma_z \begin{pmatrix} \chi_{1,1} \\ \chi_{1,0} \\ \chi_{1,-1} \end{pmatrix} = 2 \begin{pmatrix} \chi_{1,1} \\ 0 \\ -\chi_{1,-1} \end{pmatrix}. \tag{142.6}$$

Hence, we obtain by straightforward computation

$$\left.\begin{aligned} \boldsymbol{J}^2\chi_{1,1} &= (L^2+2+2L_z)\chi_{1,1}+\sqrt{2}\,L_+\chi_{1,0};\\ \boldsymbol{J}^2\chi_{1,0} &= \sqrt{2}\,L_-\chi_{1,1}+(L^2+2)\chi_{1,0}+\sqrt{2}\,L_+\chi_{1,-1};\\ \boldsymbol{J}^2\chi_{1,-1} &= \sqrt{2}\,L_-\chi_{1,0}+(L^2+2-2L_z)\chi_{1,-1}. \end{aligned}\right\} \tag{142.7}$$

Using (142.7) and the well-known relations [cf. (56.14)]

$$\left.\begin{aligned} L_+ Y_{l,m} &= -\sqrt{(l+m+1)(l-m)}\,Y_{l,m+1};\\ L_- Y_{l,m} &= -\sqrt{(l+m)(l-m+1)}\,Y_{l,m-1};\\ L_z Y_{l,m} &= m\,Y_{l,m} \end{aligned}\right\} \tag{142.8}$$

we get

$$\begin{aligned} \boldsymbol{J}^2\psi = \sum_{l=0}^{\infty} \{&[f_l(l(l+1)+2m)-g_l\sqrt{2(l+m)(l-m+1)}]\,Y_{l,m-1}\chi_{1,1}\\ &+[-f_l\sqrt{2(l+m)(l-m+1)}+g_l(l(l+1)+2)-h_l\sqrt{2(l+m+1)(l-m)}]\,Y_{l,m}\chi_{1,0}\\ &+[-g_l\sqrt{2(l+m+1)(l-m)}+h_l(l(l+1)-2m)]\,Y_{l,m+1}\chi_{1,-1}\}. \end{aligned} \tag{142.9}$$

In order to make ψ an eigenfunction of $\boldsymbol{J}^2$ this must be

$$= j(j+1)\psi,$$

which gives three independent linear equations for f_l, g_l, h_l showing that these three radial functions must be of the same form but with different amplitudes:

$$f_l = A_l F_l(r); \qquad g_l = B_l F_l(r); \qquad h_l = C_l F_l(r) \tag{142.10}$$

where the constant amplitude factors A_l, B_l, C_l may be determined from the following set of linear equations

$$\begin{aligned} [l(l+1)+2m-j(j+1)]A_l - \sqrt{2(l+m)(l-m+1)}\,B_l &= 0;\\ -\sqrt{2(l+m)(l-m+1)}\,A_l + [l(l+1)+2-j(j+1)]B_l - \sqrt{2(l+m+1)(l-m)}\,C_l &= 0;\\ -\sqrt{2(l+m+1)(l-m)}\,B_l + [l(l+1)-2m-j(j+1)]C_l &= 0. \end{aligned} \tag{142.11}$$

The determinant of these equations must vanish; if it is expanded, one finds that it becomes independent of m and has the form

$$[l(l+1)-j(j+1)]\{[l(l+1)-j(j+1)]^2+2[l(l+1)-j(j+1)]-4l(l+1)\}=0.$$

This leads to the (positive) solutions

$$j = l+1; \qquad j = l; \qquad j = l-1 \tag{142.12}$$

for which the amplitudes A_l, B_l, C_l then may be determined according to (142.11), except for a common normalization factor. Choosing arbitrarily the normalization

$$A_l^2 + B_l^2 + C_l^2 = 1 \tag{142.13}$$

we arrive at the results compiled in the table.

j	A_l	B_l	C_l
$l+1$	$-\sqrt{\frac{(l+m+1)(l+m)}{2(l+1)(2l+1)}}$	$\sqrt{\frac{(l+m+1)(l-m+1)}{(l+1)(2l+1)}}$	$-\sqrt{\frac{(l-m+1)(l-m)}{2(l+1)(2l+1)}}$
l	$\sqrt{\frac{(l-m+1)(l+m)}{2l(l+1)}}$	$\frac{m}{\sqrt{l(l+1)}}$	$-\sqrt{\frac{(l+m+1)(l-m)}{2l(l+1)}}$
$l-1$	$\sqrt{\frac{(l-m+1)(l-m)}{2l(2l+1)}}$	$\sqrt{\frac{(l+m)(l-m)}{l(2l+1)}}$	$\sqrt{\frac{(l+m+1)(l+m)}{2l(2l+1)}}$

Problem 143. Tensor force operator

The so-called tensor force between two particles 1 and 2 of spin $\frac{1}{2}$ is defined by the interaction energy

$$V = W(r) T_{12}$$

with the operator

$$T_{12} = \frac{(\boldsymbol{\sigma}_1 \cdot \boldsymbol{r})(\boldsymbol{\sigma}_2 \cdot \boldsymbol{r})}{r^2} - \frac{1}{3}(\boldsymbol{\sigma}_1 \cdot \boldsymbol{\sigma}_2). \tag{143.1}$$

To apply this operator to the spin eigenfunctions of the two-particle system.

Solution. The operator T_{12} is invariant under spin exchange. It therefore keeps the symmetry of the spin functions. Since there exists only one antisymmetrical spin function, $\chi_{0,0}$, this then must be an eigenfunction of the operator T_{12}. The three symmetrical spin functions, however, may be mixed up by its application. Since T_{12} is invariant also under exchange of the particle coordinates, i. e. under parity transformation, it will conserve parity. This means that only spherical harmonics of even order will enter the expression $T_{12}\chi$. No higher angular momenta than $l=2$ are to be expected.

In order to get details let us first apply the one-particle operator $(\boldsymbol{\sigma} \cdot \boldsymbol{r})$ to the one-particle spin functions:

$$(\boldsymbol{\sigma} \cdot \boldsymbol{r}) \begin{pmatrix} \alpha \\ \beta \end{pmatrix} = (\sigma_x x + \sigma_y y + \sigma_z z) \begin{pmatrix} \alpha \\ \beta \end{pmatrix} = \begin{pmatrix} (x+iy)\beta + z\alpha \\ (x-iy)\alpha - z\beta \end{pmatrix}. \tag{143.2}$$

It then follows directly that

$$(\boldsymbol{\sigma}_1\cdot\boldsymbol{r})(\boldsymbol{\sigma}_2\cdot\boldsymbol{r})\begin{pmatrix}\alpha_1\alpha_2\\ \alpha_1\beta_2\\ \beta_1\alpha_2\\ \beta_1\beta_2\end{pmatrix}=\begin{pmatrix}[(x+iy)\beta_1+z\alpha_1]\,[(x+iy)\beta_2+z\alpha_2]\\ [(x+iy)\beta_1+z\alpha_1]\,[(x-iy)\alpha_2-z\beta_2]\\ [(x-iy)\alpha_1-z\beta_1]\,[(x+iy)\beta_2+z\alpha_2]\\ [(x-iy)\alpha_1-z\beta_1]\,[(x-iy)\alpha_2-z\beta_2]\end{pmatrix}.$$

With

$$(x\pm iy)^2=r^2\sin^2\vartheta\,\mathrm{e}^{\pm 2i\varphi};\qquad x^2+y^2=r^2\sin^2\vartheta;$$
$$(x\pm iy)z=r^2\sin\vartheta\cos\vartheta\,\mathrm{e}^{\pm i\varphi};\qquad z^2=r^2\cos^2\vartheta,$$

this leads to

$$\frac{(\boldsymbol{\sigma}_1\cdot\boldsymbol{r})(\boldsymbol{\sigma}_2\cdot\boldsymbol{r})}{r^2}\begin{pmatrix}\alpha_1\alpha_2\\ \alpha_1\beta_2\\ \beta_1\alpha_2\\ \beta_1\beta_2\end{pmatrix}$$
$$=\begin{pmatrix}\cos^2\vartheta\,\alpha_1\alpha_2+\sin\vartheta\cos\vartheta\,\mathrm{e}^{i\varphi}(\alpha_1\beta_2+\beta_1\alpha_2)+\sin^2\vartheta\,\mathrm{e}^{2i\varphi}\beta_1\beta_2\\ \sin\vartheta\cos\vartheta\,\mathrm{e}^{-i\varphi}\alpha_1\alpha_2-\cos^2\vartheta\,\alpha_1\beta_2+\sin^2\vartheta\,\beta_1\alpha_2-\sin\vartheta\cos\vartheta\,\mathrm{e}^{i\varphi}\beta_1\beta_2\\ \sin\vartheta\cos\vartheta\,\mathrm{e}^{-i\varphi}\alpha_1\alpha_2+\sin^2\vartheta\,\alpha_1\beta_2-\cos^2\vartheta\,\beta_1\alpha_2-\sin\vartheta\cos\vartheta\,\mathrm{e}^{i\varphi}\beta_1\beta_2\\ \sin^2\vartheta\,\mathrm{e}^{-2i\varphi}\alpha_1\alpha_2-\sin\vartheta\cos\vartheta\,\mathrm{e}^{-i\varphi}(\alpha_1\beta_2+\beta_1\alpha_2)+\cos^2\vartheta\,\beta_1\beta_2\end{pmatrix}.$$

Using the notation χ_{S,m_S}, i. e. for the triplet

$$\alpha_1\alpha_2=\chi_{1,1};\qquad \frac{1}{\sqrt{2}}(\alpha_1\beta_2+\beta_1\alpha_2)=\chi_{1,0};\qquad \beta_1\beta_2=\chi_{1,-1}\tag{143.3}$$

and for the singlet

$$\chi_{0,0}=\frac{1}{\sqrt{2}}(\alpha_1\beta_2-\beta_1\alpha_2),\tag{143.4}$$

we then have for the symmetrical functions of the triplet

$$\frac{(\boldsymbol{\sigma}_1\cdot\boldsymbol{r})(\boldsymbol{\sigma}_2\cdot\boldsymbol{r})}{r^2}\begin{pmatrix}\chi_{1,1}\\ \chi_{1,0}\\ \chi_{1,-1}\end{pmatrix}\tag{143.5}$$
$$=\begin{pmatrix}\cos^2\vartheta\,\chi_{1,1}+\sqrt{2}\sin\vartheta\cos\vartheta\,\mathrm{e}^{i\varphi}\chi_{1,0}+\sin^2\vartheta\,\mathrm{e}^{2i\varphi}\chi_{1,-1}\\ \sqrt{2}\sin\vartheta\cos\vartheta\,\mathrm{e}^{-i\varphi}\chi_{1,1}+(\sin^2\vartheta-\cos^2\vartheta)\chi_{1,0}-\sqrt{2}\sin\vartheta\cos\vartheta\,\mathrm{e}^{i\varphi}\chi_{1,-1}\\ \sin^2\vartheta\,\mathrm{e}^{-2i\varphi}\chi_{1,1}-\sqrt{2}\sin\vartheta\cos\vartheta\,\mathrm{e}^{-i\varphi}\chi_{1,0}+\cos^2\vartheta\,\chi_{1,-1}\end{pmatrix}$$

and for the antisymmetrical singlet function,

$$\frac{(\boldsymbol{\sigma}_1\cdot\boldsymbol{r})(\boldsymbol{\sigma}_2\cdot\boldsymbol{r})}{r^2}\chi_{0,0}=-\chi_{0,0}.\tag{143.6}$$

The second term of T_{12} has already been discussed in Problem 140:

$$(\boldsymbol{\sigma}_1 \cdot \boldsymbol{\sigma}_2)\begin{pmatrix}\chi_{1,m_s}\\ \chi_{0,0}\end{pmatrix} = \begin{pmatrix}\chi_{1,m_s}\\ -3\chi_{0,0}\end{pmatrix}. \tag{143.7}$$

Combination of Eqs. (143.6) and (143.7) then gives at once

$$T_{12}\chi_{0,0} = 0. \tag{143.8}$$

The tensor operator therefore cannot contribute any dynamical term to a spin singlet state.

There remains further discussion of the triplet. Introducing normalized spherical harmonics according to the definitions of the table of Problem 67, we obtain from (143.5) and (143.7),

$$\begin{aligned}
T_{12}\chi_{1,1} &= \frac{2}{3}\sqrt{\frac{4\pi}{5}}\,(Y_{2,0}\chi_{1,1} + \sqrt{3}\,Y_{2,1}\chi_{1,0} + \sqrt{6}\,Y_{2,2}\chi_{1,-1});\\
T_{12}\chi_{1,0} &= \frac{2}{3}\sqrt{\frac{4\pi}{5}}\,(-\sqrt{3}\,Y_{2,-1}\chi_{1,1} - 2\,Y_{2,0}\chi_{1,0} - \sqrt{3}\,Y_{2,1}\chi_{1,-1});\\
T_{12}\chi_{1,-1} &= \frac{2}{3}\sqrt{\frac{4\pi}{5}}\,(\sqrt{6}\,Y_{2,-2}\chi_{1,1} + \sqrt{3}\,Y_{2,-1}\chi_{1,0} + Y_{2,0}\chi_{1,-1}).
\end{aligned} \tag{143.9}$$

These formulae not only show spin exchange symmetry and parity to be conserved, but also the z component of the total angular momentum. The orbital momentum, however, as well as its z component, are not good quantum numbers in a two-particle system with tensor interaction.

Problem 144. Deuteron with tensor interaction

The interaction between a proton and a neutron consists in part of a central force, and in part of a tensor force,

$$V = V_c(r) + V_t(r)\,T_{pn}. \tag{144.1}$$

The deuteron groundstate therefore is a mixture of S and D state. The eigenfunction shall be constructed, except for radial S and D factors for which a set of two coupled differential equations shall be derived, under the assumption of nuclear spin orientation in z direction.

Solution. With the nuclear spin $i=1$ (in units $\hbar$) and its component in z direction also 1, we have for the most general $S-D$ mixture,

$$\psi = f(r)\,Y_{0,0}\chi_{1,1} + g(r)\{Y_{2,0}\chi_{1,1} + \lambda\,Y_{2,1}\chi_{1,0} + \mu\,Y_{2,2}\chi_{1,-1}\} \tag{144.2}$$

with the constants λ and μ to be adjusted so that

$$\boldsymbol{I}^2\psi = 2\psi. \tag{144.3}$$

Here $\boldsymbol{I}$ is the operator of total angular momentum (nuclear spin). According to the preceding problem we have

$$\begin{aligned} \boldsymbol{I}^2 \psi = &\{\chi_{1,1}(L^2+2+2L_z)+\chi_{1,0}\sqrt{2}L_+\}(f Y_{0,0}+g Y_{2,0}) \\ &+\lambda\{\chi_{1,1}\sqrt{2}L_- +\chi_{1,0}(L^2+2)+\chi_{1,-1}\sqrt{2}L_+\} g Y_{2,1} \\ &+\mu\{\chi_{1,0}\sqrt{2}L_- +\chi_{1,-1}(L^2+2-2L_z)\} g Y_{2,2} \end{aligned}$$

and further

$$\begin{aligned} \boldsymbol{I}^2 \psi = &\chi_{1,1}[2f Y_{0,0}+(8-2\sqrt{3}\lambda) g Y_{2,0}] \\ &+\chi_{1,0}(-2\sqrt{3}+8\lambda-2\sqrt{2}\mu) g Y_{2,1}+\chi_{1,-1}(-2\sqrt{2}\lambda+4\mu) g Y_{2,2}. \end{aligned}$$

The last relation satisfies (144.3) if

$$\lambda=\sqrt{3}, \qquad \mu=\sqrt{6}, \tag{144.4}$$

so that the angular momentum eigenfunction becomes

$$\psi = f(r) Y_{0,0}\chi_{1,1}+g(r)\{Y_{2,0}\chi_{1,1}+\sqrt{3}\,Y_{2,1}\chi_{1,0}+\sqrt{6}\,Y_{2,2}\chi_{1,-1}\}. \tag{144.5}$$

The curly bracket in (144.5) is the same combination of spherical harmonics and spin functions as was obtained in the first line of Eq. (143.9) of the preceding problem, so that we may write in a more compact form:

$$\psi = \frac{1}{\sqrt{4\pi}}\{f(r)+\tfrac{3}{2}\sqrt{5}\,g(r)\,T_{pn}\}\chi_{1,1}. \tag{144.6}$$

Let us now normalize this function. From (144.5) there follows at once

$$\int_0^\infty dr\, r^2[f^2+10g^2]=1.$$

It will be suitable to put

$$f(r)=\psi_S(r)\cos\omega; \qquad g(r)=\frac{1}{\sqrt{10}}\psi_D(r)\sin\omega \tag{144.7}$$

so that

$$\int_0^\infty dr\, r^2\psi_S^2=1; \qquad \int_0^\infty dr\, r^2\psi_D^2=1 \tag{144.8}$$

and

$$\psi=\frac{1}{\sqrt{4\pi}}\left\{\psi_S(r)\cos\omega+\frac{3}{2\sqrt{2}}\psi_D(r)\sin\omega\, T_{pn}\right\}\chi_{1,1}. \tag{144.9}$$

Now, the Schrödinger equation for the relative motion (with $\hbar=1$, $m_p=m_n=1$, reduced mass $=\frac{1}{2}$, cf. Problem 150),

$$(-\nabla^2+V_c+V_t T_{pn}-E)\psi=0,$$

has to be satisfied by (144.9):

$$\left\{(-\nabla^2+V_c-E)\cos\omega\,\psi_S+\left[\frac{3}{2\sqrt{2}}(-\nabla^2+V_c-E)\sin\omega\,\psi_D+V_t\cos\omega\,\psi_S\right\}T_{pn}\right.$$
$$\left.+\frac{3}{2\sqrt{2}}V_t\sin\omega\,\psi_D\,T_{pn}^2\right\}\chi_{1,1}=0. \tag{144.10}$$

Here, if applied to a triplet spin function, χ_t, the operator T_{pn}^2 may be linearly expressed by T_{pn}, viz.

$$T_{pn}^2\chi_t=(\tfrac{8}{9}-\tfrac{2}{3}T_{pn})\chi_t. \tag{144.11}$$

This can easily be shown using the identity

$$S\begin{pmatrix}\alpha\\ \beta\end{pmatrix}=\frac{1}{r}(\boldsymbol{\sigma}\cdot\boldsymbol{r})\begin{pmatrix}\alpha\\ \beta\end{pmatrix}=\begin{pmatrix}\cos\vartheta; & \sin\vartheta\,e^{i\varphi}\\ \sin\vartheta\,e^{-i\varphi}; & -\cos\vartheta\end{pmatrix}\begin{pmatrix}\alpha\\ \beta\end{pmatrix}$$

for one-particle spin states. It then follows that the square of this operator, $S^2=1$, so that

$$S_{pn}^2=\left\{\frac{1}{r^2}(\boldsymbol{\sigma}_p\cdot\boldsymbol{r})(\boldsymbol{\sigma}_n\cdot\boldsymbol{r})\right\}^2=1\,.$$

Since we already know (p. 24) that $(\boldsymbol{\sigma}_p\cdot\boldsymbol{\sigma}_n)\chi_t=\chi_t$ we find

$$T_{pn}^2=(S_{pn}-\tfrac{1}{3})^2=1-\tfrac{2}{3}S_{pn}+\tfrac{1}{9}=\tfrac{10}{9}-\tfrac{2}{3}(T_{pn}+\tfrac{1}{3})$$

in perfect agreement with (144.11).

Eq. (144.10) may now be written

$$\left\{\cos\omega(-\nabla^2+V_c-E)\psi_S+\frac{2\sqrt{2}}{3}\sin\omega\,V_t\,\psi_D\right\}\chi_{1,1}$$
$$+\left\{\sin\omega\left[\frac{3}{2\sqrt{2}}(-\nabla^2+V_c-E)-\frac{1}{\sqrt{2}}V_t\right]\psi_D+\cos\omega\,V_t\,\psi_S\right\}T_{pn}\chi_{1,1}=0. \tag{144.12}$$

The operator T_{pn} in the second line, when applied to $\chi_{1,1}$, yields only terms with $l=2$ orthogonal to those with $l=0$ in the first line. We may therefore decompose (144.12) into two coupled radial equations, viz.

$$\cos\omega\left[\psi_S''+\frac{2}{r}\psi_S'+(E-V_c)\psi_S\right]-\sin\omega\cdot\frac{2\sqrt{2}}{3}V_t\psi_D=0 \tag{144.13}$$

and

$$\sin\omega\left[\psi_D''+\frac{2}{r}\psi_D'-\frac{6}{r^2}\psi_D+\left(E-V_c+\frac{2}{3}V_t\right)\psi_D\right]-\cos\omega\cdot\frac{2\sqrt{2}}{3}V_t\psi_S=0. \tag{144.14}$$

This is the set of differential equations required.

Problem 145. Electrical quadrupole and magnetic dipole moments of deuteron

Given the deuteron wave function determined in the preceding problem.

a) The electrical quadrupole moment of the deuteron shall be expressed in terms of the two integrals

$$A = \int_0^\infty dr\, r^4 \psi_S \psi_D; \qquad B = \int_0^\infty dr\, r^4 \psi_D^2. \tag{145.1}$$

b) The expectation value of the magnetic dipole moment shall be determined.

Solution. a) The quadrupole tensor (cf. Problem 61) can be defined by

$$Q_{ik} = \tfrac{1}{4}(3x_i x_k - r^2 \delta_{ik}).$$

In the original definition of this tensor, the factor $\frac{1}{4}$ on the right-hand side had not been used. It occurs in the present problem in consequence of $\frac{1}{2}\boldsymbol{r}$ being the proton coordinate about the centre of mass, and only the proton contributing to the quadrupole moment since the neutron carries no electric charge. The deuteron charge distribution in the state $M_I = I$ being rotationally symmetrical about the z axis, averaging of the tensor elements over the angle φ leads to the relations

$$\bar{Q}_{xy} = \bar{Q}_{yz} = \bar{Q}_{zx} = 0; \qquad \bar{Q}_{xx} = \bar{Q}_{yy} = -\tfrac{1}{2}\bar{Q}_{zz}. \tag{145.2}$$

We therefore need only evaluate the expectation value of the operator Q_{zz}, viz.

$$\begin{aligned}\langle Q_{zz}\rangle &= \sum_{\text{spin}} \int d\tau\, Q_{zz} |\psi|^2 \\ &= \frac{1}{4} \sum_{\text{spin}} \int d\tau\, r^2 (3\cos^2\vartheta - 1)|\psi|^2 \\ &= \frac{1}{2}\sqrt{\frac{4\pi}{5}} \sum_{\text{spin}} \int d\tau\, r^2\, Y_{2,0} |\psi|^2 .\end{aligned} \tag{145.3}$$

With the deuteron wave function determined in the preceding problem, viz.

$$\begin{aligned}\psi = \cos\omega\, \psi_S(r)\, Y_{0,0}\, \chi_{1,1} + \sin\omega \frac{1}{\sqrt{10}} \psi_D(r) \{ Y_{2,0}\, \chi_{1,1} + \\ + \sqrt{3}\, Y_{2,1}\, \chi_{1,0} + \sqrt{6}\, Y_{2,2}\, \chi_{1,-1} \},\end{aligned} \tag{145.4}$$

this yields by spin summation,

$$\langle Q_{zz}\rangle = \frac{1}{2}\sqrt{\frac{4\pi}{5}}\int_0^\infty dr\, r^4 \oint d\Omega\, Y_{2,0}\cdot\left\{\left|\cos\omega\, \psi_S\, Y_{0,0} + \sin\omega \frac{\psi_D}{\sqrt{10}}\, Y_{2,0}\right|^2 + \right.$$

$$\left. + \sin^2\omega \frac{\psi_D^2}{10}(3|Y_{2,1}|^2 + 6|Y_{2,2}|^2)\right\}.$$

The term with ψ_S^2 vanishes in consequence of the orthogonality of the spherical harmonics. With the product $\psi_S\,\psi_D$ the obvious integral

$$\oint d\Omega\, Y_{0,0}|Y_{2,0}|^2 = \frac{1}{\sqrt{4\pi}}$$

is coupled. It is a little more laborious to evaluate the three remaining integrals occurring with ψ_D^2, viz.

$$\left.\begin{aligned} \oint d\Omega\, Y_{2,0}|Y_{2,0}|^2 &= \frac{2}{7}\sqrt{\frac{5}{4\pi}}, \\ \oint d\Omega\, Y_{2,0}|Y_{2,1}|^2 &= \frac{1}{7}\sqrt{\frac{5}{4\pi}}, \\ \oint d\Omega\, Y_{2,0}|Y_{2,2}|^2 &= -\frac{2}{7}\sqrt{\frac{5}{4\pi}}. \end{aligned}\right\} \tag{145.5}$$

Assembling all these details, we finally arrive at the simple formula

$$\langle Q_{zz}\rangle = \frac{1}{5\sqrt{2}} A \cos\omega \sin\omega - \frac{1}{20} B \sin^2\omega\,. \tag{145.6}$$

If the admixture of D to S state is a small percentage, the parameter ω is small and the second, negative term in (145.6) represents a small correction only to the first, positive part. Therefore, $\langle Q_{zz}\rangle$ is positive so that the deuteron has an oblong shape in z direction. This is borne out by experiment.

b) The magnetic dipole operator consists of a spin part, $\mu_p\sigma_{pz} + \mu_n\sigma_{nz}$, and of an orbital momentum part to which only the proton contributes. The orbital momentum component L_z for the two-particle problem is given by

$$L_z = \frac{\hbar}{i}\left(\frac{\partial}{\partial\varphi_p} + \frac{\partial}{\partial\varphi_n}\right)$$

of which only the first part will contribute to the magnetic moment whereas both terms contribute equal parts to the orbital momentum about the center of mass. Therefore only $\frac{1}{2}L_z$ enters the orbital part of the magnetic moment,

$$\mu_{\text{orbit}} = \frac{e}{2mc} \cdot \frac{1}{2} L_z .$$

The expectation value of the z component of the magnetic moment thus becomes

$$\langle \mu \rangle = \sum_{\text{spin}} \int d\tau\, \psi^* (\mu_p \sigma_{pz} + \mu_n \sigma_{nz} + \mu_{\text{orbit}})\, \psi , \tag{145.7}$$

and the expectation values of x and y components vanish.

Application of the operators σ_{pz} and σ_{nz} to the triplet spin functions yields the relations,

$$\begin{aligned} &\sigma_{pz}\chi_{1,1} = \chi_{1,1}; &&\sigma_{nz}\chi_{1,1} = \chi_{1,1};\\ &\sigma_{pz}\chi_{1,0} = \chi_{0,0}; &&\sigma_{nz}\chi_{1,0} = -\chi_{0,0};\\ &\sigma_{pz}\chi_{1,-1} = -\chi_{1,-1}; &&\sigma_{nz}\chi_{1,-1} = -\chi_{1,-1} . \end{aligned} \tag{145.8}$$

If, therefore, we write a triplet wave function briefly in the form

$$\psi = u\chi_{1,1} + v\chi_{1,0} + w\chi_{1,-1}$$

we get

$$\begin{aligned} \langle \mu \rangle = \sum_{\text{spin}} \int d\tau & (u^*\chi_{1,1} + v^*\chi_{1,0} + w^*\chi_{1,-1}) \\ & \times \{\mu_p(u\chi_{1,1} + v\chi_{0,0} - w\chi_{1,-1}) \\ & + \mu_n(u\chi_{1,1} - v\chi_{0,0} - w\chi_{1,-1}) \\ & + \mu_{\text{orbit}}(u\chi_{1,1} + v\chi_{1,0} + w\chi_{1,-1})\} \end{aligned}$$

and, by using the orthonormality of the spin functions,

$$\langle \mu \rangle = \int d\tau \{u^*(\mu_p + \mu_n + \mu_{\text{orbit}})u + v^*\mu_{\text{orbit}}v + w^*(-\mu_p - \mu_n + \mu_{\text{orbit}})w\} .$$

Now we have (cf. (144.5))

$$u = f\,Y_{0,0} + g\,Y_{2,0}; \quad v = \sqrt{3}\,g\,Y_{2,1}; \quad w = \sqrt{6}\,g\,Y_{2,2}$$

and therefore

$$L_z u = 0; \quad L_z v = \hbar v; \quad L_z w = 2\hbar w$$

so that

$$\begin{aligned} \langle \mu \rangle &= \int d\tau \left\{ (\mu_p + \mu_n)(u^* u - w^* w) + \frac{e\hbar}{4mc}(v^* v + 2w^* w) \right\} \\ &= (\mu_p + \mu_n) \int_0^\infty dr\, r^2 (f^2 + g^2 - 6g^2) + \frac{e\hbar}{4mc} \int_0^\infty dr\, r^2 (3g^2 + 2\cdot 6g^2) . \end{aligned}$$

Measuring $\langle\mu\rangle$ in units of nuclear magnetons, $e\hbar/(2mc)$, and replacing f and g by the normalized functions ψ_S and ψ_D (144.7), we finally arrive at

$$\langle\mu\rangle=(\mu_p+\mu_n)-\tfrac{3}{2}\sin^2\omega(\mu_p+\mu_n-\tfrac{1}{2}). \tag{145.9}$$

The admixture of D state therefore only causes a second-order correction in the magnetic moment.

Problem 146. Spin functions of three particles

To construct the eigenfunctions of S_z and S^2 for a system of three particles of spin $\frac{1}{2}$.

Solution. The total spin vector operator of the system is now

$$\boldsymbol{S}=\frac{\hbar}{2}(\boldsymbol{\sigma}_1+\boldsymbol{\sigma}_2+\boldsymbol{\sigma}_3). \tag{146.1}$$

Its z component apparently has the following eigenfunctions:

$$\begin{aligned}\chi(\tfrac{3}{2})&=\alpha_1\alpha_2\alpha_3;\\ \chi(\tfrac{1}{2})&=A\alpha_1\alpha_2\beta_3+B\alpha_1\beta_2\alpha_3+C\beta_1\alpha_2\alpha_3;\\ \chi(-\tfrac{1}{2})&=A'\beta_1\beta_2\alpha_3+B'\beta_1\alpha_2\beta_3+C'\alpha_1\beta_2\beta_3;\\ \chi(-\tfrac{3}{2})&=\beta_1\beta_2\beta_3.\end{aligned} \tag{146.2}$$

The argument of χ denotes the eigenvalue of S_z in units of $\hbar$. Each of the two functions $\chi(\frac{1}{2})$ and $\chi(-\frac{1}{2})$ consists of three still degenerate functions. This degeneracy will now be dissolved by investigating the operator

$$\boldsymbol{S}^2=\left(\frac{\hbar}{2}\right)^2(\boldsymbol{\sigma}_1+\boldsymbol{\sigma}_2+\boldsymbol{\sigma}_3)^2=\left(\frac{\hbar}{2}\right)^2\{9+2(\boldsymbol{\sigma}_1\cdot\boldsymbol{\sigma}_2)+2(\boldsymbol{\sigma}_2\cdot\boldsymbol{\sigma}_3)+2(\boldsymbol{\sigma}_3\cdot\boldsymbol{\sigma}_1)\}. \tag{146.3}$$

In Problem 140 it has been shown that

$$(\boldsymbol{\sigma}_1\cdot\boldsymbol{\sigma}_2)\alpha_1\alpha_2=\alpha_1\alpha_2; \qquad (\boldsymbol{\sigma}_1\cdot\boldsymbol{\sigma}_2)\alpha_1\beta_2=2\beta_1\alpha_2-\alpha_1\beta_2;$$
$$(\boldsymbol{\sigma}_1\cdot\boldsymbol{\sigma}_2)\beta_1\alpha_2=2\alpha_1\beta_2-\beta_1\alpha_2; \qquad (\boldsymbol{\sigma}_1\cdot\boldsymbol{\sigma}_2)\beta_1\beta_2=\beta_1\beta_2;$$

or, even, more simply, that the operator

$$\Sigma_{12}=\tfrac{1}{2}(1+\boldsymbol{\sigma}_1\cdot\boldsymbol{\sigma}_2) \tag{146.4}$$

merely exchanges the spin functions of the two particles 1 and 2,

$$\Sigma_{12}\,\chi(1,2)=\chi(2,1), \tag{146.5a}$$

in detail:

$$\Sigma_{12}\alpha_1\alpha_2=\alpha_1\alpha_2; \qquad \Sigma_{12}\alpha_1\beta_2=\beta_1\alpha_2; \quad \text{etc.} \tag{146.5b}$$

The operator (146.4) is therefore called the *spin exchange operator* of particles 1 and 2.

We now can express (146.3) in terms of such exchange operators,

$$\boldsymbol{S}^2=\left(\frac{\hbar}{2}\right)^2\{3+4(\Sigma_{12}+\Sigma_{23}+\Sigma_{31})\}. \tag{146.6}$$

Application of this operator to the first and last of the four spin functions (146.2) leads to

$$\boldsymbol{S}^2\chi(\tfrac{3}{2})=\left(\frac{\hbar}{2}\right)^2\cdot 15\,\chi(\tfrac{3}{2});\quad \boldsymbol{S}^2\chi(-\tfrac{3}{2})=\left(\frac{\hbar}{2}\right)^2\cdot 15\,\chi(-\tfrac{3}{2}). \tag{146.7}$$

These two functions therefore are non-degenerate eigenfunctions of $\boldsymbol{S}^2$ already to the eigenvalue $S(S+1)=\frac{15}{4}$ or $S=\frac{3}{2}$ for two different eigenvalues of S_z. In the vector model they correspond to parallel orientation of all three spins in z or $-z$ direction.

It is not so simple to deal with the degenerate functions $\chi(\frac{1}{2})$ and $\chi(-\frac{1}{2})$. Here, application of (146.6) to $\chi(\frac{1}{2})$ yields

$$\begin{aligned}\boldsymbol{S}^2\chi(\tfrac{1}{2})=\left(\frac{\hbar}{2}\right)^2&\{3\chi(\tfrac{1}{2})+4[A\alpha_1\alpha_2\beta_3+B\beta_1\alpha_2\alpha_3+C\alpha_1\beta_2\alpha_3]\\&+4[A\alpha_1\beta_2\alpha_3+B\alpha_1\alpha_2\beta_3+C\beta_1\alpha_2\alpha_3]\\&+4[A\beta_1\alpha_2\alpha_3+B\alpha_1\beta_2\alpha_3+C\alpha_1\alpha_2\beta_3]\}\\=\left(\frac{\hbar}{2}\right)^2&\{(7A+4B+4C)\alpha_1\alpha_2\beta_3+(4A+7B+4C)\alpha_1\beta_2\alpha_3\\&+(4A+4B+7C)\beta_1\alpha_2\alpha_3\}.\end{aligned}$$

This shall become

$$=\hbar^2 S(S+1)\{A\alpha_1\alpha_2\beta_3+B\alpha_1\beta_2\alpha_3+C\beta_1\alpha_2\alpha_3\}.$$

Thus we arrive at a linear system of three homogeneous equations,

$$\begin{aligned}7A+4B+4C&=4S(S+1)\,A;\\4A+7B+4C&=4S(S+1)\,B;\\4A+4B+7C&=4S(S+1)\,C\end{aligned} \tag{146.8}$$

the determinant of which must vanish. This gives a cubic equation for the eigenvalues $S(S+1)$ possible with the solutions

$$S(S+1)=\tfrac{15}{4},\tfrac{3}{4},\tfrac{3}{4}\quad\text{or}\quad S=\tfrac{3}{2},\tfrac{1}{2},\tfrac{1}{2}. \tag{146.9}$$

The same result would be obtained by applying $\boldsymbol{S}^2$ to $\chi(-\frac{1}{2})$ if the symbols α and β are exchanged throughout.

The first eigenvalue (146.9) leads unambiguously to the solution $A=B=C$ of the system (146.8). Using a more complete notation for the spin functions, $\chi(S,S_z)$, we find a *quartet* of four completely symmetrical functions, viz.

$$\begin{aligned}
\chi\left(\frac{3}{2},\frac{3}{2}\right)&=\alpha_1\alpha_2\alpha_3;\\
\chi\left(\frac{3}{2},\frac{1}{2}\right)&=\frac{1}{\sqrt{3}}(\alpha_1\alpha_2\beta_3+\alpha_1\beta_2\alpha_3+\beta_1\alpha_2\alpha_3);\\
\chi\left(\frac{3}{2},-\frac{1}{2}\right)&=\frac{1}{\sqrt{3}}(\beta_1\beta_2\alpha_3+\beta_1\alpha_2\beta_3+\alpha_1\beta_2\beta_3);\\
\chi\left(\frac{3}{2},-\frac{3}{2}\right)&=\beta_1\beta_2\beta_3.
\end{aligned}\tag{146.10}$$

Besides this solution which corresponds to the four orientations of spin $\frac{3}{2}$ in the vector model, the double solution $S=\frac{1}{2}$ put in (146.8) leads three times to the same relation,

$$A+B+C=0;$$

we thus may express $C=-(A+B)$, but cannot then obtain separate information on A and B:

$$\begin{aligned}
\chi(\tfrac{1}{2},\tfrac{1}{2})&=A\alpha_1\alpha_2\beta_3+B\alpha_1\beta_2\alpha_3-(A+B)\beta_1\alpha_2\alpha_3;\\
\chi(\tfrac{1}{2},-\tfrac{1}{2})&=A'\beta_1\beta_2\alpha_3+B'\beta_1\alpha_2\beta_3-(A'+B')\alpha_1\beta_2\beta_3.
\end{aligned}\tag{146.11}$$

Two doublets, each with $S=\frac{1}{2}$, are still mixed up in these formulae and are still degenerate.

It is usual to decompose and normalize the doublets by the two assumptions

$$A=B=\frac{1}{\sqrt{6}}\quad\text{and}\quad A=-B=\frac{1}{\sqrt{2}}.\tag{146.12}$$

With the first assumption the doublet becomes

$$\begin{aligned}
\chi_1\left(\frac{1}{2},\frac{1}{2}\right)&=\frac{1}{\sqrt{6}}\{\alpha_1(\alpha_2\beta_3+\beta_2\alpha_3)-2\beta_1\cdot\alpha_2\alpha_3\};\\
\chi_1\left(\frac{1}{2},-\frac{1}{2}\right)&=\frac{1}{\sqrt{6}}\{\beta_1(\beta_2\alpha_3+\alpha_2\beta_3)-2\alpha_1\cdot\beta_2\beta_3\};
\end{aligned}\tag{146.13}$$

this doublet is symmetrical with respect to exchanging particles 2 and 3 (symbolic notation: $1,\overline{23}$). The other doublet will become

$$\chi_2\left(\frac{1}{2},\frac{1}{2}\right)=\frac{1}{\sqrt{2}}\alpha_1(\alpha_2\beta_3-\beta_2\alpha_3);$$

$$\chi_2\left(\frac{1}{2},-\frac{1}{2}\right)=\frac{1}{\sqrt{2}}\beta_1(\beta_2\alpha_3-\alpha_2\beta_3); \tag{146.14}$$

it is antisymmetrical in 2 and 3 (in symbols: $1,\widetilde{23}$).

Of course, it is quite arbitrary to select just particles 2 and 3 as affecting simple symmetry properties. By another choice of A and B, e.g. $B=-\frac{1}{2}A$, a function of symmetry $\overline{12},3$ would have been obtained. Only further conditions imposed on the solution in special problems can lead to the dissolution of this remaining degeneracy.

Problem 147. Neutron scattering by molecular hydrogen

Let the particles 1 and 2 of the preceding problem be the two protons of an hydrogen molecule and 3 be a slow neutron with its de Broglie wavelength large as compared to the nuclear distance. The scattering cross section shall be determined for para and orthohydrogen, separately, with the central-force *n-p* interaction (cf. Problem 140)

$$V=\tfrac{1}{4}(3V_t+V_s)+\tfrac{1}{4}(V_t-V_s)(\boldsymbol{\sigma}_n\cdot\boldsymbol{\sigma}_p)\,. \tag{147.1}$$

To connect scattering lengths with potentials, the somewhat crude assumption may be made that the scattering length is proportional to the potential well depth.

Solution. The motion of the neutron will be governed by its interaction with the two protons. If its wavelength is large, both protons are practically at the same position and we have only one relative coordinate vector $\boldsymbol{r}$. Let us denote the neutron by subscript n (instead of 3), then the neutron-molecule interaction may be written, according to (147.1),

$$V=\tfrac{1}{2}(3V_t+V_s)+\tfrac{1}{4}(V_t-V_s)(\boldsymbol{\sigma}_n,\boldsymbol{\sigma}_1+\boldsymbol{\sigma}_2) \tag{147.2}$$

with $V_t(r)$ and $V_s(r)$.

Orthohydrogen is now defined by a symmetrical, parahydrogen by an antisymmetrical spin function so that, according to the results of the preceding problem, there exist the following eight spin functions of our three-body problem:

$$
\left.\begin{aligned}
\chi\left(\frac{3}{2},+\frac{3}{2}\right) &= \alpha_1\alpha_2\alpha_n \\
\chi\left(\frac{3}{2},+\frac{1}{2}\right) &= \frac{1}{\sqrt{3}}(\alpha_1\alpha_2\beta_n+\alpha_1\beta_2\alpha_n+\beta_1\alpha_2\alpha_n) \\
\chi\left(\frac{3}{2},-\frac{1}{2}\right) &= \frac{1}{\sqrt{3}}(\beta_1\beta_2\alpha_n+\beta_1\alpha_2\beta_n+\alpha_1\beta_2\beta_n) \\
\chi\left(\frac{3}{2},-\frac{3}{2}\right) &= \beta_1\beta_2\beta_n
\end{aligned}\right\}
\begin{array}{l}\text{Quartet,}\\ \text{spin } \tfrac{3}{2},\\ o-H_2,\\ \overline{12}\; n\end{array}
\tag{147.3}
$$

$$
\left.\begin{aligned}
\chi_o\left(\frac{1}{2},+\frac{1}{2}\right) &= \frac{1}{\sqrt{6}}((\alpha_1\beta_2+\beta_1\alpha_2)\alpha_n-2\alpha_1\alpha_2\beta_n) \\
\chi_o\left(\frac{1}{2},-\frac{1}{2}\right) &= \frac{1}{\sqrt{6}}((\alpha_1\beta_2+\beta_1\alpha_2)\beta_n-2\beta_1\beta_2\alpha_n)
\end{aligned}\right\}
\begin{array}{l}\text{Doublet,}\\ \text{spin } \tfrac{1}{2},\\ o-H_2,\\ \overline{12}, n\end{array}
\tag{147.4}
$$

$$
\left.\begin{aligned}
\chi_p\left(\frac{1}{2},+\frac{1}{2}\right) &= \frac{1}{\sqrt{2}}(\alpha_1\beta_2-\beta_1\alpha_2)\alpha_n \\
\chi_p\left(\frac{1}{2},-\frac{1}{2}\right) &= \frac{1}{\sqrt{2}}(\alpha_1\beta_2-\beta_1\alpha_2)\beta_n
\end{aligned}\right\}
\begin{array}{l}\text{Doublet,}\\ \text{spin } \tfrac{1}{2},\\ p-H_2,\\ \widehat{12}, n\end{array}
\tag{147.5}
$$

These eight functions are eigenfunctions of the operator

$$\boldsymbol{S}^2=(\boldsymbol{\sigma}_1+\boldsymbol{\sigma}_2+\boldsymbol{\sigma}_n)^2$$

with the eigenvalues 15 in the quartet, and 3 in the doublet states. Since

$$\boldsymbol{S}^2=\boldsymbol{\sigma}_1^2+\boldsymbol{\sigma}_2^2+\boldsymbol{\sigma}_3^2+2(\boldsymbol{\sigma}_1\cdot\boldsymbol{\sigma}_2)+2(\boldsymbol{\sigma}_n,\boldsymbol{\sigma}_1+\boldsymbol{\sigma}_2) \tag{147.6}$$

with the three first terms all equal to 1, but the fourth contributing according to

$$(\boldsymbol{\sigma}_1\cdot\boldsymbol{\sigma}_2)=\begin{cases}+1 & \text{for orthohydrogen}\\ -3 & \text{for parahydrogen,}\end{cases} \tag{147.7}$$

we arrive at

$$
\left.\begin{aligned}
15&=9+2+2(\boldsymbol{\sigma}_n,\boldsymbol{\sigma}_1+\boldsymbol{\sigma}_2)\\
3&=9+2+2(\boldsymbol{\sigma}_n,\boldsymbol{\sigma}_1+\boldsymbol{\sigma}_2)
\end{aligned}\right.
\quad \text{for orthohydrogen} \begin{cases}\text{quartet}\\ \text{doublet}\end{cases}
$$

and

$$3=9-6+2(\boldsymbol{\sigma}_n,\boldsymbol{\sigma}_1+\boldsymbol{\sigma}_2) \quad \text{for parahydrogen doublet}$$

or

$$(\boldsymbol{\sigma}_n,\boldsymbol{\sigma}_1+\boldsymbol{\sigma}_2)=\begin{cases}2 & \text{for orthohydrogen quartet}\\ -4 & \text{for orthohydrogen doublet}\\ 0 & \text{for parahydrogen doublet.}\end{cases} \tag{147.8}$$

This, according to (147.2), leads to the following three different interactions between the neutron and the hydrogen molecule:

$$V=\tfrac{1}{2}(3V_t+V_s)+\begin{Bmatrix}\tfrac{1}{2}(V_t-V_s)\\-(V_t-V_s)\\0\end{Bmatrix}=\begin{cases}2V_t & \text{for quartet, } o-H_2\\ \tfrac{1}{2}V_t+\tfrac{3}{2}V_s & \text{for doublet, } o-H_2\\ \tfrac{3}{2}V_t+\tfrac{1}{2}V_s & \text{for doublet, } p-H_2.\end{cases} \quad (147.9)$$

Turning to the scattering question, we need only consider the limiting case of zero energy where the scattering length a is linearly connected with the potential depth constant as long as the potential hole is "small". It should, however, be noted that this is rather a crude assumption in the real neutron-proton interaction case. Making this approximation, we get the elastic scattering cross sections

$$\sigma_{\text{ortho}}=4\pi\{\tfrac{2}{3}(2a_t)^2+\tfrac{1}{3}(\tfrac{1}{2}a_t+\tfrac{3}{2}a_s)^2\};$$
$$\sigma_{\text{para}}=4\pi(\tfrac{3}{2}a_t+\tfrac{1}{2}a_s)^2$$

or

$$\sigma_{\text{ortho}}=\pi\{(3a_t+a_s)^2+2(a_t-a_s)^2\};$$
$$\sigma_{\text{para}}=\pi(3a_t+a_s)^2. \quad (147.10)$$

NB. The best values for the scattering lengths a_t and a_s of the two-nucleon problem are[1] $a_t=+5.39$ fm and $a_s=-23.7$ fm. Inserting these values into Eq. (147.10) yields $\sigma_{\text{ortho}}=55$ barn and $\sigma_{\text{para}}=1.77$ barn. The characteristic feature of this somewhat rough result is the amazingly small value of the parahydrogen cross section. This is fully borne out by experimental evidence with thermal neutrons. The para cross section would vanish entirely with $a_s=-3a_t$; its smallness shows that anyhow a_s must be large and have the opposite sign to a_t. The triplet scattering length must be positive in order to allow for a ^3S bound state, the deuteron. Hence, $a_s<0$ so that no ^1S bound state can exist. It should be noted that this sign can only be determined by interference experiments of the kind described, not by scattering of neutrons at isolated protons producing incoherent waves.

Our results apply to the limit of energy zero whereas, in experiment, the neutrons still have a few hundredths of an eV energy. Their wavelength therefore is not so very large in comparison with the molecular distance between the two protons. This causes inelastic transitions with parity change in the molecule between the rotational states $J=1$ of ortho, and $J=0$ of parahydrogen. They occur because $r_{n1}\neq r_{n2}$ so that, with the abbreviation $\frac{1}{4}(V_t-V_s)=U(r)$, we have for the spin dependent part of the interaction,

$$U(r_{n1})(\boldsymbol{\sigma}_n\cdot\boldsymbol{\sigma}_1)+U(r_{n2})(\boldsymbol{\sigma}_n\cdot\boldsymbol{\sigma}_2)=\tfrac{1}{2}(U(r_{n1})+U(r_{n2}))(\boldsymbol{\sigma}_n,\boldsymbol{\sigma}_1+\boldsymbol{\sigma}_2)$$
$$+\tfrac{1}{2}(U(r_{n1})-U(r_{n2}))(\boldsymbol{\sigma}_n,\boldsymbol{\sigma}_1-\boldsymbol{\sigma}_2).$$

In consequence of the last term, the functions (147.3) to (147.5) no longer remain eigenfunctions of the potential; it is this last term that induces ortho-para transitions.

[1] 1 fm $=10^{-13}$ cm; 1 barn $=10^{-24}$ cm^2.

IV. Many-Body Problems

A. Few Particles

Problem 148. Two repulsive particles on a circle

Two particles are fixed on a circle with a mutual repulsion given by

$$V(\varphi_1, \varphi_2) = V_0 \cos(\varphi_1 - \varphi_2) \tag{148.1}$$

to simulate e.g. the Coulomb repulsion between the two helium electrons in the ground state. The conservation of angular momentum shall be derived, and the relative motion of the particles discussed.

Solution. The Schrödinger equation

$$-\frac{\hbar^2}{2mr^2}\left(\frac{\partial^2 U}{\partial \varphi_1^2} + \frac{\partial^2 U}{\partial \varphi_2^2}\right) + V_0 \cos(\varphi_1 - \varphi_2)\, U = E \cdot U \tag{148.2}$$

permits factorization by introducing the variables

$$\alpha = \varphi_1 - \varphi_2; \qquad \beta = \tfrac{1}{2}(\varphi_1 + \varphi_2) \tag{148.3}$$

of relative and absolute motion. Then we have

$$\frac{\partial}{\partial \varphi_1} = \frac{\partial}{\partial \alpha} + \frac{1}{2}\frac{\partial}{\partial \beta}; \qquad \frac{\partial}{\partial \varphi_2} = -\frac{\partial}{\partial \alpha} + \frac{1}{2}\frac{\partial}{\partial \beta}.$$

Putting this into (148.2) we get

$$-\frac{\hbar^2}{mr^2}\left(\frac{\partial^2 U}{\partial \alpha^2} + \frac{1}{4}\frac{\partial^2 U}{\partial \beta^2}\right) + V_0 \cos\alpha \cdot U = E \cdot U.$$

Factorization now becomes possible into

$$U(\alpha, \beta) = u(\alpha)\, v(\beta) \tag{148.4}$$

and leads to the separate equations of motion

$$-\frac{\hbar^2}{mr^2}\frac{d^2u}{d\alpha^2} + V_0\cos\alpha\cdot u = E_\alpha\cdot u \tag{148.5}$$

and

$$-\frac{\hbar^2}{4mr^2}\frac{d^2v}{d\beta^2} = E_\beta\cdot v \tag{148.6}$$

with

$$E_\alpha + E_\beta = E. \tag{148.7}$$

The *absolute motion* can be determined from (148.6). The total orbital momentum operator of the two-particle system is

$$L = \frac{\hbar}{i}\left(\frac{\partial}{\partial\varphi_1} + \frac{\partial}{\partial\varphi_2}\right) = \frac{\hbar}{i}\frac{\partial}{\partial\beta}$$

so that (148.6) may as well be written

$$\frac{1}{2\Theta}L^2 v = E_\beta\cdot v \tag{148.6'}$$

with $\Theta = 2mr^2$ the total moment of inertia of both particles. Eq. (148.6′) therefore is the eigenvalue problem of the operator of rotational energy. Since (148.6) is solved by

$$v = e^{iM\beta}; \qquad M = 0, \pm 1, \pm 2, \ldots, \tag{148.8}$$

the eigenvalues of the rotational energy become

$$E_\beta = \frac{(\hbar M)^2}{2\Theta}. \tag{148.9}$$

It is much more difficult to discuss the *relative motion* determined by the differential equation (148.5) of the Mathieu type. To alleviate the discussion we transform (148.5) to the standard form by putting

$$\alpha = 2\varphi; \qquad 4\frac{mr^2}{\hbar^2}E_\alpha = \lambda; \qquad 4\frac{mr^2}{\hbar^2}V_0 = 2q \tag{148.10}$$

so that we get

$$\frac{d^2v}{d\varphi^2} + (\lambda - 2q\cos 2\varphi)v = 0. \tag{148.11}$$

We are looking for periodic solutions[1] with period 2π in the variable α, or π in the variable φ. The coefficient of v in (148.11) being an even

[1] This makes a fundamental difference to the solutions in a periodic potential of lattice theory which are not periodical, but multiply with a phase factor in each period of the lattice potential, cf. Problems 28, 29. We therefore get discrete eigenvalues λ here, but a band structure in the lattice problem.

function of φ, there are two symmetry types of solutions, even and odd. Their periodicity permits Fourier expansion so that we have

$$\begin{aligned} v_{\text{even}} &= A_0 + A_2 \cos 2\varphi + A_4 \cos 4\varphi + \cdots; \\ v_{\text{odd}} &= B_2 \sin 2\varphi + B_4 \sin 4\varphi + \cdots. \end{aligned}$$

It is usual to denote the eigenvalues of λ by $a_0, a_2, a_4, \ldots$ for even and by $b_2, b_4, \ldots$ for odd solutions[2]. It is shown in the theory of Mathieu equations that the eigenvalues may be ordered in a sequence

$$a_0 < b_2 < a_2 < b_4 < a_4 \ldots \tag{148.12}$$

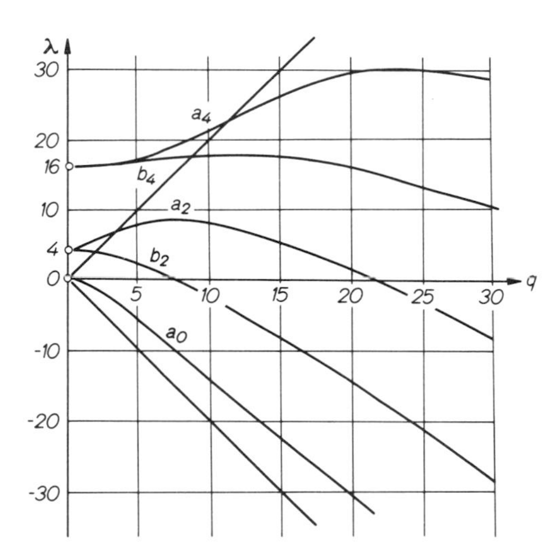

Fig. 61. Eigenvalues (λ) for different potential hole depths (q), both in dimensionless scale as defined in Eq. (148.10). The two straight lines mark the potential maximum and minimum

These eigenvalues are given, as functions of q, in Fig. 61; the relation (148.12) forbids intersections of the curves. At $q=0$ we have

$$a_0(0) = 0; \quad b_2(0) = a_2(0) = 2^2; \quad b_4(0) = a_4(0) = 4^2 \quad \text{etc.}; \tag{148.13}$$

[2] We adopt, as far as reasonably possible, the mathematical notations of Abramowitz, M., Stegun, I.A., Handbook of Mathematical Functions, Chap. 20. New York: Dover Publ., 1965. – The curves of Fig. 61 are partially constructed with the help of this Handbook.

for very large values of q there hold the asymptotic laws

$$\left.\begin{aligned} a_{2r}(q) &\to -2q+2(2r+1)\sqrt{q}-\tfrac{1}{4}[(r+1)^2+r^2]; \\ b_{2r}(q) &\to -2q+2(2r-1)\sqrt{q}-\tfrac{1}{4}[(r-1)^2+r^2]. \end{aligned}\right\} \tag{148.14}$$

The relations (148.13) and (148.14) can largely be understood by elementary considerations.

If $q=0$, Eq. (148.11) simply becomes

$$\frac{d^2 v}{d\varphi^2} + \lambda v = 0$$

with periodic solutions

$$v_{\text{even}} = A_{2r} \cos 2r\varphi; \qquad v_{\text{odd}} = B_{2r} \sin 2r\varphi$$

belonging to eigenvalues $\lambda=(2r)^2$. This exactly corresponds to Eq. (148.13).

If, on the other hand, q becomes very large, there will be a very deep potential hole around $\alpha=\pi$, almost fixing the two particles at opposite positions on the circle. It is then helpful to use, instead of φ, a variable

$$\eta = \tfrac{1}{2}(\varphi_1-\varphi_2-\pi) = \varphi - \frac{\pi}{2}. \tag{148.15}$$

From

$$\cos 2r\varphi = (-1)^r \cos 2r\eta; \qquad \sin 2r\varphi = (-1)^r \sin 2r\eta$$

it follows that the functions v_{even} and v_{odd} will be even and odd also with respect to the variable η. If the potential hole is very deep we may write in (148.11)

$$\cos 2\varphi = -\cos 2\eta \simeq -1+2\eta^2$$

and thus arrive, approximately, at the differential equation of the harmonic oscillator,

$$\frac{d^2 v}{d\eta^2} + [(\lambda+2q)-4q\eta^2]v=0. \tag{148.16}$$

Its well-known eigenvalues (cf. Problem 30) are

$$\lambda_n+2q = 2\sqrt{q}(2n+1); \qquad n=0,1,2,\ldots \tag{148.17}$$

with the eigenfunctions for even/odd n being even/odd in η. Eq. (148.17) is almost identical with (148.14) if we identify solutions at small and at large values of q as follows:

	$n=0$	1	2	3	4	5	6 …
even (a_{2r})	$r=0$		1		2		3 …
odd (b_{2r})	$r=$	1		2		3	…

The additional constant in (148.14) may be corroborated if we expand $\cos 2\eta$ one term further on; then a perturbation calculation using oscillator functions in first approximation yields instead of (148.17):

$$\lambda_n + 2q = 2\sqrt{q}(2n+1) - \tfrac{4}{3} q \langle n|\eta^4|n\rangle, \tag{148.18}$$

where the additional term turns out to bring (148.18) to perfect agreement with (148.14).

NB 1. In Fig. 61 we have drawn two diagonals, the lower one marking the depths of the potential bottoms ($-2q$), the upper one the heights of potential summits ($+2q$). Eigenvalues must, of course, always lie above the bottom line. If they lie below the summit diagonal, they describe states of libration inside the potential hole. At $q=10$, e.g., there are 4 such libration states inside the hole, the fifth (a_4) eigenvalue leading to a vibration all around the circle, including coincidence of both particles. The first four states might be called anharmonic oscillator states; from the fifth state upwards they will correspond more and more to force-free motions of independent particles.

NB 2. The relative motion of the two particles occurs under the action of a potential energy which has the same form ($V_0 \cos\alpha$) as that of a pendulum. In classical mechanics this leads to no more complicated functions than elliptical integrals, whereas in quantum mechanics we need Mathieu functions. This again, as in Problem 40, shows how much more involved is the mathematical situation in quantum than in classical mechanics.

Problem 149. Three-atomic linear molecule

The carbon dioxide molecule has a linear O═C═O form in equilibrium. Let either equal or different oxygen isotopes be used, the C=O equilibrium distance be a and the force constant of the valence vibration f. The two valence vibration frequencies shall be determined in harmonic approximation using a one-dimensional model, thus neglecting bending vibrations.

Solution. Let x_1, x_2, x_3 be the positions along the x axis and m_1, m_2, m_3 the masses of the three atoms, then the Schrödinger equation for the linear harmonic model runs

$$-\frac{\hbar^2}{2}\sum_{i=1}^{3}\frac{1}{m_i}\frac{\partial^2\Psi}{\partial x_i^2} + \frac{1}{2} f\left[(x_2-x_1-a)^2 + (x_3-x_2-a)^2\right]\Psi = E\Psi. \tag{149.1}$$

In order to factorize the solution into centre-of-mass motion and internal motion we use the variables

$$\begin{aligned} u &= x_2 - x_1 - a, \\ v &= x_3 - x_2 - a, \\ X &= \frac{1}{M}(m_1 x_1 + m_2 x_2 + m_3 x_3) \quad \text{with } M = m_1 + m_2 + m_3. \end{aligned} \tag{149.2}$$

It then follows that

$$\frac{\partial}{\partial x_1} = \frac{m_1}{M}\frac{\partial}{\partial X} - \frac{\partial}{\partial u};$$

$$\frac{\partial}{\partial x_2} = \frac{m_2}{M}\frac{\partial}{\partial X} + \frac{\partial}{\partial u} - \frac{\partial}{\partial v};$$

$$\frac{\partial}{\partial x_3} = \frac{m_3}{M}\frac{\partial}{\partial X} \qquad + \frac{\partial}{\partial v}$$

and we find by iterating these operations

$$\frac{1}{m_i}\frac{\partial^2}{\partial x_i^2} = \frac{1}{M}\frac{\partial^2}{\partial X^2} + \left(\frac{1}{m_1}+\frac{1}{m_2}\right)\frac{\partial^2}{\partial u^2} + \left(\frac{1}{m_2}+\frac{1}{m_3}\right)\frac{\partial^2}{\partial v^2} - \frac{2}{m_2}\frac{\partial^2}{\partial u\,\partial v},$$

allowing for separation of the centre-of-mass motion. Changing the definition of E in (149.1) to mean the energy of only the internal motion, we arrive at a Schrödinger equation in two variables, u and v,

$$\left\{-\frac{\hbar^2}{2}\left[\left(\frac{1}{m_1}+\frac{1}{m_2}\right)\frac{\partial^2}{\partial u^2} + \left(\frac{1}{m_2}+\frac{1}{m_3}\right)\frac{\partial^2}{\partial v^2} - \frac{2}{m_2}\frac{\partial^2}{\partial u\,\partial v}\right] + \frac{1}{2} f(u^2+v^2) - E\right\}\Psi = 0. \tag{149.3}$$

The cross term, $\partial^2/\partial u\,\partial v$, in the kinetic energy makes factorization impossible in these variables. If, however, we introduce a "rotated" system,

$$\begin{aligned} u' &= \quad u\cos\alpha + v\sin\alpha, \\ v' &= -u\sin\alpha + v\cos\alpha \end{aligned} \tag{149.4}$$

it is possible by a suitable choice of α to make the term with $\partial^2/\partial u'\,\partial v'$ vanish, whereas the potential energy remains invariant under this transformation:

$$u^2+v^2 = u'^2+v'^2.$$

The Schrödinger equation thus becomes

$$\begin{aligned} \Big\{-\frac{\hbar^2}{2}\Big[&\left(\frac{1}{m_1}+\frac{1}{m_2}\right)\left(\cos^2\alpha\frac{\partial^2}{\partial u'^2} - 2\cos\alpha\sin\alpha\frac{\partial^2}{\partial u'\,\partial v'} + \sin^2\alpha\frac{\partial^2}{\partial v'^2}\right) \\ &+\left(\frac{1}{m_2}+\frac{1}{m_3}\right)\left(\sin^2\alpha\frac{\partial^2}{\partial u'^2} + 2\sin\alpha\cos\alpha\frac{\partial^2}{\partial u'\,\partial v'} + \cos^2\alpha\frac{\partial^2}{\partial v'^2}\right) \\ &-\frac{2}{m_2}\left(\cos\alpha\sin\alpha\frac{\partial^2}{\partial u'^2} + (\cos^2\alpha-\sin^2\alpha)\frac{\partial^2}{\partial u'\,\partial v'} - \cos\alpha\sin\alpha\frac{\partial^2}{\partial v'^2}\right)\Big] \\ &+\frac{1}{2}f(u'^2+v'^2) - E\Big\}\Psi = 0. \end{aligned} \tag{149.5}$$

Here the factor of the mixed derivative will vanish if

$$\left(\frac{1}{m_3}-\frac{1}{m_1}\right)\sin 2\alpha=\frac{2}{m_2}\cos 2\alpha$$

or

$$\tan 2\alpha=\frac{2m_1 m_3}{m_2(m_1-m_3)}. \tag{149.6}$$

Using two mass constants A and B defined by

$$\left.\begin{aligned}\frac{1}{A}&=\left(\frac{1}{m_1}+\frac{1}{m_2}\right)\cos^2\alpha-\frac{2}{m_2}\cos\alpha\sin\alpha+\left(\frac{1}{m_2}+\frac{1}{m_3}\right)\sin^2\alpha\\ &\text{and}\\ \frac{1}{B}&=\left(\frac{1}{m_1}+\frac{1}{m_2}\right)\sin^2\alpha+\frac{2}{m_2}\cos\alpha\sin\alpha+\left(\frac{1}{m_2}+\frac{1}{m_3}\right)\cos^2\alpha\end{aligned}\right\} \tag{149.7}$$

the Schrödinger equation is now much simplified:

$$\left[-\frac{\hbar^2}{2A}\frac{\partial^2}{\partial u'^2}+\frac{1}{2}f u'^2\right]\Psi+\left[-\frac{\hbar^2}{2B}\frac{\partial^2}{\partial v'^2}+\frac{1}{2}f v'^2\right]\Psi=E\Psi. \tag{149.8}$$

Solution by factorization into

$$\Psi(u',v')=\psi(u')\,\varphi(v');\qquad E=E_A+E_B \tag{149.9}$$

is now possible:

$$\left.\begin{aligned}-\frac{\hbar^2}{2A}\frac{d^2\psi}{du'^2}+\frac{1}{2}f u'^2\psi=E_A\psi;\\ -\frac{\hbar^2}{2B}\frac{d^2\varphi}{dv'^2}+\frac{1}{2}f v'^2\varphi=E_B\varphi.\end{aligned}\right\} \tag{149.10}$$

The eigenvalues of these two harmonic oscillators are well known (cf. Problem 30), viz.

$$\left.\begin{aligned}E_A=\hbar\omega_A(n_A+\tfrac{1}{2});\quad \omega_A=\sqrt{f/A};\\ E_B=\hbar\omega_B(n_B+\tfrac{1}{2});\quad \omega_B=\sqrt{f/B}.\end{aligned}\right\} \tag{149.11}$$

Eqs. (149.6) and (149.7) to determine A and B from the masses and (149.11) for the energies and frequencies form the solution of the problem.

We now proceed to a closer inspection of these formulae in the normal case $m_1=m_3$ (equal oxygen isotopes) for which, according to (149.6), $\alpha=\pi/4$. The mass constants A and B then follow from (149.7)

$$\frac{1}{A}=\frac{1}{m_1};\qquad \frac{1}{B}=\frac{1}{m_1}+\frac{2}{m_2}.$$

Thence,

$$\omega_A = \sqrt{f/m_1}\,; \qquad \omega_B = \omega_A \sqrt{1 + 2\frac{m_1}{m_2}}\,. \tag{149.12}$$

Since ω_A does not depend on m_2 it may be concluded that the carbon atom stays at rest in mode A vibrations and, since the centre of mass is supposed not to move, we must have a symmetrical vibration as indicated in Fig. 62a. On the other hand, it can be shown that mode B vibrations,

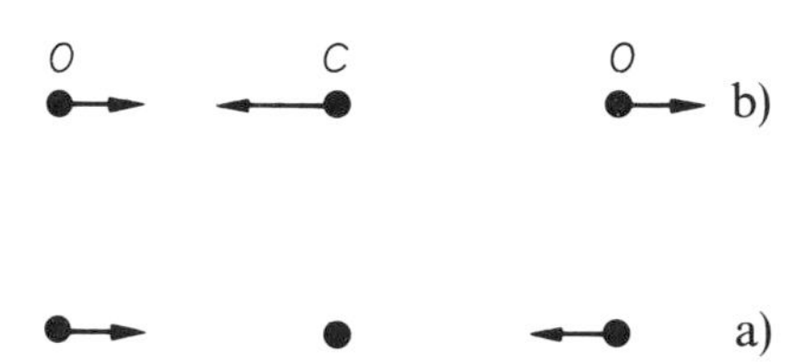

Fig. 62a and b The two valence vibrational modes of CO_2. No bending vibrations are considered in this problem

for which the carbon atom takes part in the motion, will become antisymmetric as shown in Fig. 62b. To translate this classical description of normal vibrations into quantum mechanics, we construct the wave functions according to (149.9). In the ground state we have (in arbitrary normalization)

$$\Psi_0(u', v') = \exp\left(-\frac{A\,\omega_A}{2\hbar}\,u'^2\right)\cdot\exp\left(-\frac{B\,\omega_B}{2\hbar}\,v'^2\right), \tag{149.13}$$

both factors having a sole maximum at $u'=0$ and $v'=0$, respectively, where

$$u' = \frac{1}{\sqrt{2}}(x_3 - x_1 - 2a)\,; \qquad v' = \frac{1}{\sqrt{2}}(x_1 + x_3 - 2x_2)\,. \tag{149.14}$$

The zero-point vibration therefore occurs about the positions $x_1 + x_3 = 2x_2$ with the carbon atom halfway between the two oxygen atoms, and $x_3 - x_1 = 2a$, i.e. both oxygen atoms a distance $2a$ apart from each other. The most probable position of the ground state therefore is just the classical equilibrium position.

If the A mode is excited to, say, its first excited state, a factor u' is to be added to Ψ_0. Since a function

$$\psi(u') = u'\,e^{-\frac{1}{2}\lambda u'^2}\,; \qquad \lambda = \frac{A\,\omega_A}{\hbar}$$

has two maxima of opposite signs at $u' = \pm\lambda^{-\frac{1}{2}}$, we now have the most probable positions shifted to

$$x_3 - x_1 - 2a = \pm\sqrt{\frac{2}{\lambda}}; \qquad x_1 + x_3 - 2x_2 = 0. \tag{149.15}$$

The most probable position in which to find a vibrating particle has its classical equivalent in the two turning points where its time of stay is longest. The two maximum values of u' in (149.15) therefore mark something like the classical vibration amplitudes. The condition $v'=0$ shows that the carbon atom most probably still lies halfway between the two oxygen atoms, but these are now alternately at a smaller or at a larger distance apart from each other, just as indicated on Fig. 62a.

If, on the other hand, the B mode is excited we have inversely for the maximum

$$x_3 - x_1 - 2a = 0; \quad x_1 + x_3 - 2x_2 = \pm\sqrt{\frac{2}{\mu}}; \quad \mu = \frac{B\omega_B}{\hbar} \tag{149.16}$$

so that the most probable distance between the two oxygen atoms $(x_3 - x_1)$ remains $2a$ as in equilibrium, both being shifted to and fro with respect to the carbon atom as indicated on Fig. 62b.

Problem 150. Centre-of-mass motion

In classical mechanics the motion of the centre of mass in a many-body problem with only internal forces acting can be separated from the relative motion of the particles. It shall be shown that the same holds for quantum mechanics. Special attention shall be given to the case of only two particles.

Solution. We start with the hamiltonian of a system of N particles not subjected to external forces,

$$H = -\frac{\hbar^2}{2}\sum_{i=1}^{N}\frac{1}{m_i}\nabla_i^2 + \frac{1}{2}\sum_{i=1}^{N}\sum_{k=1}^{N}{}' V_{ik}(x_i - x_k, y_i - y_k, z_i - z_k) \tag{150.1}$$

and replace the $3N$ coordinates x_i, y_i, z_i by the position coordinates X, Y, Z of the centre of mass and the coordinates $\xi_\lambda, \eta_\lambda, \zeta_\lambda$ defining the position of particle λ ($\lambda = 1, 2, \ldots, N-1$) relative to particle N:

$$\left.\begin{aligned} X &= \frac{1}{M}\sum_{i=1}^{N} m_i x_i; \qquad M = \sum_{i=1}^{N} m_i; \\ \xi_\lambda &= x_\lambda - x_N \qquad (\lambda = 1, 2, \ldots, N-1) \end{aligned}\right\} \tag{150.2}$$

and correspondingly for Y, Z, η_λ, ζ_λ. The use of these coordinates, of course, breaks up the natural symmetry of (150.1) by artificially distinguishing particle N from the rest.

We easily obtain from (150.2) the operators

$$\frac{\partial}{\partial x_\nu}=\frac{m_\nu}{M}\frac{\partial}{\partial X}+\frac{\partial}{\partial \xi_\nu} \qquad (\nu=1,2,\ldots,N-1);$$

$$\frac{\partial}{\partial x_N}=\frac{m_N}{M}\frac{\partial}{\partial X}-\sum_\lambda \frac{\partial}{\partial \xi_\lambda}$$

and

$$\sum_{i=1}^{N}\frac{1}{m_i}\frac{\partial^2}{\partial x_i^2}=\sum_{\lambda=1}^{N-1}\frac{1}{m_\lambda}\left(\frac{m_\lambda^2}{M^2}\frac{\partial^2}{\partial X^2}+2\frac{m_\lambda}{M}\frac{\partial^2}{\partial X\,\partial \xi_\lambda}+\frac{\partial^2}{\partial \xi_\lambda^2}\right)$$

$$+\frac{1}{m_N}\left(\frac{m_N^2}{M^2}\frac{\partial^2}{\partial X^2}-2\frac{m_N}{M}\sum_\lambda\frac{\partial^2}{\partial X\,\partial \xi_\lambda}+\sum_\mu\sum_\lambda\frac{\partial^2}{\partial \xi_\mu \partial \xi_\lambda}\right)$$

$$=\frac{1}{M}\frac{\partial^2}{\partial X^2}+\left\{\sum_\lambda\frac{1}{m_\lambda}\frac{\partial^2}{\partial \xi_\lambda^2}+\frac{1}{m_N}\sum_\mu\sum_\lambda\frac{\partial^2}{\partial \xi_\mu \partial \xi_\lambda}\right\},$$

where the sums over Greek subscripts run from 1 to $N-1$. The essential feature of this result is the cancelling of all mixed derivatives $\partial^2/\partial X\,\partial \xi_\lambda$ permitting separation of the hamiltonian,

$$H=H_0+H_r, \tag{150.3}$$

in a centre-of-mass part,

$$H_0=-\frac{\hbar^2}{2M}\left(\frac{\partial^2}{\partial X^2}+\frac{\partial^2}{\partial Y^2}+\frac{\partial^2}{\partial Z^2}\right), \tag{150.4}$$

and a part describing the relative motion of the particles,

$$H_r=-\frac{\hbar^2}{2}\left\{\sum_\lambda\frac{1}{m_\lambda}\nabla_\lambda^2+\frac{1}{m_N}\nabla_\lambda\cdot\nabla_\mu\right\}+V, \tag{150.5}$$

with the potential energy

$$V=\tfrac{1}{2}\sum_\lambda\sum_\mu{}' V_{\lambda\mu}(\xi_\lambda-\xi_\mu,\eta_\lambda-\eta_\mu,\zeta_\lambda-\zeta_\mu)+\sum_\lambda V_{\lambda N}(\xi_\lambda,\eta_\lambda,\zeta_\lambda) \tag{150.6}$$

independent of the centre-of-mass coordinates. The solution of the Schrödinger equation,

$$(H_0+H_r)\,U=E\cdot U \tag{150.7}$$

then permits factorization into

$$U=\varphi(X,Y,Z)\,u(\xi_\lambda,\eta_\lambda,\zeta_\lambda) \tag{150.8}$$

with

$$-\frac{\hbar^2}{2M}\nabla^2\varphi = E_0\varphi\,; \tag{150.9}$$

$$(H_r+V)\,u = E_r u\,; \tag{150.10}$$

$$E_0+E_r = E\,. \tag{150.11}$$

Eq. (150.9) is solved by the plane wave

$$\varphi = e^{i\boldsymbol{K}\cdot\boldsymbol{R}} \qquad E_0 = \frac{\hbar^2 K^2}{2M} \tag{150.12}$$

with $\boldsymbol{R}=(X,Y,Z)$. This is the centre-of-mass law as in classical mechanics: the total mass M of the system moves with constant momentum $\hbar\boldsymbol{K}$. The relative motion of the particles about the centre of mass, governed by (150.10), is quite independent thereof.

The third term in Eq. (150.5) prevents further factorization of $u(\xi_\lambda,\eta_\lambda,\zeta_\lambda)$. Only in the case of the two-body problem, $N=2$ with $\lambda=\mu=1$ only, the hamiltonian of the relative motion simplifies to

$$H_r = -\frac{\hbar^2}{2}\left\{\frac{1}{m_1}\nabla_1^2+\frac{1}{m_2}\nabla_1^2\right\}+V_{12}(\xi_1,\eta_1,\zeta_1)\,. \tag{150.13}$$

Introducing, as in classical mechanics, the reduced mass m^* by putting

$$\frac{1}{m_1}+\frac{1}{m_2}=\frac{1}{m^*} \tag{150.14}$$

and omitting the subscripts of the relative coordinates (and of V_{12}), we arrive at

$$-\frac{\hbar^2}{2m}\nabla^2 u+V(\xi,\eta,\zeta)\,u=E_r u\,, \tag{150.15}$$

i.e. at the Schrödinger equation of an equivalent one-body problem.

NB. In Problem 67, the hydrogen atom has been treated as a one-body problem with its nucleus at rest. According to Eq. (150.15), we should more correctly introduce the reduced mass m^* of nucleus and electron, instead of m, the mass of the electron. No other change is required to take account of the participation of the nucleus in the relative motion about the centre of mass. Since the nuclear mass, say M, is very large as compared with m, Eq. (150.14) leads to

$$m^*=m\left(1-\frac{m}{M}\right),$$

approximately. Comparing e.g. the red spectral line $H_\alpha\,(n=3\rightarrow n=2)$ of the hydrogen atom with the frequency

$$\nu(H_\alpha)=\frac{5}{36}\,\frac{m_H^* e^4}{2\hbar^2 h}$$

and the corresponding line of the deuterium atom,

$$\nu(D_\alpha) = \frac{5}{36} \frac{m_D^* e^4}{2\hbar^2 h},$$

we find a line shift of

$$\nu(D_\alpha) - \nu(H_\alpha) = \frac{m_D^* - m_H^*}{m_H^*} \nu(H_\alpha) \simeq \frac{m}{2M_H} \nu(H_\alpha)$$

because $M_D \simeq 2M_H$. This difference is not very difficult to observe. It amounts to $4.12\,\mathrm{cm}^{-1}$ at a wavelength of 6563 Å. Heavy hydrogen was discovered in 1931 by Urey, Brickwedde and Murphy who observed this weak D_α satellite of the H_α line in natural hydrogen [Phys. Rev. **40**, 1 (1932)].

Problem 151. Virial theorem

To prove that the virial theorem

$$2E_{\mathrm{kin}} + E_{\mathrm{pot}} = 0$$

holds for any quantum mechanical system kept together by Coulomb forces only. The proof shall be performed by a scale transformation of the wave function of the system keeping normalization constant.

Solution. A system of N particles of masses m_i and electric charges e_i satisfies the Schrödinger equation

$$-\frac{\hbar^2}{2} \sum_{i=1}^{N} \frac{1}{m_i} \nabla_i^2 \Psi + \frac{1}{2} \sum_{i=1}^{N} \sum_{k=1}^{N} \frac{e_i e_k}{r_{ik}} \Psi = E\Psi \tag{151.1}$$

with Ψ being normalized according to

$$\int d\tau_1 \int d\tau_2 \ldots \int d\tau_N \, \Psi^* \Psi = 1 . \tag{151.2}$$

Kinetic and potential energy of the system in a state Ψ may be computed from the formulae

$$E_{\mathrm{kin}} = -\frac{\hbar^2}{2} \sum_{i=1}^{N} \frac{1}{m_i} \int d\tau_1 \int d\tau_2 \ldots \int d\tau_N \, \Psi^* \nabla_i^2 \Psi \tag{151.3a}$$

and

$$E_{\mathrm{pot}} = \frac{1}{2} \sum_{i=1}^{N} {\sum_{k=1}^{N}}' e_i e_k \int d\tau_1 \int d\tau_2 \ldots \int d\tau_N \, \Psi^* \frac{1}{r_{ik}} \Psi . \tag{151.3b}$$

A scale transformation,

$$\boldsymbol{r}_i' = \lambda \boldsymbol{r}_i \tag{151.4}$$

keeping (151.2) intact, means that the wave function

$$\Psi(\boldsymbol{r}_1, \boldsymbol{r}_2, \ldots \boldsymbol{r}_N)$$

is replaced by

$$\Psi_\lambda = \lambda^{3N/2}\, \Psi(\lambda \boldsymbol{r}_1, \lambda \boldsymbol{r}_2, \ldots \lambda \boldsymbol{r}_N)\,. \tag{151.5}$$

When introducing (151.5) into the energy expressions (151.3a, b) and passing over to the new variables (151.4), we find

$$\nabla_i^2 = \lambda^2\, \nabla_i'^2 \qquad \frac{1}{r_{ik}} = \lambda \frac{1}{r'_{ik}}$$

so that instead of the true energy of the system

$$E = E_{\rm kin} + E_{\rm pot}\,,$$

we obtain

$$E(\lambda) = \lambda^2\, E_{\rm kin} + \lambda\, E_{\rm pot}\,. \tag{151.6}$$

This function of λ apparently must be a minimum when we select from the set (151.5) of functions the correct solution of the Schrödinger equation, i.e. for $\lambda = 1$. Therefore,

$$\frac{\partial E(\lambda)}{\partial \lambda} = 2\lambda\, E_{\rm kin} + E_{\rm pot}$$

must vanish with $\lambda = 1$, i.e.

$$2\, E_{\rm kin} + E_{\rm pot} = 0\,. \tag{151.7}$$

This is the virial theorem which was to be proved.

NB. The theorem need not hold in approximate solutions. It is, therefore, remarkable that it can be proved for a Thomas-Fermi atom, cf. Problem 175.

Problem 152. Slater determinant

Let the wave function of a many-particle problem with N equal particles be factorized into a product of single-particle wave functions and antisymmetrized according to the Pauli principle. The expectation value of an operator describing the action of an external force field shall be reduced to single-particle integrals.

Solution. Let $u_i(\nu)$ be a single-particle wave function of the ν-th particle in state i, depending on its space coordinates and spin variable, all contracted into the symbol ν. The antisymmetrized product for a system of N equal particles may then be written as the Slater determinant,

$$\psi = C \begin{vmatrix} u_1(1)\, u_1(2) \ldots u_1(N) \\ u_2(1)\, u_2(2) \ldots u_2(N) \\ \ldots\ldots\ldots\ldots\ldots\ldots \\ u_N(1)\, u_N(2) \ldots u_N(N) \end{vmatrix} \tag{152.1}$$

or, by expansion of the determinant,

$$\psi = C \sum_{P} (-1)^P P(u_1, u_2, \dots u_N), \tag{152.2}$$

where P means any permutation of the functions u_i with their arguments ν in standard ordering $1, 2, \dots, N$. If P is an even (odd) permutation, the corresponding term in the sum over all permutations is positive (negative).

An operator Ω describing an external force will act on all particles in the same way, i.e.

$$\Omega = \sum_{\nu=1}^{N} \Omega_\nu . \tag{152.3}$$

Its expectation value then is

$$\langle \psi | \Omega | \psi \rangle = |C|^2 \sum_{P,P'} (-1)^{P+P'} \langle P'(u_1 \dots u_N) | \sum_{\nu} \Omega_\nu | P(u_1, \dots u_N) \rangle . \tag{152.4}$$

Let us now single out of (152.4) one term, Ω_ν, acting only upon functions of the ν-th particle coordinates and spin. Any other coordinate set, say μ, will then occur in some other function, say u_j, in both permutations P and P', because in any other combination the term would vanish by the orthogonality of the single-particle functions,

$$\langle u_j | u_k \rangle = \delta_{jk} . \tag{152.5}$$

This means identity of permutations P and P', the signature in (152.4) always being $+1$, and the term Ω_ν thus contributing only one-particle integrals,

$$\langle P' | \Omega_\nu | P \rangle = \delta_{PP'} \sum_{i=1}^{N} \langle u_i(\nu) | \Omega_\nu | u_i(\nu) \rangle . \tag{152.6}$$

Now, in the wave function ψ a factor $u_i(\nu)$, with fixed i and ν, is combined with a determinant of rank $N-1$. Therefore there still remain, among a total of $N!$ possible permutations, $(N-1)!$ permutations of the remaining $N-1$ functions except u_i over the remaining $N-1$ particles except ν. Hence,

$$\langle \psi | \Omega_\nu | \psi \rangle = |C|^2 (N-1)! \sum_{i=1}^{N} \langle u_i(\nu) | \Omega_\nu | u_i(\nu) \rangle . \tag{152.7}$$

This result, of course, will hold for whatever term Ω_ν we pick out of the sum (152.3), so that in $\langle \psi | \Omega | \psi \rangle$ we have a total of N equal expressions of the form (152.7). Hence,

$$\langle \psi | \Omega | \psi \rangle = |C|^2 N! \sum_{i=1}^{N} \langle u_i(\nu) | \Omega_\nu | u_i(\nu) \rangle . \tag{152.8}$$

It remains to determine the normalization constant C so that

$$\langle\psi|\psi\rangle=1 \tag{152.9}$$

is the probability (viz. certainty) of finding N particles anywhere. This can be formally achieved by putting $\Omega_\nu=1/N$ in Eq. (152.3) thus making $\Omega=1$. Making use of the single-particle normalization, according to (152.5), Eq. (152.8) then yields

$$\langle\psi|\psi\rangle=|C|^2 N!\sum_{i=1}^{N}\frac{1}{N}=|C|^2 N!$$

or, using (152.9),

$$C=N!^{-\frac{1}{2}} . \tag{152.10}$$

Eq. (152.8) may then finally be written

$$\langle\psi|\Omega|\psi\rangle=\sum_{i=1}^{N}\langle u_i(\nu)|\Omega_\nu|u_i(\nu)\rangle \tag{152.11}$$

as the simple sum of the expectation values of the single-particle states.

NB. If we neglect symmetrization and replace (152.1) by the simple product

$$\mathring{\psi}=u_1(1)u_2(2)\ldots u_N(N) \tag{152.12}$$

we get

$$\langle\mathring{\psi}|\Omega|\mathring{\psi}\rangle=\sum_{\nu=1}^{N}\langle u_\nu|\Omega_\nu|u_\nu\rangle \tag{152.13}$$

and

$$\langle\mathring{\psi}|\mathring{\psi}\rangle=1, \tag{152.14}$$

i.e. essentially the same results as we found in (152.11) and (152.9) for the antisymmetrized wave function. Neither for interaction between particles, contradicting the structure of (152.3), nor for the use of non-orthogonal single-particle functions, contradicting (152.5), do these relations hold.

Problem 153. Exchange in interaction terms with Slater determinant

For the factorized, antisymmetrized wave function of the preceding problem the expectation value of a particle-pair interaction,

$$\Omega=\tfrac{1}{2}\sum_{\mu\nu}{}'\Omega_{\mu\nu} \tag{153.1}$$

shall be determined.

Solution. Using the same notation and normalization as in the preceding problem, the expectation value of one term of (153.1) may be written,

$$\langle\psi|\Omega_{\mu\nu}|\psi\rangle = \frac{1}{N!}\sum_{P,P'}(-1)^{P+P'}\langle P'(u_1,\ldots,u_N)|\Omega_{\mu\nu}|P(u_1,\ldots,u_N)\rangle\,. \tag{153.2}$$

All functions u_n with arguments neither μ nor ν must be identical pairs in P and P' if the corresponding term in the sum is not to vanish. There being $N-2$ such pairs of functions of as many arguments, there still remain $(N-2)!$ permutations among them. If these all are identical in P and P', only one pair of functions, say $u_i u_j$, will remain for arguments μ and ν in each non-vanishing term of (153.2),

$$\langle\psi|\Omega_{\mu\nu}|\psi\rangle = \frac{(N-2)!}{N!}\sum_{ij}{}' \{\langle u_i(\mu)u_j(\nu)|\Omega_{\mu\nu}|u_i(\mu)u_j(\nu)\rangle - \langle u_j(\mu)u_i(\nu)|\Omega_{\mu\nu}|u_i(\mu)u_j(\nu)\rangle\}\,. \tag{153.3}$$

In the first, classical term of the curly bracket, the permutations P and P' coincide completely, even with respect to μ and ν, in the second, exchange term one different permutation $(ij\to ji)$ just changes the sign.

Let us now consider the sum (153.1) of such operators. Then,

$$\langle\psi|\Omega|\psi\rangle = \frac{1}{2N(N-1)}\sum_{\mu\nu}{}'\sum_{ij}{}' \{\langle u_i(\mu)u_j(\nu)|\Omega_{\mu\nu}|u_i(\mu)u_j(\nu)\rangle - \langle u_j(\mu)u_i(\nu)|\Omega_{\mu\nu}|u_i(\mu)u_j(\nu)\rangle\}\,. \tag{153.4}$$

Here μ and ν are dummies so that the sum $\sum_{\mu\nu}'$ consists of $N(N-1)$ equal terms (μ, ν and ν, μ here being counted as different terms). Thence, the expectation value wanted, becomes

$$\langle\psi|\Omega|\psi\rangle = \tfrac{1}{2}\sum_{ij}{}' \{\langle u_i(1)u_j(2)|\Omega_{12}|u_i(1)u_j(2)\rangle - \langle u_j(1)u_i(2)|\Omega_{12}|u_i(1)u_j(2)\rangle\}\,. \tag{153.5}$$

Use of the symbols 1 and 2 is, of course, quite arbitrary.

It should be emphasized again that here and in the preceding problem each single-particle wave function u_i comprises space as well as spin state.

Problem 154. Two electrons in the atomic ground state

The K shell of an atom is composed of two electrons in the $1s$ state. Its energy shall be approximated by using screened hydrogen wave functions in the field of a nucleus of charge Ze and infinitely large mass.

Solution. The hamiltonian of the problem is (in atomic units, $e=\hbar=m=1$):

$$H=-\frac{1}{2}(\nabla_1^2+\nabla_2^2)-Z\left(\frac{1}{r_1}+\frac{1}{r_2}\right)+\frac{1}{r_{12}}; \tag{154.1}$$

the approximate wave function is according to the table of Problem 67

$$U=u(r_1)\,u(r_2)=\frac{\alpha^3}{\pi}\,\mathrm{e}^{-\alpha(r_1+r_2)} \tag{154.2}$$

with

$$\alpha=Z-\sigma \tag{154.3}$$

and σ the screening constant. It is to be expected that $0<\sigma<1$, because the nuclear charge, in its effect on each electron, is only partly screened by the other electron. The two factors of (154.2) satisfy the wave equations

$$\left(-\frac{1}{2}\nabla_1^2-\frac{\alpha}{r_1}\right)u(r_1)=-\frac{1}{2}\alpha^2\,u(r_1);$$

$$\left(-\frac{1}{2}\nabla_2^2-\frac{\alpha}{r_2}\right)u(r_2)=-\frac{1}{2}\alpha^2\,u(r_2)$$

so that

$$HU=\left\{\left(-\frac{1}{2}\alpha^2-\frac{Z-\alpha}{r_1}\right)+\left(-\frac{1}{2}\alpha^2-\frac{Z-\alpha}{r_2}\right)+\frac{1}{r_{12}}\right\}U$$

and the energy becomes

$$E=\iint d\tau_1\,d\tau_2\left\{-\alpha^2-\frac{\sigma}{r_1}-\frac{\sigma}{r_2}+\frac{1}{r_{12}}\right\}U^2. \tag{154.4}$$

With U according to (154.2) and using the normalization of each factor u, this yields

$$E=-\alpha^2-2\sigma\frac{\alpha^3}{\pi}\int d\tau_1\frac{\mathrm{e}^{-2\alpha r_1}}{r_1}+\frac{\alpha^6}{\pi^2}\iint d\tau_1\,d\tau_2\frac{\mathrm{e}^{-2\alpha(r_1+r_2)}}{r_{12}}. \tag{154.5}$$

The first integral in (154.5) can be evaluated in an elementary way:

$$\int d\tau_1\frac{\mathrm{e}^{-2\alpha r_1}}{r_1}=4\pi\int_0^\infty dr_1\,r_1\,\mathrm{e}^{-2\alpha r_1}=\frac{\pi}{\alpha^2}. \tag{154.6}$$

To calculate the double integral

$$J=\iint d\tau_1\,d\tau_2\frac{\mathrm{e}^{-2\alpha(r_1+r_2)}}{r_{12}}$$

we expand $1/r_{12}$ into Legendre polynomials of the angle Θ between the vectors $\boldsymbol{r}_1$ and $\boldsymbol{r}_2$:

$$\frac{1}{r_{12}}=\begin{cases}\dfrac{1}{r_2}\sum\limits_{n=0}^{\infty}\left(\dfrac{r_1}{r_2}\right)^n P_n(\cos\Theta) & 0\le r_1\le r_2;\\[2ex] \dfrac{1}{r_1}\sum\limits_{n=0}^{\infty}\left(\dfrac{r_2}{r_1}\right)^n P_n(\cos\Theta) & r_2\le r_1<\infty.\end{cases}$$

Then only the first term of the expansion ($n=0$) contributes to J, and we obtain

$$J=\int d\tau_2\,\mathrm{e}^{-2\alpha r_2}\cdot 4\pi\left\{\frac{1}{r_2}\int_0^{r_2} dr_1\,r_1^2\,\mathrm{e}^{-2\alpha r_1}+\int_{r_2}^{\infty} dr_1\,r_1\,\mathrm{e}^{-2\alpha r_1}\right\}.$$

With $d\tau_2=4\pi r_2^2\,dr_2$ and $0\le r_2<\infty$, all the integrations become elementary again and lead to

$$J=\frac{5\pi^2}{8\alpha^5}. \tag{154.7}$$

The energy expression (154.5) with the integrals (154.6) and (154.7) then becomes

$$E=-\alpha^2-2(Z-\alpha)\alpha+\tfrac{5}{8}\alpha. \tag{154.8}$$

Up to this point, we have not yet disposed of the value of α that we now choose optimally in the sense of variational calculus putting

$$\frac{dE}{d\alpha}=0. \tag{154.9}$$

That leads to

$$\alpha=Z-\tfrac{5}{16} \tag{154.10}$$

and

$$E=-(Z-\tfrac{5}{16})^2. \tag{154.11}$$

It should be remarked that this value of α makes (154.2) the exact solution of the hamiltonian

$$H^0=-\frac{1}{2}(\nabla_1^2+\nabla_2^2)-\alpha\left(\frac{1}{r_1}+\frac{1}{r_2}\right) \tag{154.12}$$

which permits factorization. Comparing (154.12) with (154.2) we find

$$H'=H-H^0=-\sigma\left(\frac{1}{r_1}+\frac{1}{r_2}\right)+\frac{1}{r_{12}}.$$

Defining H' as perturbation, the energy shift of a first-order perturbation theory,

$$\Delta E=\iint d\tau_1\,d\tau_2\,U\,H'\,U,$$

would vanish if the screening constant σ is chosen according to (154.10).

We add a few numerical remarks. The theory describes atoms stripped of all electrons except the two in the K shell. There exist experimental values for

$$\begin{array}{llllll} Z= & 2 & 3 & 4 & 6 & 8 \\ & \mathrm{He} & \mathrm{Li}^{+} & \mathrm{Be}^{++} & \mathrm{C}^{4+} & 0^{6+}. \end{array}$$

In all these cases, it is not E itself which has been observed, but the ionization energy I necessary to strip the ion of only one of the two K electrons. The remaining ion, keeping only its last electron, then has the energy

$$E' = -\tfrac{1}{2} Z^2$$

so that the ionization energy becomes

$$I = (Z - \tfrac{5}{16})^2 - \tfrac{1}{2} Z^2 . \tag{154.13}$$

The accompanying table shows that the agreement of Eq. (154.13) with experiment improves continuously with increasing Z. This is reasonable, because the role of the interaction term $1/r_{12}$ becomes less important as the coupling of each electron to the centre becomes stronger with increasing Z.

Z	I in eV theor.	I in eV exper.
2	23,2	24,5
3	74,1	75,6
4	152,2	153,6
6	390	393
8	737	738

Problem 155. Excited states of the helium atom

In a neutral helium atom let one electron be in the $1s$ ground state and the other in an n, l excited state ($n \geqq 2$, $l \geqq 1$). The ionization energy for the (n, l) electron shall then be determined for both, ortho and parahelium using hydrogen-like wave functions with screening of one nuclear charge by the $1s$ electron. The method shall be applied numerically to the $2p$ state ($n=2$, $l=1$).

Solution. If the $1s$ electron is exposed to the full nuclear charge $2e$, but the (n, l) electron only to the screened charge e, we may describe the two one-electron states by solution of the differential equations

$$\left(-\frac{1}{2}\nabla^2 - \frac{2}{r}\right) u = E_1 u ; \quad \left(-\frac{1}{2}\nabla^2 - \frac{1}{r}\right) v_{nl} = E_n v_{nl} \tag{155.1}$$

with

$$u \equiv |1\rangle = \sqrt{\frac{8}{\pi}}\, e^{-2r}; \qquad E_1 = -2;$$

$$v_{nl} \equiv |n\rangle = R_{nl}(r)\, Y_{l,m}(\vartheta, \varphi); \qquad E_n = -\frac{1}{2n^2} \tag{155.2}$$

and R_{nl} the normalized radial functions of the hydrogen atom (see Problem 67).

The Schrödinger equation of the two-electron problem

$$\left\{-\frac{1}{2}\nabla_1^2 - \frac{1}{2}\nabla_2^2 - \frac{2}{r_1} - \frac{2}{r_2} + \frac{1}{r_{12}}\right\}\psi = E\psi \tag{155.3}$$

shall be approximately solved by the symmetrized product wave function

$$\psi = u(1)\, v_n(2) + \varepsilon\, v_n(1)\, u(2) = |1\,n\rangle + \varepsilon |n\,1\rangle \tag{155.4}$$

with $\varepsilon = +1$ for parahelium (spins antiparallel) and $\varepsilon = -1$ for orthohelium (spins parallel). The function ψ is normalized according to $\langle\psi|\psi\rangle = 2$.

In order to satisfy (155.3) as well as possible by (155.4) we apply $\langle 1\,n|$ to (155.3):

$$\langle 1\,n| -\frac{1}{2}\nabla_1^2 - \frac{1}{2}\nabla_2^2 - \frac{2}{r_1} - \frac{2}{r_2} + \frac{1}{r_{12}}|\psi\rangle = E\langle 1\,n|\psi\rangle\,. \tag{155.5}$$

Since

$$\langle 1|1\rangle = 1; \qquad \langle n|n\rangle = 1; \qquad \langle 1|n\rangle = 0$$

(if $l \neq 0$) the integrals will partially split up and partially vanish. For instance we have

$$\langle 1\,n| -\frac{1}{2}\nabla_1^2 - \frac{2}{r_1}|1\,n\rangle = E_1; \qquad \langle 1\,n| -\frac{1}{2}\nabla_1^2 - \frac{2}{r_1}|n\,1\rangle = 0;$$

$$\langle 1\,n| -\frac{1}{2}\nabla_2^2 - \frac{2}{r_2}|1\,n\rangle = E_n - \langle n|\frac{1}{r}|n\rangle \qquad \text{etc.}$$

Since for all one-electron states the virial theorem (cf. Problem 151) leads to $E_{\text{pot}} = 2E$, we obtain

$$\langle n|\frac{1}{r}|n\rangle = -E_{\text{pot}}^{(n)} = -2E_n = \frac{1}{n^2}\,.$$

Thus, finally, Eq. (155.5) leads to the energy expression

$$E = -2 - \frac{3}{2n^2} + \mathscr{C} + \varepsilon\mathscr{E} \tag{155.6}$$

with the abbreviations

$$\mathscr{C} = \langle 1\,n|\frac{1}{r_{12}}|1\,n\rangle \tag{155.7}$$

for the classical, and

$$\mathscr{E} = \langle 1\,n|\frac{1}{r_{12}}|n\,1\rangle \tag{155.8}$$

for the exchange integral of the electron-electron interaction. It remains to evaluate these two integrals.

In both cases we expand $1/r_{12}$ into spherical harmonics of the angle ϑ_{12} between the position vectors $\boldsymbol{r}_1$ and $\boldsymbol{r}_2$ of the electrons:

$$\frac{1}{r_{12}} = \begin{cases} \dfrac{1}{r_2}\sum\limits_{\lambda=0}^{\infty}\left(\dfrac{r_1}{r_2}\right)^{\lambda} P_\lambda(\cos\vartheta_{12}) & \text{if } r_1 < r_2; \\ \dfrac{1}{r_1}\sum\limits_{\lambda=0}^{\infty}\left(\dfrac{r_2}{r_1}\right)^{\lambda} P_\lambda(\cos\vartheta_{12}) & \text{if } r_1 > r_2. \end{cases} \tag{155.9}$$

When first performing the integrations over the polar angles, we have to calculate the integrals

$$\mathscr{C}_{\text{ang}} = \oint d\Omega_2 |\,Y_{l,m}(2)|^2 \oint d\Omega_1\, P_\lambda(\cos\vartheta_{12}) \tag{155.10}$$

and

$$\mathscr{E}_{\text{ang}} = \oint d\Omega_2\, Y^*_{l,m}(2) \oint d\Omega_1\, Y_{l,m}(1)\, P_\lambda(\cos\vartheta_{12})\,. \tag{155.11}$$

In $\mathscr{C}_{\text{ang}}$, the inner integral becomes $4\pi\delta_{\lambda,0}$ so that of the series (155.9) there remains only a contribution from the term $\lambda=0$. The classical interaction therefore is

$$\mathscr{C} = 4\pi \int_0^\infty dr_2\, r_2^2 |u(r_2)|^2 \left\{ \frac{1}{r_2}\int_0^{r_2} dr_1\, r_1^2 |R_{nl}(r_1)|^2 + \int_{r_2}^\infty dr_1\, r_1 |R_{nl}(r_1)|^2 \right\}. \tag{155.12}$$

In order to calculate the inner integral of (155.11) we use the spherical harmonics addition theorem,

$$P_\lambda(\cos\vartheta_{12}) = \frac{4\pi}{2\lambda+1}\sum_{\mu=-\lambda}^{+\lambda} Y^*_{\lambda,\mu}(1)\, Y_{\lambda,\mu}(2)\,; \tag{155.13}$$

then

$$\oint d\Omega_1\, Y_{l,m}(1)\, P_\lambda(\cos\vartheta_{12}) = \frac{4\pi}{2l+1}\, Y_{l,m}(2)\, \delta_{l,\lambda}$$

and

$$\mathscr{E}_{\text{ang}} = \frac{4\pi}{2l+1}\,\delta_{l,\lambda}. \tag{155.14}$$

The exchange integral therefore receives a contribution only from the term $\lambda = l$ of the series (155.9) so that

$$\mathscr{E} = \frac{4\pi}{2l+1} \int_0^\infty dr_2\, r_2^2\, u(r_2)\, R_{nl}(r_2) \left\{ \frac{1}{r_2} \int_0^{r_2} dr_1\, r_1^2 \left(\frac{r_1}{r_2}\right)^l u(r_1)\, R_{nl}(r_1) + \int_{r_2}^\infty dr_1\, r_1 \left(\frac{r_2}{r_1}\right)^l u(r_1)\, R_{nl}(r_1) \right\}. \tag{155.15}$$

The higher the values of n and l of the excited electron, the more will our approximation improve because of decreasing overlap of the two one-electron wave functions. Except therefore for S states, the method will be worst for $n=2$ and $l=1$, but may be well trusted elsewhere if in that special case it produces reasonable results. We therefore now proceed to calculate the energy for this special excited state of the helium atom and compare the result with experimental evidence.

The normalized radial function R_{nl} then becomes

$$R_{2,1} = \frac{1}{\sqrt{24}}\, r e^{-\frac{r}{2}}, \tag{155.16}$$

and the radial integrals (155.12) and (155.15) can easily, though in a somewhat cumbersome way, be evaluated using (155.2) for u and (155.16) for $R_{2,1}$. The results are

$$\mathscr{C} = \tfrac{1}{4}\left(1 - \tfrac{13}{3125}\right) = 0.24896$$

and

$$\mathscr{E} = \tfrac{14}{3}\left(\tfrac{4}{5}\right)^3 \left(\tfrac{1}{5}\right)^4 = 0.00382.$$

This leads to

$$E = -2.12604 + \varepsilon \cdot 0.00382$$

in atomic units. The ionization energy is the difference of E and the energy $E^+ = -2$ of He^+ in the ground state (i.e. with one electron still in the $1s$ state and the other removed),

$$I = E^+ - E = 0.12604 - \varepsilon \cdot 0.00382$$

or

$$I = (3.429 - \varepsilon \cdot 0.104)\ \mathrm{eV}.$$

This result may be compared with experiment as shown in the table. The results fit quite nicely, and even the splitting between para and ortho states is not as bad as might be expected from its being rather sensitive to the overlap and mutual polarization of the two one-electron functions. It should be noted that the para state with the space symmetric wave function lies *above* the antisymmetric ortho state, the

	ε	ionization energy in eV	
		theory	experiment
para	$+1$	3.325	3.368
ortho	-1	3.533	3.623
difference		0.208	0.255

situation thus being the opposite of the one in the H_2 molecule (Problem 163). This can easily be verified in our calculation where the integral $\mathscr{E}$, Eq. (155.8), gives the only contribution depending on the sign ε; and since it derives from the mutual repulsion of the two electrons, $\mathscr{E}$ is positive, thus raising the energy level for $\varepsilon=+1$.

Problem 156. Excited S states of the helium atom

The method of the preceding problem shall be extended to the configuration $1s$, ns using again the undisturbed wave function for the $1s$ state, but making no specializing assumptions on the ns wave function. It shall be shown that, if overlap and exchange integrals are small, an effective potential field can be constructed in which the ns electron moves.

Solution. The wave function will be written as a symmetrized product of one-electron states,

$$\psi=u(1)\,v_n(2)+\varepsilon\,v_n(1)\,u(2)=|1\,n\rangle+\varepsilon|n\,1\rangle \tag{156.1}$$

with $\varepsilon=\pm 1$ and the $1s$ state function in atomic units

$$u\equiv|1\rangle=\sqrt{\frac{8}{\pi}}\,e^{-2r};\quad \left(-\frac{1}{2}\nabla^2-\frac{2}{r}\right)u=-2u;\quad \langle 1|1\rangle=1\,. \tag{156.2}$$

Of the function for the ns state we know only that it does not depend on angles, and that it too is supposed to be normalized,

$$v_n(r)\equiv|n\rangle;\quad \langle n|n\rangle=1\,. \tag{156.3}$$

No further specialization of $|n\rangle$ will be attempted.

The wave function ψ is an approximate solution of the Schrödinger equation

$$(H-E)\psi=0 \tag{156.4}$$

with the atomic hamiltonian

$$H=-\frac{1}{2}\nabla_1^2-\frac{2}{r_1}-\frac{1}{2}\nabla_2^2-\frac{2}{r_2}+\frac{1}{r_{12}}. \tag{156.5}$$

These are the basic equations of our problem. We start by forming the Hilbert product of (156.4) with $\langle 1\,n|$,

$$\langle 1\,n|H-E|1\,n\rangle+\varepsilon\langle 1\,n|H-E|n\,1\rangle=0 \tag{156.6}$$

and determining the integrals occurring with H given by (156.5). So far the formulae do not yet appreciably differ from those of the preceding problem. One main difference, however, is seen immediately since the two functions $|1\rangle$ and $|n\rangle$ are no longer orthogonal because they both belong to $l=0$ but have different potential fields. Hence we have to introduce the overlap integral

$$S=\langle 1|n\rangle=\langle n|1\rangle\,. \tag{156.7}$$

Further, let us again use the abbreviations

$$\langle 1\,n|\frac{1}{r_{12}}|1\,n\rangle=\mathscr{C}\,;\qquad \langle 1\,n|\frac{1}{r_{12}}|n\,1\rangle=\mathscr{E}\,. \tag{156.8}$$

It then remains to evaluate the following integrals

$$\begin{aligned}
\langle 1\,n|-\frac{1}{2}\nabla_1^2-\frac{2}{r_1}|1\,n\rangle &= \langle 1|-\frac{1}{2}\nabla^2-\frac{2}{r}|1\rangle=-2\,;\\
\langle 1\,n|-\frac{1}{2}\nabla_2^2-\frac{2}{r_2}|1\,n\rangle &= \langle n|-\frac{1}{2}\nabla^2-\frac{2}{r}|n\rangle=K_n;\\
\langle 1\,n|-\frac{1}{2}\nabla_1^2-\frac{2}{r_1}|n\,1\rangle &= \langle 1\,n|-\frac{1}{2}\nabla_2^2-\frac{2}{r_2}|n\,1\rangle=-2S^2\,.
\end{aligned} \tag{156.9}$$

In the last line, the identity

$$\langle 1|\nabla^2|n\rangle=\langle n|\nabla^2|1\rangle$$

has been used. Eq. (156.6) then may be written

$$-2+K_n+\mathscr{C}-E+\varepsilon(-4S^2+\mathscr{E}-E\,S^2)=0$$

or

$$E=-2+\frac{K_n+\mathscr{C}+\varepsilon(\mathscr{E}-2S^2)}{1+\varepsilon S^2}. \tag{156.10}$$

Since $E^+ = -2$ is the energy of the He^+ ion ground state, the ionization energy $I = E^+ - E$ becomes

$$I = -\frac{K_n + \mathscr{C} + \varepsilon(\mathscr{E} - 2S^2)}{1 + \varepsilon S^2}. \tag{156.11}$$

Either E or I may then be determined by evaluating the integrals S, K_n, $\mathscr{C}$, $\mathscr{E}$ for a set of sufficiently pliable functions $|n\rangle$ defined by Ritz parameters and extremizing E or I by their suitable choice.

If the overlap integral S and the exchange integral $\mathscr{E}$ are very small, Eqs. (156.10) and (156.11) simplify to

$$E = -2 + K_n + \mathscr{C}; \qquad I = -(K_n + \mathscr{C}). \tag{156.12}$$

The same expressions would be achieved by neglecting symmetrization (i.e. with $\varepsilon = 0$), and it is in this sense only that symmetrization in many-body problems may occasionally be omitted.

Falling back upon the definitions (156.8) and (156.9) of the integrals K_n and $\mathscr{C}$, (156.12) may be written in more detail

$$E = -2 + \langle n|\Omega|n\rangle \tag{156.13}$$

with the operator

$$\Omega = -\frac{1}{2}\nabla^2 - \frac{2}{r} + \int d\tau' \frac{u(r')^2}{|\boldsymbol{r} - \boldsymbol{r}'|}. \tag{156.14}$$

The choice of such a normalized function $|n\rangle$ as makes E a minimum is performed by variation,

$$\delta(\langle n|\Omega|n\rangle + \lambda\langle n|n\rangle) = 0$$

with a Lagrange multiplicator λ. Since

$$\delta\langle n|\Omega|n\rangle = 2\langle \delta n|\Omega|n\rangle; \quad \delta\langle n|n\rangle = 2\langle \delta n|n\rangle$$

we arrive at

$$\langle \delta n|\Omega + \lambda|n\rangle = 0$$

or, $|\delta n\rangle$ being an arbitrary function, at the differential equation

$$(\Omega + \lambda)|n\rangle = 0.$$

Rewriting (156.13) in the form

$$\langle n|\Omega - E - 2|n\rangle = 0$$

we see that $\lambda = -E - 2$. Hence, $|n\rangle$ is to satisfy the one-electron Schrödinger equation

$$-\tfrac{1}{2}\nabla^2|n\rangle + V_{\text{eff}}(r)|n\rangle = (E+2)|n\rangle \tag{156.15}$$

with the effective potential

$$V_{\text{eff}}(r) = -\frac{2}{r} + \int d\tau' \frac{u(r')^2}{|\boldsymbol{r}-\boldsymbol{r}'|}. \tag{156.16}$$

This is exactly the electrostatic potential to be derived from the Poisson equation $\nabla^2 V_{\text{eff}} = +4\pi\rho$ with the charge density ρ composed of the nuclear point charge $+2$ and the negative space charge $-u^2$ of the $1s$ electron. Evaluation of (156.16) with the wave function u of Eq. (156.2) renders

$$V_{\text{eff}} = -\frac{1}{r} - \left(\frac{1}{r} + 2\right) e^{-4r}. \tag{156.17}$$

If $|n\rangle$ is determined from (156.15), this may be considered a sufficient approximation to *all* integrals of the energy expression (156.10) as long as $S \ll 1$ and $\mathscr{E} \ll \mathscr{C}$ are no more than slight corrections.

Appendix. Numerical calculations along the general outline sketched before may become rather involved. As an example, we give numerical results for an abridged variational procedure making (156.12) extremal, which we have performed for the $2s$ state using the set of trial functions

$$|2\rangle = A\left(e^{-2r} - p r e^{-\frac{1}{2}r}\right) \tag{156.18}$$

normalized according to $\langle 2|2\rangle = 1$ and defined by the Ritz parameter p. These functions have finite value at $r=0$, they have a zero as necessary for a $2s$ state, and show the correct asymptotic behaviour determined by the second term. The first term describes deviations from hydrogen-like behaviour at small distances where the nuclear charge becomes less and less completely screened; since the $1s$ distribution is described by e^{-2r}, this effect should show approximately the same dependence on distance.

The approximate function (156.18) leads to the following numerical results.

$$\pi A^2 = \left(\tfrac{1}{8} - \tfrac{768}{625}p + 96p^2\right)^{-1};$$

$$K_2 = 4\pi A^2\left(-\tfrac{1}{16} + \tfrac{384}{625}p - 11p^2\right);$$

$$\mathscr{C} = \pi A^2\left(\tfrac{5}{32} - \tfrac{512\cdot 6359}{125\cdot 28561}p + \tfrac{74688}{3125}p^2\right);$$

$$S = 8\sqrt{2\pi}\, A\left(\tfrac{1}{32} - \tfrac{96}{625}p\right);$$

$$\mathscr{E} = 128\pi A^2\left(\tfrac{5}{4096} - \tfrac{25436}{3570125}p + \tfrac{1056}{78125}p^2\right).$$

The energy expression (156.12) has then been minimized by suitable choice of p. This leads to a quadratic equation in p with one positive solution $p=0.1105$. Without symmetrization this yields an ionization energy $I=-(K_2+\mathscr{C})=0.145$ atomic units or $I=3.94$ eV. If, with the same value of p, the full energy expression (156.11) is evaluated we find

$$I=\frac{0{,}145-0{,}021\,\varepsilon}{1+0{,}0225\,\varepsilon}\ \text{atomic units}$$

and arrive at the tabulated results. The theoretical approximation,

symmetry	ionization energy for $2s$ in eV	
	theory	experiment
para, $\varepsilon=+1$	3.30	3.97
ortho, $\varepsilon=-1$	4.62	4.76
difference	1.32	0.79

as always in variational procedures, leads to somewhat higher energy term values than the correct ones. The rather large ortho-para splitting is reproduced with an accuracy of about 35% even by this very simplified approximation.

Problem 157. Lithium ground state

To calculate the binding energy of a lithium atom ($Z=3$) in its ground state. For the two $1s$ electrons the screened hydrogen-like functions of Problem 154 may be used. Exchange is to be neglected.

Solution. The hamiltonian of the problem is

$$H=\left\{-\frac{1}{2}(\nabla_1^2+\nabla_2^2)-3\left(\frac{1}{r_1}+\frac{1}{r_2}\right)+\frac{1}{r_{12}}\right\}$$
$$+\left\{-\frac{1}{2}\nabla_3^2-\frac{1}{r_3}\right\}+\left\{\frac{1}{r_{13}}+\frac{1}{r_{23}}-\frac{2}{r_3}\right\} \tag{157.1}$$

where the first curly bracket corresponds to the two-electron problem of Li^+, the second bracket leads to the $2s$ function of the third electron in the field of the screened nucleus of rest charge 1, and in the third term the remaining interactions are assembled. The treatment of the third electron thus indicated would be quite correct, were the radius of the K shell very small compared to the extension of the $2s$ wave function; since it is not, the use of a hydrogen function with central charge 1 for the third electron is an approximation.

We write the eigenfunction in product form

$$U(1,2,3)=u(1)\,u(2)\,v(3). \tag{157.2}$$

Here $u(r)$ means the $1s$ function

$$u(r)=\frac{\alpha^{\frac{3}{2}}}{\sqrt{\pi}}\,e^{-\alpha r} \tag{157.3}$$

and

$$\alpha=Z-\tfrac{5}{16}=2.6875 \tag{157.4}$$

the effective nuclear charge originated by the mutual screening effect of the two $1s$ electrons in Li^+ (cf. Problem 154). From Problem 154 we further gather that the energy of the $(1s)^2$ state,

$$E^+=\iint d\tau_1\,d\tau_2\,u(1)\,u(2)\left\{-\frac{1}{2}(\nabla_1^2+\nabla_2^2)-3\left(\frac{1}{r_1}+\frac{1}{r_2}\right)+\frac{1}{r_{12}}\right\}u(1)\,u(2)\,, \tag{157.5}$$

becomes

$$E^+=-\alpha^2\,. \tag{157.6}$$

For the third electron we take the eigenfunction from the table of Problem 67 (hydrogen problem); in its lowest state, $2s$, it is

$$v(r)=\frac{1}{\sqrt{8\pi}}\left(1-\tfrac{1}{2}r\right)e^{-\frac{1}{2}r} \tag{157.7}$$

and satisfies the differential equation

$$\left(-\frac{1}{2}\nabla_3^2-\frac{1}{r_3}\right)v(r_3)=-\tfrac{1}{8}v(r_3)\,. \tag{157.8}$$

If we put the functions (157.3) and (157.7) into the energy expression,

$$E=\iiint d\tau_1\,d\tau_2\,d\tau_3\,u(1)\,u(2)\,v(3)\,H\,u(1)\,u(2)\,v(3) \tag{157.9}$$

with the hamiltonian (157.1), the first bracket of (157.1) will contribute E^+, Eq. (157.6), and the second bracket $-\frac{1}{8}$, according to (157.8), so that

$$E=E^+-\frac{1}{8}+\iiint d\tau_1\,d\tau_2\,d\tau_3\,u(1)^2\,u(2)^2\,v(3)^2\left(\frac{1}{r_{13}}+\frac{1}{r_{23}}-\frac{2}{r_3}\right). \tag{157.10}$$

The last integral has still to be evaluated. It may be simplified into

$$J=2\int d\tau_3\,v(3)^2\left\{\int d\tau_1\,\frac{u(1)^2}{r_{13}}-\frac{1}{r_3}\right\}. \tag{157.11}$$

Using [cf. Problem 44, Eq. (44.19)]

$$\int d\tau_1\,\frac{u(1)^2}{r_{13}}=\frac{1}{r_3}\left\{1-(1+\alpha r_3)\,e^{-2\alpha r_3}\right\}$$

we may combine

$$J = -2 \int d\tau_3 \frac{v(3)^2}{r_3} (1+\alpha r_3)\, e^{-2\alpha r_3}.$$

With v according to (157.7) this integral is elementary though cumbersome and yields

$$J = -\frac{\frac{1}{2}+3\alpha+16\alpha^3}{(1+2\alpha)^5}. \qquad (157.12)$$

The energy of the ground state of the lithium atom then becomes

$$E = -\alpha^2 - \frac{1}{8} - \frac{\frac{1}{2}+3\alpha+16\alpha^3}{(1+2\alpha)^5}. \qquad (157.13)$$

Numerical computation with the value (157.4) for α gives the ionization energy

$$I = E^+ - E = \frac{1}{8} - \frac{\frac{1}{2}+3\alpha+16\alpha^3}{(1+2\alpha)^5} \qquad (157.14)$$

as

$$I = 0.1553 = 4.23 \text{ eV}.$$

This is to be compared with the experimental value of 5.37 eV. The approximation, of course, is not very good. The reason for the difference is to be sought neither in the use of complete screening of the third electron by the K shell, nor in neglecting any small difference of the α values between atom and ion. Both these corrections are far too small to account for a discrepany of more than 1 eV. There remain two features of the wave function which may still account for it: its product form, and the neglect of symmetrization and hence of exchange binding.

Problem 158. Exchange correction to lithium ground state

To correct the energy of the lithium ground state, found in the preceding problem, by taking account of the correct symmetry of the eigenfunction.

Solution. The eigenfunction must again describe a state with two electrons in the $1s$ state $u(r)$ and one electron in the $2s$ state $v(r)$, as defined in the preceding problem. Symmetrization requires the inclusion of spins. A totally antisymmetrical eigenfunction is then obtained by the Slater determinant (see Problem 152)

$$\psi = \frac{1}{\sqrt{6}} \begin{vmatrix} u(1)\,\alpha(1); & u(2)\,\alpha(2); & u(3)\,\alpha(3) \\ u(1)\,\beta(1); & u(2)\,\beta(2); & u(3)\,\beta(3) \\ v(1)\,\alpha(1); & v(2)\,\alpha(2); & v(3)\,\alpha(3) \end{vmatrix} \qquad (158.1)$$

in which the spin functions α and β describe opposite spin directions. The determinant (158.1) is an approximate solution of the Schrödinger equation

$$(H-E)\psi=0 \tag{158.2}$$

with H the hamiltonian of the preceding problem. From

$$\langle\psi|H-E|\psi\rangle=0\,,$$

where the scalar Hilbert product includes spin summation, we then are led on, by performing the sum over spins, to

$$\iiint d\tau_1\,d\tau_2\,d\tau_3\{u(1)u(2)v(3)-v(1)u(2)u(3)\}(H-E)u(1)u(2)v(3)=0 \tag{158.3}$$

or, with the abbreviations

$$\tilde{E}=\iiint d\tau_1\,d\tau_2\,d\tau_3\,u(1)u(2)v(3)\,H\,u(1)u(2)v(3)\,; \tag{158.4}$$

$$\mathscr{E}=\iiint d\tau_1\,d\tau_2\,d\tau_3\,v(1)u(2)u(3)\,H\,u(1)u(2)v(3)\,; \tag{158.5}$$

$$S=\int d\tau_1\,v(1)u(1)\,, \tag{158.6}$$

to the corrected energy formula

$$E=\frac{\tilde{E}-\mathscr{E}}{1-S^2}. \tag{158.7}$$

Here, by $\tilde{E}$ we denote the uncorrected energy as determined in Eq. (157.13) of the preceding problem, viz.

$$\tilde{E}=-(\alpha^2+\tfrac{1}{8})+J;\qquad J=-\frac{\tfrac{1}{2}+3\alpha+16\alpha^3}{(1+2\alpha)^5}\,. \tag{158.8}$$

$\mathscr{E}$ is the exchange energy, and S the overlap integral of the functions u and v which, as we know, are not orthogonal. The main problem still remaining will then be the evaluation of the exchange energy (158.5). If we write the hamiltonian in the form

$$\begin{aligned}H=\left(-\frac{1}{2}\nabla_1^2-\frac{\alpha}{r_1}\right)+\left(-\frac{1}{2}\nabla_2^2-\frac{\alpha}{r_2}\right)+\left(-\frac{1}{2}\nabla_3^2-\frac{1}{r_3}\right)\\ -\frac{3-\alpha}{r_1}-\frac{3-\alpha}{r_2}-\frac{2}{r_3}+\frac{1}{r_{12}}+\frac{1}{r_{13}}+\frac{1}{r_{23}}\end{aligned} \tag{158.9}$$

then its application to $u(1)u(2)v(3)$ leads in the first line of (158.9) simply to multiplication by $-\frac{1}{2}\alpha^2$, $-\frac{1}{2}\alpha^2$, $-\frac{1}{8}$, respectively, so that we get

$$\mathscr{E} = -(\alpha^2+\tfrac{1}{8})\,S^2 - (3-\alpha)\,S\int d\tau_1 \frac{u(1)\,v(1)}{r_1} - (3-\alpha)\,S^2\int d\tau_2 \frac{u(2)^2}{r_2}$$
$$-2S\int d\tau_3 \frac{u(3)\,v(3)}{r_3} + S\int\int d\tau_1\,d\tau_2 \frac{u(1)\,v(1)\cdot u(2)^2}{r_{12}}$$
$$+\int\int d\tau_1\,d\tau_3 \frac{u(1)\,v(1)\cdot u(3)\,v(3)}{r_{13}} + S\int\int d\tau_2\,d\tau_3 \frac{u(2)^2\cdot u(3)\,v(3)}{r_{23}}.$$

Using the further abbreviations

$$U=\int d\tau \frac{u^2}{r};\qquad V=\int d\tau \frac{u\,v}{r}; \tag{158.10}$$

$$X=\int\int d\tau\,d\tau' \frac{u(r)\,v(r)\cdot u(r')\,v(r')}{|\boldsymbol{r}-\boldsymbol{r}'|}; \tag{158.11}$$

$$Y=\int\int d\tau\,d\tau' \frac{u(r)^2\cdot u(r')\,v(r')}{|\boldsymbol{r}-\boldsymbol{r}'|}, \tag{158.12}$$

the exchange energy becomes

$$\mathscr{E} = -(\alpha^2+\tfrac{1}{8})\,S^2 - (5-\alpha)\,S\,V - (3-\alpha)\,S^2\,U + 2\,S\,Y + X \tag{158.13}$$

and the corrected energy (158.7),

$$E=\tilde{E}+\frac{[J+(3-\alpha)\,U]\,S^2+[(5-\alpha)\,V-2\,Y]\,S-X}{1-S^2}. \tag{158.14}$$

As a last step, we may now proceed to evaluate the integrals S, Eq. (158.6), U, V, X, Y, Eqs. (158.10–12), with the functions

$$u(r)=\frac{\alpha^{\frac{3}{2}}}{\sqrt{\pi}}\,\mathrm{e}^{-\alpha r};\qquad v(r)=\frac{1}{\sqrt{8\pi}}(1-\tfrac{1}{2}r)\,\mathrm{e}^{-\frac{1}{2}r}.$$

A little difficulty arises only in the two-particle integrals X and Y where $1/|\boldsymbol{r}-\boldsymbol{r}'|$ can be expanded into Legendre polynomials for $\cos(\boldsymbol{r},\boldsymbol{r}')$. Only the term P_0 of this series contributes, since neither u nor v depend upon polar angles, hence the inner integral in X and Y becomes

$$\int d\tau' \frac{u(r')\,v(r')}{|\boldsymbol{r}-\boldsymbol{r}'|} = 4\pi\left\{\frac{1}{r}\int_0^r dr'\,r'^2\,u(r')\,v(r') + \int_r^\infty dr'\,r'\,u(r')\,v(r')\right\}.$$

All other computations are elementary and yield the following results:

$$S=\frac{2\sqrt{2}\,\alpha^{\frac{3}{2}}(\alpha-1)}{(\alpha+\frac{1}{2})^4};\qquad U=\alpha;\qquad V=\frac{\alpha^2-\frac{1}{4}}{2(\alpha-1)}S;$$

$$X=\frac{\alpha^3(\frac{5}{2}\alpha^2-\frac{15}{4}\alpha+\frac{13}{8})}{(\alpha+\frac{1}{2})^7};$$

$$Y=\frac{66\alpha^6+26\alpha^5-25\alpha^4-16\alpha^3-\frac{23}{8}\alpha^2-\frac{1}{8}\alpha}{2(\alpha-1)(3\alpha+\frac{1}{2})^4}S.$$

Numerical values with $\alpha=2.6875$ are

$$\begin{aligned}
&S=0.203;\\
&J=-0.030;\\
&U=2.6875 \qquad [J+(3-\alpha)U]\,S^2=+0.0334;\\
&V=0.419 \qquad [(5-\alpha)V-2Y]\,S=+0.0735;\\
&X=0.0558;\\
&Y=0.303.
\end{aligned}$$

The two positive contributions to the numerator in (158.14) therefore exceed the negative term $-X$, so that the exchange correction yields a smaller binding energy of the lithium atom. The reason for this rather unhappy result lies in the choice of $v(r)$ which is far too small in the overlap zone, as the increasing effective charge to which the $2s$ electron is subjected when penetrating into the $1s$ core has been neglected. This causes a small error only in the uncorrected energy of the preceding problem, but will have a large effect upon the exchange correction. Since V and Y are linear, but X is quadratic in the overlap product uv, the third (negative) term in the numerator of (158.14) should be much larger whereas the second (positive) term would be only moderately increased by using a better approximation for $v(r)$, so that the entire expression may easily change its sign.

Problem 159. Dielectric susceptibility

Let the states of an atom be described by the Schrödinger equation $H|n\rangle=E_n\cdot|n\rangle$, its ground state being $|0\rangle$. The dielectric polarizability α of the atom (or the susceptibility χ of a substance consisting of N atoms per cm^3) shall be determined. What can, in general, be said on the polarizability of alkali atoms?

Solution. If an electrical field $\mathscr{E}$ is applied to the atom in z direction, it causes a perturbation energy

$$W = e\mathscr{E} \sum_\lambda z_\lambda \tag{159.1}$$

with $-e$ the electron charge and the subscript λ numbering the atomic electrons. The equation

$$(H+W)\,\psi = E\psi \tag{159.2}$$

is then approximately solved by

$$\psi = |0\rangle + \sum_n{}' \frac{\langle n|W|0\rangle}{E_0 - E_n} |n\rangle$$

or

$$\psi = |0\rangle + e\mathscr{E} \sum_n{}' \frac{\langle n|\sum z_\lambda|0\rangle}{E_0 - E_n} |n\rangle\,. \tag{159.3}$$

The component in field direction of the dipole moment of this state has the expectation value

$$p_z = -e\langle\psi|\sum_\lambda z_\lambda|\psi\rangle\,. \tag{159.4}$$

In first order this is composed of two terms:

$$p_z = -e\left\{\langle 0|\sum_\lambda z_\lambda|0\rangle \right.$$
$$\left. + \sum_n{}' \left[\frac{\langle n|W|0\rangle}{E_0 - E_n} \langle 0|\sum_\lambda z_\lambda|n\rangle + \frac{\langle 0|W|n\rangle}{E_0 - E_n} \langle n|\sum_\lambda z_\lambda|0\rangle\right]\right\}.$$

The first term is the moment of the undisturbed state (if any). The second term describes the moment induced by the field. Denoting this latter moment by p_{ind} the polarizability α is defined by

$$p_{\text{ind}} = \alpha\mathscr{E}\,. \tag{159.5}$$

We thus find

$$\alpha = 2e^2 \sum_n{}' \frac{|\langle n|\sum z_\lambda|0\rangle|^2}{E_n - E_0}\,. \tag{159.6}$$

Since E_0 is supposed to be the ground state, the denominator will be positive. So therefore will be the polarizability, too.

The susceptibility χ connects the dielectric polarization $P = N p_{\text{ind}}$ with the field,

$$P = \chi\mathscr{E} \tag{159.7}$$

so that

$$\chi = 2Ne^2 \sum_n{}' \frac{|\langle n|\sum z_\lambda|0\rangle|^2}{E_n - E_0} \tag{159.8}$$

and $\chi > 0$.

For alkali atoms consisting of a core plus one single outer electron, the excitation of a core electron requires much energy, thus leading to a large denominator in (159.8). For a rough orientation it therefore suffices to study excitations of the outer electron only, moving in the field of an unaffected core. Its eigenfunctions may be written

$$|0\rangle = u(r); \qquad |n\rangle = v_n(r)\, Y_{l,m}(\vartheta,\varphi)$$

because the ground state $|0\rangle$ is an s state not depending on polar angles. With

$$z = r\cos\vartheta = r\sqrt{\frac{4\pi}{3}}\, Y_{1,0}$$

the matrix element then becomes

$$\langle n|z|0\rangle = \int_0^\infty dr\, r^3 v_n(r)\, u(r)\cdot \oint d\Omega\, Y^*_{l,m}\cos\vartheta\,.$$

This vanishes, except for excitation of states with $l=1$ and $m=0$, when it becomes

$$\langle n|z|0\rangle = \sqrt{\frac{4\pi}{3}}\int_0^\infty dr\, r^3 v_n(r)\, u(r)\,.$$

Further computations imply detailed knowledge of the radial parts of the eigenfunctions.

If dimensionless units are not being used, it is easily seen that the polarizability α has the dimension of a volume so that it will, roughly speaking, be of the order of $(\hbar^2/me^2)^3$.

NB. One can as well calculate the second-order Stark effect with an energy shift

$$\Delta_2 E = \sum_n{}' \frac{|\langle 0|W|n\rangle|^2}{E_0 - E_n}$$

which must be $= -\frac{1}{2}\alpha\mathscr{E}^2$. The result for α is the same as above.

Problem 160. Diamagnetic susceptibility of neon

To compute the diamagnetic susceptibility of neon ($Z=10$) using hydrogen-like wave functions with different screening constants $\sigma_{n,l}$. The following screening constants may be used:

$$\sigma_{1,0}=0.23 \qquad \sigma_{2,0}=3.26 \qquad \sigma_{2,1}=4.11\,.$$

Solution. The diamagnetic susceptibility per mole is given by the formula [cf. (128.14)]

$$\chi = -\frac{e^2}{6mc^2} N \sum \langle r^2 \rangle \tag{160.1}$$

where N is the Loschmidt number ($N = 6.02 \times 10^{23}$) and the sum is to be extended over all electrons of one atom (or molecule). The expectation values of r^2 in states with eigenfunctions

$$u_{n,l,m} = \frac{1}{r} \chi_{n,l}(r)\, Y_{l,m}(\vartheta, \varphi)$$

are given by the integrals

$$\langle r^2 \rangle = \int_0^\infty dr\, r^2 |\chi_{n,l}|^2 . \tag{160.2}$$

The radial parts of hydrogen-like wave functions may be taken from the table of Problem 67, with Z replaced by $Z - \sigma$; the resulting values of the integrals (160.2), in units of $(\hbar^2/me^2)^2$ then become[3]

for $(n,l) =$	$(1,0)$	$(2,0)$	$(2,1)$
$(Z-\sigma)^2 \langle r^2 \rangle =$	3	42	30

as may be checked by elementary integrations. The respective numbers of electrons in the three (n,l) states are 2, 2, 6. The order of magnitude of the susceptibilities will be determined by the factor

$$\chi_0 = \frac{e^2}{6mc^2} N \left(\frac{\hbar^2}{me^2} \right)^2 = 0.790 \times 10^{-6}\ \mathrm{cm^3/mole} . \tag{160.3}$$

We thus obtain for the susceptibility of neon:

$$\begin{aligned} \chi_{\mathrm{Ne}} &= -\chi_0 \left\{ \frac{2 \cdot 3}{(10-\sigma_{1,0})^2} + \frac{2 \cdot 42}{(10-\sigma_{2,0})^2} + \frac{6 \cdot 30}{(10-\sigma_{2,1})^2} \right\} \\ &= -5.61 \times 10^{-6}\ \mathrm{cm^3/mole} . \end{aligned} \tag{160.4}$$

This result may be compared with the experimental value of

$$\chi_{\mathrm{Ne}} = -6.7 \times 10^{-6}\ \mathrm{cm^3/mole} .$$

---

[3] These are special cases of the general relation

$$\langle r^2 \rangle = \frac{n^2}{2Z^2} \{5n^2 + 1 - 3l(l+1)\}$$

the deduction of which is cumbersome and not very interesting. For details see Bethe, H. A., Salpeter, E. E., *Encyclopedia of Physics*, vol. **35** (1957), p. 103.

It should be remarked that the contribution of the outermost subshell (n,l) is by far the biggest of the three terms in (160.4) contributing $\chi(1s)=-0.05$, $\chi(2s)=-1.46$ and $\chi(2p)=-4.10\,\mathrm{cm}^3/\mathrm{mole}$. Unfortunately the screening effect is not only very big for the outermost electrons but also rather uncertain experimentally.

Problem 161. Van der Waals attraction

Two hydrogen atoms in their *ground states* are at a distance R from each other. The nuclei may be supposed at rest. It shall be shown that the interaction between the two atoms vanishes in first order of a perturbation calculation, and that a second-order approximation leads to Van der Waals attraction. Of the interaction part of the hamiltonian, only the leading term proportional to the lowest negative power of R is to be used (large distance approximation).

Solution. Let us denote the position of electron 1 relative to nucleus a by $\boldsymbol{r}_1$ with components $x_1\ y_1\ z_1$ and the position of electron 2 relative to nucleus b by $\boldsymbol{r}_2$ with components $x_2\ y_2\ z_2$. The z direction shall

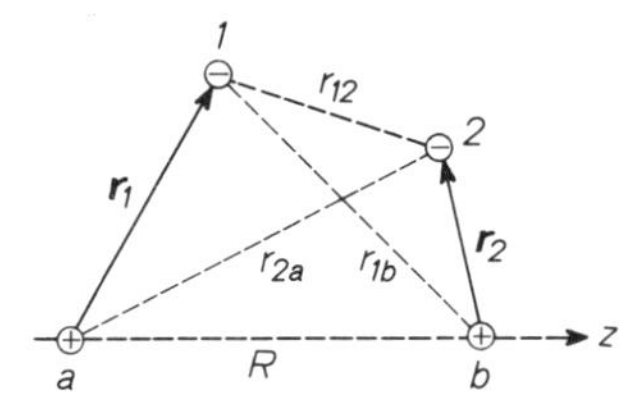

Fig. 63. Notations. The distances marked by broken lines enter the interaction (161.3)

coincide with the nuclear axis (Fig. 63). With the nuclei at rest, i.e. in Born-Oppenheimer approximation, we then have the hamiltonian

$$H=H^0+H' \tag{161.1}$$

with

$$H^0=-\frac{\hbar^2}{2m}(\nabla_1^2+\nabla_2^2)-\frac{e^2}{r_1}-\frac{e^2}{r_2} \tag{161.2}$$

describing two independent atoms, and

$$H'=\frac{e^2}{R}+\frac{e^2}{r_{12}}-\frac{e^2}{r_{1\mathrm{b}}}-\frac{e^2}{r_{2\mathrm{a}}} \tag{161.3}$$

their mutual interaction. We shall use H' as perturbation energy. If H' is expanded in a series of negative powers of R (i.e. for $r_1 \ll R$ and $r_2 \ll R$), the large-distance main term comes from the interaction of the two dipoles a1 and b2 with dipole moments $\boldsymbol{p}_1 = -e\boldsymbol{r}_1$ and $\boldsymbol{p}_2 = -e\boldsymbol{r}_2$ respectively, viz.

$$H' = \frac{\boldsymbol{p}_1 \boldsymbol{p}_2}{R^3} - 3\frac{(\boldsymbol{p}_1 \boldsymbol{R})(\boldsymbol{p}_2 \boldsymbol{R})}{R^5}. \tag{161.4}$$

In coordinate formulation, (161.4) becomes

$$H' = \frac{e^2}{R^3}(x_1 x_2 + y_1 y_2 - 2 z_1 z_2). \tag{161.5}$$

This expression shall be used in the following calculations.

Let $u_0(r)$ be the wave function of the atomic ground state. The zero-order wave function of the entire system then is the product

$$U(1,2) = u_0(r_1)\, u_0(r_2) \tag{161.6}$$

where symmetrization has been omitted since exchange contributions tend exponentially towards zero at large distances R and may thus be neglected.

In zero-order approximation the sum of the two atom energies is the energy of the system. In first order we have to add

$$E' = \langle U|H'|U\rangle = 0.$$

It can easily be seen that this term needs must vanish. Taking, e.g., the first term of (161.5), we have

$$\frac{e^2}{R^3}\langle U|x_1 x_2|U\rangle = \frac{e^2}{R^3}\langle u_0|x|u_0\rangle^2 = \frac{e^2}{R^3}\{\int d\tau\, u_0^2(r)\, x\}^2,$$

and these integrals describing dipole moments of atomic states with spherical symmetry indeed vanish[4].

The second-order energy perturbation is

$$E'' = \sum_n{}' \frac{|\langle 0|H'|n\rangle|^2}{E_0 - E_n}, \tag{161.7}$$

[4] They vanish not only for S states but always, if both atoms are in the same state, see the following problem. Even for two excited states, $|u_0|^2$ depends upon polar angles as the square of a spherical harmonic. This can be decomposed into a sum of spherical harmonics of *even* orders only. In the integrand it is multiplied by a coordinate x, y, or z, i.e. with a spherical harmonic of first, hence of *odd* order. Orthogonality properties of spherical harmonics thus cause the product integral to vanish.

the sum being extended over all excited states, and 0 referring to the ground state. As $E_n > E_0$, all denominators of the sum are negative so that $E'' < 0$ and there arises an attraction. The matrix elements depend upon R only through the constant factor R^{-3} in Eq. (161.5) which stands in front of each integral. E'' therefore must be of the form

$$E'' = -\frac{C}{R^6}$$

with C a positive constant. This is, however, the well-known distance dependence of the Van der Waals attraction.

Literature. Schiff, L. I.: *Quantum Mechanics.* New York 1949. pp. 174—178.

Problem 162. Excitation degeneracy

Two hydrogen atoms at rest, a distance R apart, shall be in *different* quantum states, one in the ground state, the other in a P state. Dipole-dipole interaction exists between them, as deduced in the preceding problem. It shall be shown that now there is a non-vanishing first-order contribution to the energy of the system, even at large distances where overlap of eigenfunctions may again be neglected. This first-order energy perturbation shall be calculated.

Solution. Let $|l\,m\rangle$ be written for an atomic wave function with quantum numbers l and m. The ground state may then be described by $|00\rangle$, and the three possible P states by $|1\,m\rangle$ with $m=1,\ 0,\ -1$. Zero-order wave functions of the system, in product form, are then

$$|00, 1\,m\rangle \quad \text{and} \quad |1\,m, 00\rangle, \tag{162.1}$$

the first pair of quantum numbers referring to the first, the second pair to the second atom.

The perturbation energy (161.5) may be used. Introducing the symbols

$$\xi = x + i\,y, \qquad \xi^\dagger = x - i\,y \tag{162.2}$$

it may be reshaped into

$$H' = \frac{e^2}{2R^3}(\xi_1\,\xi_2^\dagger + \xi_1^\dagger\,\xi_2 - 4\,z_1\,z_2)\,. \tag{162.3}$$

This operator, being linear in the coordinates of either electron, has matrix elements with functions of type (162.1) different from zero only if for each of the two electrons an S and a P state are combined[5]. These are the matrix elements

$$\langle 1\,m_1,00|H'|00,1\,m_2\rangle \quad \text{and} \quad \langle 00,1\,m_2|H'|1\,m_1,00\rangle\,.$$

The bilinear structure of H', Eq. (162.3), allows decomposition into atomic matrix elements:

$$\begin{aligned}\langle 1\,m_1,00|H'|00,1\,m_2\rangle = \frac{e^2}{2R^3}\{&\langle 1\,m_1|\xi|00\rangle\langle 00|\xi^\dagger|1\,m_2\rangle\\ &+\langle 1\,m_1|\xi^\dagger|00\rangle\langle 00|\xi|1\,m_2\rangle\\ &-4\langle 1\,m_1|z|00\rangle\langle 00|z|1\,m_2\rangle\}\,. \qquad (162.4)\end{aligned}$$

Denoting the radial parts of the atomic wave functions by $f_l(r)$ and using the abbreviation

$$r_0 = \int_0^\infty dr\, r^3 f_0(r) f_1(r)\,, \qquad (162.5)$$

the following results are obtained for the non-vanishing integrals occurring in (162.4):

$$\left.\begin{aligned}\langle 1\,1|\xi|00\rangle &= \sqrt{\tfrac{2}{3}}\,r_0\,; & \langle 00|\xi|1-1\rangle &= -\sqrt{\tfrac{2}{3}}\,r_0\,;\\ \langle 1-1|\xi^\dagger|00\rangle &= -\sqrt{\tfrac{2}{3}}\,r_0\,; & \langle 00|\xi^\dagger|1\,1\rangle &= \sqrt{\tfrac{2}{3}}\,r_0\,;\\ \langle 1\,0|z|00\rangle &= \sqrt{\tfrac{1}{3}}\,r_0\,; & \langle 00|z|1\,0\rangle &= \sqrt{\tfrac{1}{3}}\,r_0\,.\end{aligned}\right\} \qquad (162.6)$$

All other combinations of quantum numbers lead to vanishing matrix elements. Eq. (162.4) then becomes

$$\begin{aligned}\langle 1\,m_1,00|H'|00,1\,m_2\rangle = \frac{e^2 r_0^2}{2R^3}\{&\tfrac{2}{3}\delta_{1,m_1}\delta_{1,m_2}+\tfrac{2}{3}\delta_{-1,m_1}\delta_{-1,m_2}\\ &-\tfrac{4}{3}\delta_{0,m_1}\delta_{0,m_2}\}\,. \qquad (162.7)\end{aligned}$$

Thus, only with $m_1 = m_2$ the matrix element (162.7) does not vanish.

We are now prepared to write down the secular determinant of the six degenerate zero-order wave functions (162.1). If E' is the first-order perturbation and the six functions are used in the order

$$|00,11\rangle\,;\quad |11,00\rangle\,;\quad |00,10\rangle\,;\quad |10,00\rangle\,;\quad |00,1-1\rangle\,;\quad |1-1,00\rangle$$

[5] Coupling of any states with even l and odd $l' = l \pm 1$ would do, e.g. $P-D$ coupling. This however exceeds the limitations set by Eq. (162.1).

the secular equation becomes

$$\begin{vmatrix} -E' & \frac{2}{3}\varepsilon & 0 & 0 & 0 & 0 \\ \frac{2}{3}\varepsilon & -E' & 0 & 0 & 0 & 0 \\ 0 & 0 & -E' & -\frac{4}{3}\varepsilon & 0 & 0 \\ 0 & 0 & -\frac{4}{3}\varepsilon & -E' & 0 & 0 \\ 0 & 0 & 0 & 0 & -E' & \frac{2}{3}\varepsilon \\ 0 & 0 & 0 & 0 & \frac{2}{3}\varepsilon & -E' \end{vmatrix} = 0 \qquad (162.8)$$

with

$$\varepsilon = \frac{e^2 r_0^2}{2R^3}. \qquad (162.9)$$

The determinant can be decomposed into three 2×2 ones, thus extremely simplifying its evaluation. The results are therefore immediately shown in the accompanying table. Here Λ stands for the sum of the m values of both atoms, i.e. for the component of the total electron orbital momentum along the nuclear axis. The classification symbols used are those of molecular spectroscopy, the signs Σ and Π referring to $\Lambda=0$ and ± 1, respectively, and the subscripts g and u to wave functions even and odd. The two Π_g states having the same energy are still degenerate; so are the two Π_u states. The last column gives the interaction energy E' in multiples of ε.

State	Λ	Wave function (unnormalized)	E'
Π_g	1	$\lvert 00, 11\rangle + \lvert 11, 00\rangle$	$+\frac{2}{3}\varepsilon$
Π_u	1	$\lvert 00, 11\rangle - \lvert 11, 00\rangle$	$-\frac{2}{3}\varepsilon$
Σ_g	0	$\lvert 00, 10\rangle + \lvert 10, 00\rangle$	$-\frac{4}{3}\varepsilon$
Σ_u	0	$\lvert 00, 10\rangle - \lvert 10, 00\rangle$	$+\frac{4}{3}\varepsilon$
Π_g	-1	$\lvert 00, 1-1\rangle + \lvert 1-1, 00\rangle$	$+\frac{2}{3}\varepsilon$
Π_u	-1	$\lvert 00, 1-1\rangle - \lvert 1-1, 00\rangle$	$-\frac{2}{3}\varepsilon$

Introducing the hydrogen wave functions (see Problem 67)

$$f_0 = 2\,\mathrm{e}^{-r}; \qquad f_1 = \frac{\sqrt{6}}{12}\, r\, \mathrm{e}^{-r/2}$$

in atomic units, the integral r_0, Eq. (162.5), can be evaluated:

$$r_0 = \sqrt{6} \cdot \tfrac{128}{243} \qquad (162.10)$$

and, from Eq. (162.9),

$$\varepsilon = \frac{16384}{19683} \frac{e^2 a_0^2}{R^3} \tag{162.11}$$

with a_0 the Bohr radius. In all states, therefore, the interaction energy E' becomes proportional to R^{-3}, decreasing more slowly than, and thus dominating, the Van der Waals energy ($\propto R^{-6}$) at large distances. The sign of the interaction still depends upon the state of the system: in the states Σ_u and Π_g there is repulsion between the two atoms and attraction only in the states Σ_g and Π_u.

Literature. Herzberg, G.: *Spectra of diatomic molecules*, p. 378. — Landau-Lifschitz: *Quantum Mechanics*, p. 302. — Margenau, H.: Rev. Modern Phys. **11**, 1 (1939). — King, G. W., Van Vleck, J. H.: Phys. Rev. **55**, 1165 (1939).

Problem 163. Neutral hydrogen molecule

To find the binding energy and equilibrium distance of the neutral hydrogen molecule by a method analogous to that of the H_2^+ treatment given in Problem 44.

Solution. This is a two-body problem in the Born-Oppenheimer approximation of fixed nuclei. Let the two nuclei (protons) be denoted by a and b, and the two electrons by 1 and 2, then we have the hamiltonian, in atomic units,

$$H = -\tfrac{1}{2}(\nabla_1^2 + \nabla_2^2) + \frac{1}{r_{12}} - \left(\frac{1}{r_{a1}} + \frac{1}{r_{b1}} + \frac{1}{r_{a2}} + \frac{1}{r_{b2}}\right) + \frac{1}{R} \tag{163.1}$$

with R the distance between the nuclei. At large distances R, the wave function should pass over into the product of the separate atoms, either becoming of the form $f(\boldsymbol{r}_{a1})\, f(\boldsymbol{r}_{b2})$ if electron 1 forms an atom with nucleus a, and 2 with b, or of the form $f(\boldsymbol{r}_{b1})\, f(\boldsymbol{r}_{a2})$ if the two electrons are exchanged. A reasonable approach at finite distances R will be a linear combination of two such products, and symmetry considerations lead to the choice of the symmetrical solution

$$U(1,2) = \alpha[f(\boldsymbol{r}_{a1})\, f(\boldsymbol{r}_{b2}) + f(\boldsymbol{r}_{b1})\, f(\boldsymbol{r}_{a2})] \tag{163.2}$$

for the ground state (with antiparallel electron spins, according to the Pauli principle). The antisymmetric combination, which is an eigenfunction as well, would lead to a larger energy with no attraction and no formation of a molecule at all.

If we put (163.2) into the Schrödinger equation,

$$H U = E \cdot U \tag{163.3}$$

with H the hamiltonian (163.1) we get

$$\begin{aligned} F(\boldsymbol{r}_{a1}) f(\boldsymbol{r}_{b2}) + f(\boldsymbol{r}_{a1}) F(\boldsymbol{r}_{b2}) + \left[\frac{1}{r_{12}} - \frac{1}{r_{b1}} - \frac{1}{r_{a2}} - E + \frac{1}{R}\right] f(\boldsymbol{r}_{a1}) f(\boldsymbol{r}_{b2}) \\ + F(\boldsymbol{r}_{b1}) f(\boldsymbol{r}_{a2}) + f(\boldsymbol{r}_{b1}) F(\boldsymbol{r}_{a2}) \\ + \left[\frac{1}{r_{12}} - \frac{1}{r_{a1}} - \frac{1}{r_{b2}} - E + \frac{1}{R}\right] f(\boldsymbol{r}_{b1}) f(\boldsymbol{r}_{a2}) = 0 \end{aligned} \tag{163.4}$$

where F stands as an abbreviation:

$$F(\boldsymbol{r}_{a1}) = \left(-\frac{1}{2}\nabla_1^2 - \frac{1}{r_{a1}}\right) f(\boldsymbol{r}_{a1}). \tag{163.5}$$

We now apply the operator

$$\int d\tau_1 \int d\tau_2\, f^*(\boldsymbol{r}_{a1}) f^*(\boldsymbol{r}_{b2}) \ldots$$

to Eq. (163.4). The function f is supposed to be normalized. Using the following abbreviations:

$$S = \int d\tau_1\, f^*(\boldsymbol{r}_{a1}) f(\boldsymbol{r}_{b2}) \tag{163.6}$$

for the overlap integral,

$$\mathscr{C} = \int d\tau_1 \frac{1}{r_{b1}} |f(\boldsymbol{r}_{a1})|^2 \tag{163.7}$$

and

$$\mathscr{C}' = \int\int d\tau_1\, d\tau_2 \frac{1}{r_{12}} |f(\boldsymbol{r}_{a1})|^2 |f(\boldsymbol{r}_{b2})|^2 \tag{163.8}$$

for the classical interaction integrals,

$$\mathscr{E} = \int d\tau_1 \frac{1}{r_{a1}} f^*(\boldsymbol{r}_{a1}) f(\boldsymbol{r}_{b1}) \tag{163.9}$$

and

$$\mathscr{E}' = \int\int d\tau_1\, d\tau_2 \frac{1}{r_{12}} f^*(\boldsymbol{r}_{a1}) f(\boldsymbol{r}_{b1}) f(\boldsymbol{r}_{a2}) f^*(\boldsymbol{r}_{b2}) \tag{163.10}$$

for the exchange integrals, and finally

$$A = \int d\tau_1 f^*(\boldsymbol{r}_{a1}) F(\boldsymbol{r}_{a1}) \tag{163.11}$$

and

$$A' = \int d\tau_1 f^*(\boldsymbol{r}_{a1}) F(\boldsymbol{r}_{b1}) \tag{163.12}$$

for the two remaining integrals, we find by this procedure

$$2(A+A'S)-2(\mathscr{C}+\mathscr{E}S)+(\mathscr{C}'+\mathscr{E}') = \left(E-\frac{1}{R}\right)(1+S^2) \quad (163.13)$$

or

$$E=2\frac{A+A'S}{1+S^2}-\frac{2(\mathscr{C}+\mathscr{E}S)-(\mathscr{C}'+\mathscr{E}')}{1+S^2}+\frac{1}{R}. \quad (163.14)$$

In a way analogous to that used in Problem 44 for H_2^+ we now put

$$f(\boldsymbol{r}) = \sqrt{\frac{\gamma^3}{\pi}}\, e^{-\gamma r}. \quad (163.15)$$

For $\gamma=1$ this would be the wave function of the atomic ground state; with γ a variational Ritz parameter we may still get a better approximation. Specializing to (163.15) we find, according to (163.5),

$$F(r_{a1}) = \left(-\frac{1}{2}\gamma^2+\frac{\gamma-1}{r_{a1}}\right) f(r_{a1})$$

so that (163.11) and (163.12) yield

$$A=-\tfrac{1}{2}\gamma^2+\gamma(\gamma-1); \qquad A'=-\tfrac{1}{2}\gamma^2 S+(\gamma-1)\mathscr{E}. \quad (163.16)$$

It can further be shown that the overlap integral S depends only upon the combination

$$\rho=\gamma R \quad (163.17)$$

which we may use as a second Ritz parameter besides γ, and that the four remaining interaction integrals $\mathscr{C}$, $\mathscr{C}'$, $\mathscr{E}$, $\mathscr{E}'$ all become proportional to γ so that we may write

$$\mathscr{C}=\gamma\overline{\mathscr{C}}(\rho); \quad \mathscr{C}'=\gamma\overline{\mathscr{C}}'(\rho); \quad \mathscr{E}=\gamma\overline{\mathscr{E}}(\rho); \quad \mathscr{E}'=\gamma\overline{\mathscr{E}}'(\rho). \quad (163.18)$$

The energy expression (163.14) thus becomes

$$E=-a\gamma+b\gamma^2 \quad (163.19)$$

with

$$a(\rho) = \frac{2(1+\overline{\mathscr{C}})+4S\overline{\mathscr{E}}-(\overline{\mathscr{C}}'+\overline{\mathscr{E}}')}{1+S^2}-\frac{1}{\rho} \quad (163.20)$$

and

$$b(\rho) = \frac{1-S^2+2S\overline{\mathscr{E}}}{1+S^2} \quad (163.21)$$

depending only on ρ. We then get an energy minimum if

$$\frac{\partial E}{\partial \gamma} = -a + 2b\gamma = 0$$

or

$$\gamma = \frac{a}{2b}; \tag{163.22}$$

it amounts to

$$E = -\frac{a^2}{4b}. \tag{163.23}$$

It remains to evaluate the five integrals (163.6–10). Three of them have been computed in the H_2^+, Problem 44, viz.

$$S = (1 + \rho + \tfrac{1}{3}\rho^2)\mathrm{e}^{-\rho},$$

$$\overline{\mathscr{C}} = \frac{1}{\rho}[1 - (1+\rho)\mathrm{e}^{-2\rho}],$$

$$\overline{\mathscr{E}} = (1+\rho)\mathrm{e}^{-\rho}.$$

The integral $\mathscr{C}'$ is a little more difficult but can still be evaluated in an elementary way if integration over the coordinates of the electron 2 is first performed using spherical polar coordinates with the origin in b and $b1$ the polar axis:

$$\frac{1}{\pi}\int d\tau_2 \frac{1}{r_{12}} \mathrm{e}^{-2r_{b2}} = \overline{\mathscr{C}}(r_{b1}).$$

The second integration, over electron 1, then leads to integrals already evaluated, only some having factors 2 in the exponential. The result is

$$\overline{\mathscr{C}}' = \frac{1}{\rho}[1 - (1 + \tfrac{11}{8}\rho + \tfrac{3}{4}\rho^2 + \tfrac{1}{6}\rho^3)\mathrm{e}^{-2\rho}].$$

Real difficulties, however, are encountered in the last integral, $\mathscr{E}'$, which can no longer be reduced to elementary integrations. The result, first obtained by Sugiura, is

$$\overline{\mathscr{E}}' = [\tfrac{5}{8} - \tfrac{23}{20}\rho - \tfrac{3}{5}\rho^2 - \tfrac{1}{15}\rho^3]\mathrm{e}^{-2\rho} + \frac{6}{5}\frac{\varphi(\rho)}{\rho}$$

with

$$\varphi(\rho) = S(\rho)^2(\log\rho + C) - S(-\rho)^2 E_1(4\rho) + 2S(\rho)S(-\rho)\,E_1(2\rho),$$

where

$$E_1(z) = \int_z^\infty \frac{dt}{t}\mathrm{e}^{-t}$$

is the exponential integral.

It is easily seen that, for large r, this integral tends to zero as $e^{-2\rho}$ and that, at $\rho=0$, it has a limit $\bar{\mathscr{E}}'(0)=\frac{5}{8}$ in agreement with the theory of the helium ground state emerging for nuclear distance zero (cf. Problem 154).

Numerical computation yields the following table. The binding energy of the molecule has a maximum at about $R=1.46$ a.u. corresponding to an equilibrium distance of $R_0=0.77$ Å (instead of the experimental value 0.742 Å). The energy then is $E=-1.139$ which is to be compared with the binding energy of two separate hydrogen atoms in the ground state, $2E_0=-1$. If the zero-point energy of the molecular vibrations is denoted by $\frac{1}{2}\hbar\omega$, we therefore have a dissociation energy of

$$D=2E_0-(E+\tfrac{1}{2}\hbar\omega)=0.139-\tfrac{1}{2}\hbar\omega.$$

The zero-point energy can be determined by the same procedure as in the case of H_2^+ (Problem 44), but the energy parabola in the neighbourhood of the minimum is much less accurately determined by the table.

ρ	γ	$-E$	R
1.3	1.145	1.120	1.133
1.4	1.152	1.127	1.214
1.5	1.160	1.131	1.293
1.6	1.164	1.137	1.374
1.7	1.166	1.139	1.458
1.8	1.164	1.137	1.546
1.9	1.161	1.134	1.635
2.0	1.156	1.129	1.730

However, we find a value close to 0.010 a.u. or 0.27 eV, with an error of about $\pm 5\%$, in perfect agreement with experiment ($\hbar\omega=0.54$ eV). Thus, the dissociation energy turns out to be

$$D=0.138 \text{ a.u.}=3.75 \text{ eV}$$

whereas its experimental value is $D=4.45$ eV. The agreement is not so bad, as we have explained in the similar situation obtaining in the case of H_2^+.

NB. It should be mentioned that, for large distances R or ρ, the parameter γ tends towards 1 and the function f to the atomic ground state wave function. In the original Heitler-London method, this value was used throughout, so that there was no Ritz parameter γ. In that rougher picture the dissociation energy becomes only 2.90 eV and the equilibrium distance 0.88 Å. The increasing values of γ which, according to the table, obtain during an adiabatic approach of the two atoms describe a contraction of their electron wave functions in binding.

Literature. Heitler, W., London, F.: Z. Physik **44**, 455 (1927). — *Evaluation of the integral* $\mathscr{E}'$*:* Sugiura, Y., Z. Physik **45**, 484 (1927). — *Variation of* γ*:* Wang, S. C., Phys. Rev. **31**, 579 (1928); Rosen, N.: Phys. Rev. **38**, 2099 (1931). — *Better variational approximations:* James, H. M., Coolidge, A. S., J. chem. Phys. **1**, 825 (1933); **3**, 129 (1935).

Problem 164. Scattering of equal particles

A beam of particles of charge e collides with a target consisting of particles of the same kind at rest. How does the angular distribution by scattering compare with that expected of classical physics, if allowance is made for the correct symmetry of the wave function? This shall be discussed for unpolarized particle beams of spins 0, $\frac{1}{2}$, and 1.

Solution. The Rutherford amplitude has been derived in Problem 110 in the centre-of-mass system. It is

$$f(\vartheta) = -\frac{\varkappa^*}{2k^*} e^{2i\eta_0} \frac{e^{-i\kappa^* \log \sin^2 \frac{\vartheta}{2}}}{\sin^2 \frac{\vartheta}{2}}; \qquad \eta_0 = \Gamma(1 + i\varkappa^*); \tag{164.1}$$

here the abbrevations κ^* and k^* refer to centre-of-mass values,

$$\varkappa^* = \frac{e^2}{\hbar v}; \qquad k^* = \frac{m^* v}{\hbar}; \qquad E^* = \frac{\hbar^2 k^{*2}}{2m^*} \tag{164.2}$$

and m^* is the reduced mass. For two equal particles, $m^* = \frac{1}{2}m$. The relative velocity of the two particles, v, is independent of the frame of reference. Therefore, if κ, k and E refer to the laboratory frame, we get

$$\varkappa^* = \varkappa; \qquad k^* = \tfrac{1}{2}k; \qquad E^* = \tfrac{1}{2}E \tag{164.3}$$

so that

$$f(\vartheta) = -\frac{\varkappa}{k} e^{2i\eta_0} \frac{e^{-i\kappa \log \sin^2 \frac{\vartheta}{2}}}{\sin^2 \frac{\vartheta}{2}}; \qquad \eta_0 = \Gamma(1 + i\varkappa). \tag{164.4}$$

Further, the angle of deflection in the laboratory frame, Θ, for equal masses becomes

$$\Theta = \tfrac{1}{2}\vartheta \tag{164.5}$$

with ϑ the angle in the centre-of-mass system. The solid angle element therefore transforms according to

$$d\omega = 2\pi \sin\vartheta\, d\vartheta = 2\pi \cdot 4\cos\Theta \sin\Theta\, d\Theta = 4\cos\Theta\, d\Omega, \qquad (164.6)$$

and the differential Rutherford cross section in the laboratory frame becomes

$$\frac{d\sigma_R}{d\Omega} = 4\cos\Theta \left(\frac{\varkappa}{k}\right)^2 \frac{1}{\sin^4\Theta} \quad \text{with} \quad \frac{\varkappa}{k} = \frac{e^2}{m v^2} = \frac{e^2}{2E}. \qquad (164.7)$$

Even in classical mechanics, this formula has to undergo an essential correction. Since it is impossible to distinguish a scattered particle from a recoil particle, if both are of equal kind, both have to be added in the cross section. As, according to (164.5), the two paths will be perpendicular to each other, the recoil particle emerges from the target under the angle $\frac{\pi}{2} - \Theta$ and we get, instead of (164.7),

$$\frac{d\sigma_{\text{class}}}{d\Omega} = 4\cos\Theta \left(\frac{\varkappa}{k}\right)^2 \left\{\frac{1}{\sin^4\Theta} + \frac{1}{\cos^4\Theta}\right\}. \qquad (164.8)$$

This is the classical expression with which we have to compare the quantum mechanical result now to be derived.

According to quantum theory, we have not to superimpose intensities (i.e. cross sections) but amplitudes. Let $u(\boldsymbol{r})$ be the unsymmetrized wave function in the centre-of-mass frame with $\boldsymbol{r}$ the relative coordinate vector. Its asymptotic behaviour, apart from logarithmic phases, is

$$u(\boldsymbol{r}) \to e^{i k^* z} + f(\vartheta) \frac{e^{i k^* r}}{r}.$$

The plane wave part can be written

$$e^{i k^* (z_1 - z_2)},$$

describing one particle of velocity $+\frac{1}{2}v$ and the other of velocity $-\frac{1}{2}v$ in z direction. If centre-of-mass motion is superimposed by a factor

$$e^{i k^* (z_1 + z_2)},$$

particle 2 is brought to rest (target particle), and there remains the motion of particle 1 according to $e^{2 i k^* z_1} = e^{i k z_1}$ (colliding particle). This holds for the unsymmetrized wave function of the two-particle system. Its symmetrization means replacing $u(\boldsymbol{r})$ by

$$u(\boldsymbol{r}) + \varepsilon u(-\boldsymbol{r}) \quad \text{with } \varepsilon = \pm 1.$$

For the asymptotic spherical wave part this means replacing $f(\vartheta)$ by

$$f(\vartheta)+f(\pi-\vartheta)$$

leaving r unchanged, so that, with (164.4), instead of $f(\vartheta)$ we get

$$f(\vartheta)+\varepsilon f(\pi-\vartheta)=-\frac{e^2}{m v^2}\,e^{2i\eta_0}\left\{\frac{e^{-i\kappa\log\sin^2\frac{\vartheta}{2}}}{\sin^2\frac{\vartheta}{2}}+\varepsilon\frac{e^{-i\kappa\log\sin^2\frac{\vartheta}{2}}}{\cos^2\frac{\vartheta}{2}}\right\} \tag{164.9}$$

and, instead of the classical cross section formula (164.8),

$$\frac{d\sigma}{d\Omega}=4\cos\Theta\left(\frac{e^2}{m v^2}\right)^2\left|\frac{e^{-i\varkappa\log\sin^2\Theta}}{\sin^2\Theta}+\frac{e^{-i\varkappa\log\cos^2\Theta}}{\cos^2\Theta}\right|^2. \tag{164.10}$$

Evaluation of the last expression leads to

$$\frac{d\sigma}{d\Omega}=4\cos\Theta\left(\frac{e^2}{m v^2}\right)^2\left\{\frac{1}{\sin^4\Theta}+\frac{1}{\cos^4\Theta}+\varepsilon\frac{2\cos(\frac{e^2}{\hbar v}\log\tan^2\Theta)}{\sin^2\Theta\cos^2\Theta}\right\}, \tag{164.11}$$

with an interference term added to the classical expression. For comparing both, quantum theoretical and classical cross sections, the ratio

$$\frac{d\sigma}{d\sigma_{\text{class}}}=1+\varepsilon\frac{2\tan^2\Theta\cos(\frac{e^2}{\hbar v}\log\tan^2\Theta)}{1+\tan^4\Theta} \tag{164.12}$$

may be useful.

As a last step we now have to decide which part of an unpolarized beam will be symmetrical, and which antisymmetrical. If the scattering particles are *fermions* of spin $\frac{1}{2}$ (two protons or two electrons) whose total wave function must be antisymmetrical, we have weights $\frac{3}{4}$ for the spin-symmetrical, space-antisymmetrical triplet and $\frac{1}{4}$ for the opposite symmetry of the singlet state so that we get

$$d\sigma=\tfrac{3}{4}d\sigma_-+\tfrac{1}{4}d\sigma_+$$

with the subscript signs referring to the two signs of ε in (164.11). Therefore, unpolarized beam experiments in this case yield

$$\varepsilon_{\text{eff}}=-\tfrac{3}{4}+\tfrac{1}{4}=-\tfrac{1}{2}$$

and

$$\frac{d\sigma}{d\sigma_{\text{class}}}=1-\frac{\tan^2\Theta\cos(\frac{e^2}{\hbar v}\log\tan^2\Theta)}{1+\tan^4\Theta}. \tag{164.13}$$

This function is shown in Fig. 64 for proton-proton scattering at an energy $E = 100$ keV. This is about the highest energy for which no very appreciable scattering anomaly occurs due to the short-range nuclear attraction between the protons (cf. the following problem). At this energy, we have $\varkappa = 0.50$ for protons. The function (164.13) is invariant under the transformation $\Theta \to \frac{\pi}{2} - \Theta$ so that it need only be computed for values $0 < \Theta \leqq 45°$. Passing on to much lower energies, $\varkappa$ becomes so large that $\cos(\varkappa \log \tan^2 \Theta)$ has several oscillations in this interval.

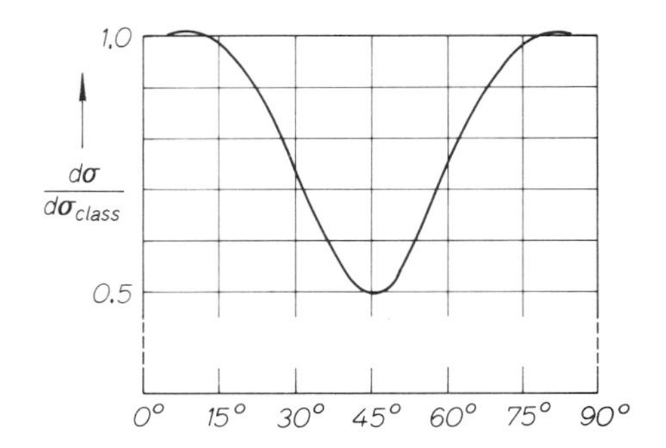

Fig. 64. Scattering of two equal fermions. The figure shows the ratio of quantum theoretical to classical scattering intensity as a function of scattering angle. The curve has an infinite number of oscillations in the vicinities of 0° and 90°, with decreasing amplitudes

Finally, for very large $\varkappa$, these oscillations will be so rapid that they can no longer be resolved experimentally so that the classical expression remains.

If the particles are *bosons* without spin (e.g. two α particles or two pions), only the space symmetrical state with $\varepsilon = +1$ occurs. Of course, for α particles e^2 has to be replaced by $4e^2$. If the particles are bosons of spin 1 (e.g. two deuterons), the total spins possible are 2 (weight $\frac{5}{9}$), 1 (weight $\frac{3}{9}$), and 0 (weight $\frac{1}{9}$) with space symmetry for total spins 2 and 0. Then we get

$$\varepsilon_{\text{eff}} = \left(\tfrac{5}{9} + \tfrac{1}{9}\right) - \tfrac{3}{9} = +\tfrac{1}{3}$$

and

$$\frac{d\sigma}{d\sigma_{\text{class}}} = 1 + \frac{2}{3}\,\frac{\tan^2\Theta \cos\left(\frac{e^2}{\hbar v}\log\tan^2\Theta\right)}{1 + \tan^4\Theta}. \tag{164.14}$$

Literature. The classical symmetrized expression (164.8) was first used by Darwin, C. G.: Proc. Roy. Soc. London, **A 120,** 631 (1928). The quantum mechanical formula was derived by Mott, N. F., ibid: **126,** 259 (1930). This formula was corroborated experimentally with α particles, as described in the papers by Chadwick, J.: Proc. Roy. Soc. London, **A 128,** 114 (1930) and by Blackett, P. M. S., Champion, F. C., ibid.: **130,** 380 (1931), and with protons by Gerthsen, C.: Ann. Physik **9,** 769 (1931).

Problem 165. Anomalous proton-proton scattering

The short-range nuclear force between two protons gives rise to an attraction which causes a scattering anomaly above energies of about 100 keV. This anomaly shall be described by an additional phase shift δ_0 in the partial wave $l=0$.

Solution. In Problem 112 we have treated anomalous scattering for charged particles without symmetrization and derived Eq. (112.5), in the centre-of-mass frame,

$$f(\vartheta) = -\frac{\varkappa}{2k^* \sin^2 \frac{\vartheta}{2}} e^{-i\varkappa \log \sin^2 \frac{\vartheta}{2}} + \frac{1}{2ik^*}(e^{2i\delta_0} - 1). \tag{165.1}$$

Let us instead for what follows use the abbreviation

$$f(\vartheta) = f_R(\vartheta) + f_a \tag{165.2}$$

where the first term describes the Rutherford scattering of the Coulomb field and the second,

$$f_a = \frac{1}{2ik^*}(e^{2i\delta_0} - 1) = \frac{1}{k^*} e^{i\delta_0} \sin\delta_0, \tag{165.3}$$

is the anomaly amplitude. Symmetrization, then, according to the preceding problem, with

$$k^* = \tfrac{1}{2}k; \quad \tfrac{1}{2}\vartheta = \Theta, \tag{165.4}$$

leads to the relation

$$\frac{d\sigma}{d\Omega} = 4\cos\Theta\,\{\tfrac{3}{4}|f(\vartheta) - f(\pi-\vartheta)|^2 + \tfrac{1}{4}|f(\vartheta) + f(\pi-\vartheta)|^2\}, \tag{165.5}$$

or

$$\frac{d\sigma}{d\Omega} = 4\cos\Theta\,\{\tfrac{3}{4}|f_R(\vartheta) - f_R(\pi-\vartheta)|^2 + \tfrac{1}{4}|f_R(\vartheta) + f_R(\pi-\vartheta) + 2f_a|^2\}. \tag{165.6}$$

In the first square of (165.6), the anomaly not depending upon ϑ cancels out. It is remarkable that even for higher energies, where higher angular momenta contribute to f_a so that

$$f_a = \frac{1}{2ik^*} \sum_{l=0}^{\infty} (2l+1)\, e^{2i(\eta_l - \eta_0)} (e^{2i\delta_l} - 1)\, P_l(\cos\vartheta),$$

in consequence of

$$P_l(\cos(\pi - \vartheta)) = (-1)^l P_l(\cos\vartheta),$$

all contributions with even l would cancel out in the triplet, and all with odd l in the singlet term. Thus, contributions would arise only from the terms

$$^1S,\ ^3P,\ ^1D,\ ^3F,\ ^1G, \ldots$$

This is in complete agreement with the Pauli principle applied separately to each of these partial waves (forbidding ^3S, ^1P, ...).

If the absolute squares are evaluated, Eq. (165.6) yields

$$\frac{d\sigma}{d\Omega} = 4\cos\Theta\,\{|f_R(\vartheta)|^2 + |f_R(\pi-\vartheta)|^2 - \mathrm{Re}[f_R(\vartheta) f_R^*(\pi-\vartheta)]$$
$$+\mathrm{Re}[f_a f_R^*(\vartheta)] + \mathrm{Re}[f_a f_R^*(\pi-\vartheta)] + |f_a|^2\}. \tag{165.7}$$

In the first line of this equation there stand the terms contributing to Coulomb scattering as discussed in the preceding problem; in the second line there stand two interference terms of Coulomb and anomalous scattering, and the anomaly itself. With the explanations of the symbols f_R and f_a given above, these last three terms can easily be computed:

$$\mathrm{Re}[f_a f_R^*(\vartheta)] = -\frac{2\varkappa}{k^2}\frac{\sin\delta_0}{\sin^2\Theta}\cos(\delta_0 + \varkappa\log\sin^2\Theta);$$

$$\mathrm{Re}[f_a f_R^*(\pi-\vartheta)] = -\frac{2\varkappa}{k^2}\frac{\sin\delta_0}{\cos^2\Theta}\cos(\delta_0 + \varkappa\log\cos^2\Theta);$$

$$|f_a|^2 = \frac{4}{k^2}\sin^2\delta_0\,.$$

It is usual to give the so-called scattering ratio,

$$R = \frac{d\sigma}{d\sigma_C}, \tag{165.8}$$

where $d\sigma_C$ means the Coulomb term only (as derived in the preceding problem or from the first line only of (165.7)). This becomes

$$R = 1 + \frac{-\frac{2}{\varkappa}\sin\delta_0\left[\frac{\cos(\delta_0+\varkappa\log\sin^2\Theta)}{\sin^2\Theta} + \frac{\cos(\delta_0+\varkappa\log\cos^2\Theta)}{\cos^2\Theta}\right] + \frac{4}{\varkappa^2}\sin^2\delta_0}{\frac{1}{\sin^4\Theta} + \frac{1}{\cos^4\Theta} - \frac{\cos(\varkappa\log\tan^2\Theta)}{\sin^2\Theta\cos^2\Theta}} \tag{165.9}$$

This formula holds upwards to such energies where ^3P scattering begins to play a role, i.e. upwards to a few MeV proton energy.

Eq. (165.9) gives $R=1$ for $\Theta=0°$ and for $\Theta=90°$ where the singularity of Coulomb scattering outgrows the finite anomaly.

More significance attaches to the value of R at the angle $\Theta=45°$ where we get

$$R(45°)=1-\frac{2}{\varkappa}\sin\delta_0\cos(\delta_0-\log 2)+\frac{1}{\varkappa^2}\sin^2\delta_0. \qquad (165.10)$$

Let us first discuss this expression for a rather small proton energy, say, of 250 keV. Then $\varkappa=0.316$ is still rather large and δ_0 very small, so that the second term in (165.10) by far outweighs the third one. Observation at 45° then easily decides on the sign of the additional force: If it is an attraction, $\delta_0>0$ and $R(45°)<1$; if it is a repulsion, $\delta_0<0$ and $R(45°)>1$. Experiment shows that the nuclear force is attractive.

Let us now go on to higher energies, say, to 1 MeV ($\varkappa=0.158$). Fig. 65 shows the value of $R(45°)$ then to be expected for different positive and negative values of δ_0, according to (165.10). Since we

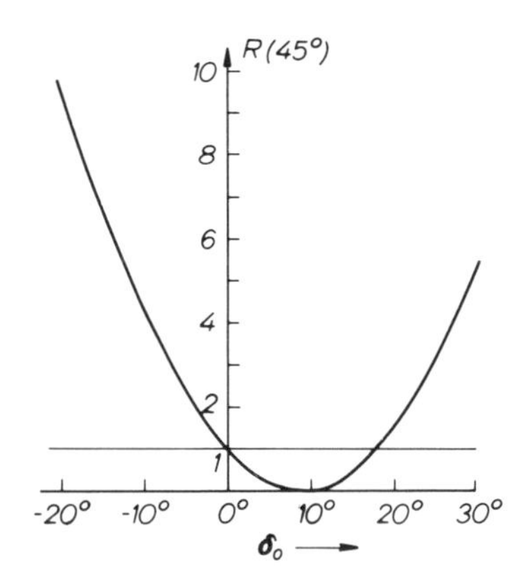

Fig. 65. Anomalous proton-proton scattering at 1 MeV. The ratio of actual to pure Coulomb scattering at $\Theta=45°$ is shown as a function of the phase angle δ_0. Positive (negative) values of δ_0 correspond to short-range attraction (repulsion)

have already decided in favour of $\delta_0>0$ (attraction), there is a unique determination of δ_0 if $R(45°)>1$. This really is observed at 1 MeV, viz. $R(45°)=4.6$. The phase angle then turns out to be $\delta_0=32°$. In this way δ_0 can be uniquely determined as a function of energy.

The ratios R at other angles of deflection may then be computed from (165.9) with the value of δ_0 so determined, and the whole curve of angular distribution for each energy be compared with experiment. Thus the theory can be corroborated in more detail. Perfect agreement has thus been obtained.

Literature. Blatt, J. M., Jackson, J. D.: Rev. Modern Phys. **22,** 77 (1950). — Flügge, S.: Ergebn. exakt. Naturwiss. **26,** 165 (1952).

Problem 166. Inelastic scattering

A beam of protons is passing a target consisting of alkali atoms. Excitation cross sections for the outermost atomic electron, originally in its ground state, shall be determined by treating the proton-atom interaction as a perturbation. Momentum transfer to the atomic core shall be neglected (infinitely heavy nucleus).

Solution. Let us use atomic units throughout ($\hbar=1$, $e=1$, $m=1$) and let $\boldsymbol{r}_1$ be the position vector of the proton, $\boldsymbol{r}_2$ of the electron. The hamiltonian may then be decomposed into three parts,

$$H = H_1 + H_2 + H_{12} \tag{166.1}$$

with

$$H_1 = -\frac{1}{2M}\nabla_1^2 \tag{166.2}$$

describing force-free motion of a proton of mass M,

$$H_2 = -\tfrac{1}{2}\nabla_2^2 + V(r_2) \tag{166.3}$$

describing the motion of the outermost electron in the field of the atomic core, and

$$H_{12} = -V(r_1) - \frac{1}{r_{12}} \tag{166.4}$$

describing the interaction of the proton with the atomic core and the electron to be excited. This last term of the hamiltonian shall be regarded as a perturbation. Such a formulation of the problem is reasonable as long as the energy of the proton is not big enough to excite any one of the core electrons, so that we may simply deal with only the outermost electron and a rigid core.

Let the eigenfunctions of the operator H_2 be u_ν (where the subscript ν comprises the three quantum numbers n, l, m, and $\nu=0$ denotes the ground state) with eigenvalues W_ν,

$$H_2 u_\nu(\boldsymbol{r}_2) = W_\nu(\boldsymbol{r}_2), \tag{166.5}$$

and let $\hbar \boldsymbol{k}$ be the momentum of an undeflected beam proton. Then the zero-order solution of the Schrödinger equation, neglecting H_{12}, is

$$U^0(\boldsymbol{r}_1, \boldsymbol{r}_2) = \mathrm{e}^{i\boldsymbol{k}\cdot\boldsymbol{r}_1} u_0(\boldsymbol{r}_2). \tag{166.6}$$

The first-order solution may be written as an expansion with respect to the complete orthogonal set $\{u_\nu\}$, viz.

$$U(\boldsymbol{r}_1, \boldsymbol{r}_2) = \mathrm{e}^{i\boldsymbol{k}\cdot\boldsymbol{r}_1} u_0(\boldsymbol{r}_2) + \sum_\mu{}' F_\mu(\boldsymbol{r}_1) u_\mu(\boldsymbol{r}_2) \tag{166.7}$$

where $\mu=0$ is excluded from, and the integral over the continuum functions included in, the sum.

Putting (166.7) in the Schrödinger equation we arrive at

$$\sum_\mu{}' \{\nabla_1^2 F_\mu + [k^2 - 2M(W_\mu - W_0)] F_\mu - 2MH_{12} F_\mu\} u_\mu(\boldsymbol{r}_2) = 2MH_{12} \mathrm{e}^{i\boldsymbol{k}\cdot\boldsymbol{r}_1} u_0(\boldsymbol{r}_2).$$

In the first order of perturbation we neglect H_{12} on the left-hand side. With the notation

$$k_\mu^2 = k^2 - 2M(W_\mu - W_0) \tag{166.8}$$

we then have

$$\sum_\mu{}' \{\nabla_1^2 F_\mu(\boldsymbol{r}_1) + k_\mu^2 F_\mu(\boldsymbol{r}_1)\} u_\mu(\boldsymbol{r}_2) = 2MH_{12} \mathrm{e}^{i\boldsymbol{k}\cdot\boldsymbol{r}_1} u_0(\boldsymbol{r}_2). \tag{166.9}$$

Multiplying (166.9) by $u_\nu^*(\boldsymbol{r}_2)$ and integrating over $\boldsymbol{r}_2$ we obtain a set of independent differential equations for the F_ν's:

$$\nabla_1^2 F_\nu + k_\nu^2 F_\nu = \Phi_\nu(\boldsymbol{r}_1) \tag{166.10}$$

with

$$\Phi_\nu(\boldsymbol{r}_1) = 2M \mathrm{e}^{i\boldsymbol{k}\cdot\boldsymbol{r}_1} \int d^3 r_2\, u_\nu^*(\boldsymbol{r}_2) H_{12} u_0(\boldsymbol{r}_2). \tag{166.11}$$

Eq. (166.10) is an inhomogeneous equation and can be solved by using a Green's function,

$$F_\nu(\boldsymbol{r}_1) = -\frac{1}{4\pi} \int d^3 r' \frac{\mathrm{e}^{ik_\nu|\boldsymbol{r}_1 - \boldsymbol{r}'|}}{|\boldsymbol{r}_1 - \boldsymbol{r}'|} \Phi_\nu(\boldsymbol{r}'). \tag{166.12}$$

In order to derive cross sections from the solution, we next have to study the asymptotic behaviour of (166.12) for $r_1 \to \infty$. The integral (166.11) in $\Phi_\nu(\boldsymbol{r}_1)$ decreases as $1/r_1^2$ with large r_1 since H_{12} is the interaction of a proton with the neutral atom, which according to (166.4), for $r_1 \gg r_2$, tends towards—$(\boldsymbol{r}_1 \cdot \boldsymbol{r}_2)/r_1^3$. The factor $\Phi_\nu(\boldsymbol{r}')$ in (166.12) thus practically limits the integration domain in (166.12) to atomic dimensions. For $r_1 \to \infty$ therefore $r_1 \gg r'$ may be supposed so that

$$F_\nu(\boldsymbol{r}_1) \to -\frac{\mathrm{e}^{ik_\nu r_1}}{4\pi r_1} \int d^3 r'\, \mathrm{e}^{-ik_\nu r' \cos(\boldsymbol{r}_1, \boldsymbol{r}')} \Phi_\nu(\boldsymbol{r}'). \tag{166.13}$$

This is an outgoing spherical wave,

$$F_\nu(\boldsymbol{r}_1) \to f(\vartheta_1)\frac{e^{ik_\nu r_1}}{r_1} \tag{166.14}$$

with a scattering amplitude

$$f(\vartheta_1) = -\frac{1}{4\pi}\int d^3r'\, e^{-ik_\nu r'\cos(\boldsymbol{r}_1,\boldsymbol{r}')}\,\Phi_\nu(\boldsymbol{r}') \tag{166.15}$$

still depending upon the direction in which the proton is finally deflected. It follows that the differential inelastic cross section for the proton, under excitation of the outermost alkali electron to the state ν, then is

$$d\sigma = \frac{k_\nu}{k}|f(\vartheta_1)|^2\, d\Omega_1 \tag{166.16}$$

because the velocity of the outgoing protons, and thence their current, are lowered by the factor k_ν/k. This is an immediate consequence of interpreting (166.8) as energy conservation law.

Let us now go into some more details with respect to the angular distribution of the inelastically scattered protons. In the exponent of (166.15) we introduce a vector $\boldsymbol{k}_\nu$ in the direction of $\boldsymbol{r}_1$ so that

$$k_\nu r'\cos(\boldsymbol{r}_1,\boldsymbol{r}') = \boldsymbol{k}_\nu\cdot\boldsymbol{r}'$$

and, using (166.11) and (166.4), we get

$$f(\vartheta_1) = -\frac{M}{2\pi}\int d^3r'\, e^{i(\boldsymbol{k}-\boldsymbol{k}_\nu)\cdot\boldsymbol{r}'}\int d^3r_2\, u_\nu^*(\boldsymbol{r}_2)\left\{-V(r') - \frac{1}{|\boldsymbol{r}'-\boldsymbol{r}_2|}\right\}u_0(\boldsymbol{r}_2).$$

It is obvious that the term with $V(r')$ does not contribute to inelastic collisions ($\nu \neq 0$). This follows from orthogonality as well as from the physical fact that interaction of the proton with the core cannot excite an electron which forms no part of the core. Further introducing a vector $\boldsymbol{K}_\nu = \boldsymbol{k}-\boldsymbol{k}_\nu$ for the change of momentum we then arrive at

$$f(\vartheta_1) = \frac{M}{2\pi}\int d^3r_2\, u_\nu^*(\boldsymbol{r}_2)u_0(\boldsymbol{r}_2)\int d^3r'\, e^{i\boldsymbol{K}_\nu\cdot\boldsymbol{r}'}\frac{1}{|\boldsymbol{r}'-\boldsymbol{r}_2|}. \tag{166.17}$$

The integrand of the inner integral in this formula consists of two factors, both of which may be expanded into spherical harmonics of the angles ϑ' between $\boldsymbol{K}_\nu$ and $\boldsymbol{r}'$, and ϑ_2' between $\boldsymbol{r}'$ and $\boldsymbol{r}_2$:

$$e^{i\boldsymbol{K}_\nu\cdot\boldsymbol{r}'} = \frac{1}{K_\nu r'}\sum_{l=0}^{\infty}\sqrt{4\pi(2l+1)}\,i^l j_l(K_\nu r')\,Y_{l,0}(\vartheta')$$

and

$$\frac{1}{|\boldsymbol{r}'-\boldsymbol{r}_2|} = \sum_{n=0}^{\infty}\sqrt{\frac{4\pi}{2n+1}}\,R_n\,Y_{n,0}(\vartheta_2')$$

with

$$\left.\begin{aligned} R_n &= \frac{1}{r_2}\left(\frac{r'}{r_2}\right)^n \quad \text{if } r'<r_2, \\ R_n &= \frac{1}{r'}\left(\frac{r_2}{r'}\right)^n \quad \text{if } r'>r_2. \end{aligned}\right\} \tag{166.18}$$

Referring all angles to the direction of $\boldsymbol{K}_\nu$ as polar axis, we may apply the addition theorem of spherical harmonics to $Y_{n,0}(\vartheta_2')$, viz.

$$Y_{n,0}(\vartheta_2') = \sqrt{\frac{4\pi}{2n+1}}\sum_{m=-n}^{+n} Y^*_{n,m}(\vartheta',\varphi')\,Y_{n,m}(\vartheta_2,\varphi_2)$$

with ϑ', φ' defining the direction of $\boldsymbol{r}'$ and ϑ_2, φ_2 that of $\boldsymbol{r}_2$ with respect to $\boldsymbol{K}_\nu$. Then, in the inner integral of (166.17), the angular integrations may be performed and we get

$$\int d^3r'\, e^{i\boldsymbol{K}_\nu\cdot\boldsymbol{r}'}\frac{1}{|\boldsymbol{r}'-\boldsymbol{r}_2|} = \sum_{l=0}^{\infty}\int_0^{\infty} dr'\, r'^2\,\frac{4\pi}{K_\nu r'}\sqrt{\frac{4\pi}{2l+1}}\,i^l j_l(K_\nu r')\,R_l\,Y_{l,0}(\vartheta_2).$$

Using the abbreviation

$$g_l(r_2) = \sqrt{\frac{4\pi}{2l+1}}\,i^l\int_0^{\infty} dr'\,r'^2\,\frac{j_l(K_\nu r')}{K_\nu r'}\,R_l \tag{166.19}$$

Eq. (166.17) becomes

$$f(\vartheta_1) = 2M\int d^3r_2\,u_\nu^*(\boldsymbol{r}_2)\,u_0(\boldsymbol{r}_2)\sum_{l=0}^{\infty} g_l(r_2)\,Y_{l,0}(\vartheta_2). \tag{166.20}$$

The integral (166.19) can be explicitly evaluated. Using $y=K_\nu r'$ as integration variable and putting $x=K_\nu r_2$, we find with (166.18),

$$g_l(r_2) = \sqrt{\frac{4\pi}{2l+1}}\,i^l\,\frac{1}{K_\nu^2}\left\{x^{-l-1}\int_0^x dy\,y^{l+1} j_l(y) + x^l\int_x^{\infty} dy\,y^{-l} j_l(y)\right\}.$$

These integrals are well known from the theory of Bessel functions:

$$\int_0^x dy\, y^{l+1} j_l(y) = x^{l+1} j_{l+1}(x);$$

$$\int_x^\infty dy\, y^{-l} j_l(y) = x^{-l} j_{l-1}(x)$$

so that the curly bracket above becomes

$$j_{l+1}(x) + j_{l-1}(x) = (2l+1)\frac{j_l(x)}{x}.$$

Hence,

$$g_l(r_2) = \sqrt{4\pi(2l+1)}\, i^l \frac{j_l(K_\nu r_2)}{K_\nu^3 r_2}. \tag{166.21}$$

In the evaluation of the scattering amplitude (166.20), we can go one step farther. We know that the states u_ν may be factorized with a spherical harmonic. Since u_0, the ground state, does not depend on angles, there are, therefore, in (166.20) products of two spherical harmonics to be integrated. Orthogonality then selects only one term of the sum (166.20). With

$$u_\nu \equiv \frac{1}{r_2} \chi_{n,l}(r_2)\, Y_{l,m}(\vartheta_2, \varphi_2) \tag{166.22a}$$

where $\boldsymbol{K}_\nu$ has been used as axis of quantization, and

$$u_0 = \frac{1}{r_2} \chi_0(r_2) \tag{166.22b}$$

as the ground state, the scattering amplitude becomes

$$f(\vartheta_1) = \frac{2M}{K_\nu^2} \delta_{m0} \sqrt{4\pi(2l+1)}\, i^l \int_0^\infty dr_2\, \chi_{n,l}(r_2)\, \chi_0(r_2) \frac{j_l(K_\nu r_2)}{K_\nu r_2}. \tag{166.23}$$

The states of different m are degenerate. Only such a linear combination of them can be excited that its angular momentum has no component about the direction of the momentum transfer vector K_ν. No selection rules exist for l. It should further be remarked that the expression (166.23) still depends upon the angle of deflection, ϑ_1, of the proton since K_ν does,

$$K_\nu^2 = k^2 + k_\nu^2 - 2k k_\nu \cos\vartheta_1 \tag{166.24}$$

k_ν being derived from the energy law (166.8) and independent of ϑ_1.

NB. The same results can be obtained by applying the Golden Rule (Problem 183) to this process.

B. Very Many Particles: Quantum Statistics

Problem 167. Electron gas in a metal

In a rough approximation, the conduction electrons in a metal may be treated as freely moving with potential walls at the surface which prevent their leaving the metal. In a cube of silver (density $\rho = 10.5\,\text{gm/cm}^3$, atomic weight 108, one conduction electron per silver ion) there shall be determined

a) the highest electron energy occurring in the ground state,
b) the average kinetic electron energy,
c) the pressure of the electron gas.

Temperature excitation may be neglected.

Solution. In a silver cube of volume L^3 the possible electron energies, according to Problem 18, are given by

$$E = \frac{\hbar^2 \pi^2}{2mL^2}(n_1^2 + n_2^2 + n_3^2) \tag{167.1}$$

with n_1, n_2, n_3 positive integers $(=1,2,3,\ldots)$. In each state described by a triple of quantum numbers (n_1, n_2, n_3) there are two electrons of opposite spin directions, according to the Pauli principle. Since in our metal cube a great many electrons have to be distributed, we shall essentially have large values of the quantum numbers.

Let us consider a space with coordinates n_1, n_2, n_3. Each lattice point with integer coordinates in its first octant corresponds to a state of energy (167.1). If we denote the radius from the origin in this space by n, so that

$$n_1^2 + n_2^2 + n_3^2 = n^2, \tag{167.2}$$

a spherical shell between radii n and $n + dn$ in this octant will contain

$$\frac{1}{8} \cdot 4\pi n^2\, dn = \frac{\pi}{2} n^2\, dn$$

integer lattice points. Occupying each of them with two electrons of opposite spins, we have $\pi n^2\, dn$ electrons between n and $n + dn$. As, on the other hand, the energy (167.1) depends only upon n,

$$E = \frac{\hbar^2 \pi^2}{2mL^2} n^2\,; \qquad dE = \frac{\hbar^2 \pi^2}{mL^2} n\, dn \tag{167.3}$$

there will be, in the interval dn,

$$dN = \pi n^2\, dn = \pi \sqrt{\frac{2mL^2 E}{\hbar^2 \pi^2}} \cdot \frac{mL^2}{\hbar^2 \pi^2}\, dE$$

or

$$dN = \sqrt{2m}\,\frac{mL^3}{\pi^2 \hbar^3}\sqrt{E}\, dE \tag{167.4}$$

electrons with energies between E and $E + dE$.

a) The highest electron energy in the ground state of this electron gas, ζ, follows from the total number of electrons, N, in the cube to be a given constant:

$$N = \sqrt{2m}\,\frac{mL^3}{\pi^2 \hbar^3}\int_0^{\zeta}\sqrt{E}\, dE. \tag{167.5}$$

Let us denote the electron density by

$$\mathcal{N} = \frac{N}{L^3}, \tag{167.6}$$

then we arrive at a formula independent of the volume of the metal considered:

$$\mathcal{N} = \frac{1}{3\pi^2}\left(\frac{2m\zeta}{\hbar^2}\right)^{\frac{3}{2}} \tag{167.7a}$$

or

$$\zeta = \frac{\hbar^2}{2m}(3\pi^2 \mathcal{N})^{\frac{2}{3}}. \tag{167.7b}$$

As $\mathcal{N} = \rho/M$ with M the mass of a silver atom ($M = 1.80 \times 10^{-22}$ gm) we have

$$\mathcal{N} = 5.85 \times 10^{22}\,\mathrm{cm}^{-3}$$

so that Eq. (167.7b) leads to an upper energy limit of

$$\zeta = 8.80 \times 10^{-12}\,\mathrm{erg} = 5.55\,\mathrm{eV}.$$

This is an energy so large compared to thermal energies ($kT = 0.026$ eV at 300 °K) that thermal excitation may indeed only slightly change the distribution of electrons over the energy states. The reason for this effect, called degeneracy of the electron gas (Fermi gas), is, of course, the very small mass of the electron in the denominator of (167.7b).

The energy limit ζ is, in general, called the *Fermi energy* of the electron gas.

b) The average energy of a conduction electron is determined by

$$\bar{E} = \int dN\, E / \int dN \tag{167.8}$$

or, according to (167.4),

$$\bar{E} = \int_0^\zeta dE \sqrt{E}\, E \Big/ \int_0^\zeta dE \sqrt{E} = \tfrac{3}{5}\zeta . \tag{167.9}$$

c) Pressure may always be defined, without using any thermodynamics, by the work of compression if a volume V is diminished by dV:

$$dW = p\, dV .$$

This work is used to increase the total energy content U of the gas by dU,

$$dW = dU .$$

This total energy (at $T=0$) is

$$U = N\bar{E} = \tfrac{3}{5} N \zeta . \tag{167.10}$$

According to (167.7b), the Fermi energy ζ depends on $\mathcal{N} = N/V$ and therefore on V:

$$U \propto V^{-\frac{2}{3}}; \qquad \frac{dU}{U} = -\frac{2}{3}\frac{dV}{V} .$$

Thus we find for the pressure,

$$p = -\frac{dU}{dV} = \frac{2}{3}\frac{U}{V} = \frac{2}{5}\mathcal{N}\zeta . \tag{167.11}$$

With the numerical values of $\mathcal{N}$ and ζ determined above, this leads to a pressure of

$$p = 2.06 \times 10^{11}\ \text{dyn/cm}^2$$

or about 200,000 atmospheres. This immense pressure is counterbalanced by the strong Coulomb attraction between the conduction electrons and the lattice ions.

Problem 168. Paramagnetic susceptibility of a metal

To determine the paramagnetic susceptibility of a metal, for zero temperature, by treating the conduction electrons as a Fermi gas, neglecting the polarizability of the lattice ions.

Solution. According to the preceding problem the conduction electrons form a Fermi gas whose energy limit (the Fermi energy) is

$$\zeta = \frac{\hbar^2}{2m}(3\pi^2 \mathcal{N})^{\frac{2}{3}}, \tag{168.1}$$

with $\mathcal{N}$ the number of conduction electrons per unit volume. The energy difference, ΔE, between two consecutive electron levels follows from

$$4\pi p^2 \Delta p = 4\pi\sqrt{2mE}\, m\Delta E = \frac{(2\pi\hbar)^3}{V}. \tag{168.2}$$

In the vicinity of the Fermi energy this yields (writing $\Delta E_0 \propto \zeta^{-\frac{1}{2}} = \zeta/\zeta^{\frac{3}{2}}$ and putting (168.1) for $\zeta^{\frac{3}{2}}$),

$$\Delta E_0 = \frac{4}{3}\,\frac{\zeta}{\mathcal{N}V}. \tag{168.3}$$

All levels $E<\zeta$ are occupied by pairs of electrons of opposite spin directions, all levels $E>\zeta$ are unoccupied, at zero temperature.

If now a magnetic field is applied to the metal, energy can be gained by separating pairs of electrons and directing the spins of each pair so that both are parallel to the field strength $\mathcal{H}$. If ν pairs are separated, the energy gain apparently becomes

$$2\nu\cdot\mu\mathcal{H} \quad \text{with } \mu = \frac{e\hbar}{2mc}. \tag{168.4}$$

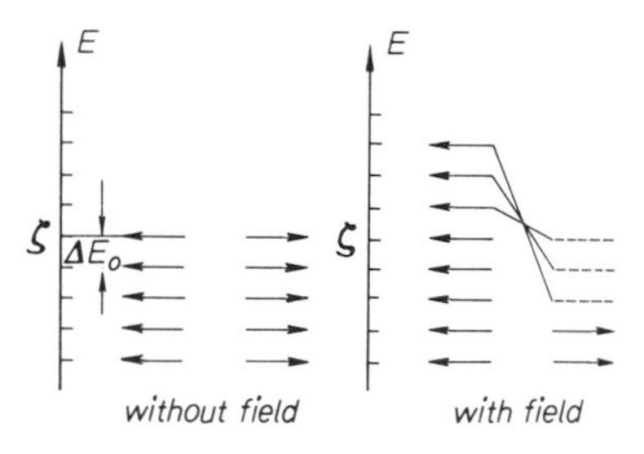

Fig. 66. Spin-flip near Fermi energy under action of a magnetic field

Such a separation is, of course, possible only by raising one electron of the pair to an unoccupied level beyond $E=\zeta$. This means, however, an expenditure of kinetic energy counteracting the gain (168.4). Fig. 66

shows that the amount of energy expended in separating the first pair (by transferring an electron from the uppermost occupied to the lowest unoccupied level) is ΔE_0, for the second pair it is $3\Delta E_0$, for the third pair $5\Delta E_0$, etc. In general, the separation of ν pairs requires an expenditure of

$$[1+3+5+\cdots+(2\nu-1)]\Delta E_0=\nu^2\Delta E_0. \tag{168.5}$$

Equilibrium will be reached if the total energy change W of the gas affected by the magnetic field,

$$W=-2\nu\cdot\mu\mathscr{H}+\nu^2\Delta E_0, \tag{168.6}$$

is a minimum:

$$\frac{dW}{d\nu}=-2\mu\mathscr{H}+2\nu\Delta E_0=0,$$

i.e. for

$$\nu=\frac{\mu\mathscr{H}}{\Delta E_0} \tag{168.7}$$

with

$$W_{\min}=-\frac{(\mu\mathscr{H})^2}{\Delta E_0}. \tag{168.8}$$

If more pairs were separated, the total energy of the gas would increase again. In equilibrium, the total magnetic moment of the metal becomes

$$\mathscr{M}=2\nu\cdot\mu=\frac{2\mu^2\mathscr{H}}{\Delta E_0}$$

and, by definition, its paramagnetic susceptibility per unit volume,

$$\chi=\frac{\mathscr{M}}{\mathscr{H}V}=\frac{2\mu^2}{V\Delta E_0}, \tag{168.9}$$

according to (168.3) and (168.1),

$$\chi=\frac{e^2}{4\pi mc^2}\left(\frac{3\mathscr{N}}{\pi}\right)^{\frac{1}{3}}. \tag{168.10}$$

To evaluate this expression numerically we may write

$$\mathscr{N}=\frac{z\rho}{m_H A},$$

expressing the electron density by the mass density ρ of the metal, the mass of one of its atoms $m_H A$ (with A the atomic weight) and its valence z. Then we find,

$$\chi_{\text{para}}=1.86\times10^{-6}\left(\frac{z\rho}{A}\right)^{\frac{1}{3}}. \tag{168.11}$$

To compare this result with experimental values, the diamagnetic susceptibility of the lattice ions has to be subtracted.

NB. In problem 160 we have calculated the diamagnetic susceptibility of neon which must be practically identical with that of Na^+. It turned out to be

$$\chi_{\text{dia}} = -5.61 \times 10^{-6}\ \text{cm}^3/\text{mole}$$

which, with a density of sodium metal of about $1\ \text{gm/cm}^3 = \frac{1}{23}\ \text{mole/cm}^3$ is the same as

$$\chi_{\text{dia}} = -0.25 \times 10^{-6}$$

in the dimensionless scale used in the present problem. From (168.11), on the other hand, we obtain the electron contribution

$$\chi_{\text{para}} = +0.66 \times 10^{-6}.$$

The expressions are of the same order of magnitude so that in some metals (e.g. in caesium) even a resultant diamagnetism is observed.

Literature. Frenkel, J.: Z. Physik **49,** 31 (1928).

Problem 169. Field emission, uncorrected for image force

To determine the electron current emitted from a metallic surface under the action of a high electric field strength $\mathscr{E}$. The temperature can be supposed to be low; image force and lattice structure shall be neglected.

Solution. Let $z=0$ be the metallic surface. The interior ($z<0$) may have constant potential energy, $V=0$, whereas the exterior ($z>0$) has potential V_0. Inside the metal the conduction electrons form the ground state of a Fermi gas, occupying all levels up to the Fermi energy ζ. Outside, there holds the potential

$$V(z) = V_0 - e\mathscr{E}z. \tag{169.1}$$

As shown in Fig. 67 a potential barrier is formed beyond the surface. Let E_z be the part of the electron energy corresponding to its velocity

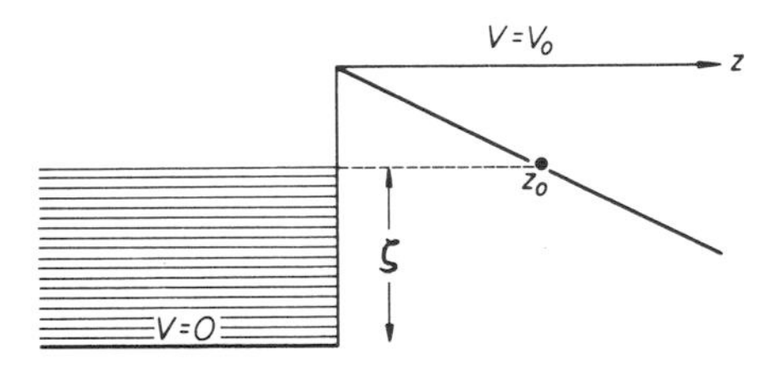

Fig. 67. Field emission of electrons from a metal surface. Left-hand side: densely lying electron levels inside the metal up to the Fermi energy ζ. Right-hand side: Potential outside the metal under action of an electrical field

component in z direction; then the potential barrier has a transmission coefficient T, to be calculated in WKB approximation,

$$T=\exp\left\{-2\frac{\sqrt{2m}}{\hbar}\int_0^{z_0} dz\sqrt{V(z)-E_z}\right\}, \tag{169.2}$$

rapidly decreasing with decreasing E_z. Here, $V(z)$ is the expression (169.1) and

$$E_z=\frac{m}{2}v_z^2; \qquad z_0=\frac{V_0-E_z}{e\mathcal{E}}. \tag{169.3}$$

The integration in (169.2), performed in an elementary way, yields

$$T=\exp\left\{-\frac{4}{3}\frac{\sqrt{2m}}{\hbar e\mathcal{E}}(V_0-E_z)^{\frac{3}{2}}\right\}. \tag{169.4}$$

The electric current density (per cm^2) is

$$j=e\int dn\, v_z T \tag{169.5}$$

with dn the number of conduction electrons per cm^3 and per momentum space element $dp_x dp_y dp_z$. For the Fermi gas, there is

$$dn=2\frac{dp_x dp_y dp_z}{h^3} \qquad (h=2\pi\hbar)$$

inside the Fermi sphere, i.e. for

$$p_x^2+p_y^2+p_z^2\le 2m\zeta, \tag{169.6}$$

and, outside it, $dn=0$. With cylindrical coordinates ρ, φ, p_z in momentum space according to

$$p_x=\rho\cos\varphi, \qquad p_y=\rho\sin\varphi, \qquad \rho^2+p_z^2\le 2m\zeta,$$

the integral (169.5) then may be written

$$j=\frac{2e}{h^3}2\pi\int_0^{\sqrt{2m\zeta}} dp_z \int_0^{\sqrt{2m\zeta-p_z^2}} d\rho\,\rho(p_z/m)\,T,$$

the integration being extended over all electrons with $v_z>0$. If we use the auxiliary variable

$$\varepsilon=\zeta-E_z \tag{169.7}$$

the integral simplifies to

$$j=\frac{4\pi e m}{h^3}\int_0^{\zeta} d\varepsilon\, \varepsilon\, T(\varepsilon) \tag{169.8}$$

with

$$T=\exp\left\{-\frac{4}{3}\frac{\sqrt{2m}}{\hbar e\mathscr{E}}(V_0-\zeta+\varepsilon)^{\frac{3}{2}}\right\}. \tag{169.9}$$

To evaluate the integral (169.8) we use the fact that, starting from $\varepsilon=0$ (maximum energy $E_z=\zeta$), the transmission coefficient $T(\varepsilon)$ decreases rapidly with increasing ε. Therefore, mainly the electrons with small values of ε contribute to the integral (169.8) and we may expand

$$(V_0-\zeta+\varepsilon)^{\frac{3}{2}}=(V_0-\zeta)^{\frac{3}{2}}+\tfrac{3}{2}\varepsilon(V_0-\zeta)^{\frac{1}{2}}+\cdots$$

With the abbreviation

$$2\frac{\sqrt{2m}}{\hbar e\mathscr{E}}(V_0-\zeta)^{\frac{3}{2}}=q \tag{169.10}$$

we then obtain

$$T=\mathrm{e}^{-\frac{2}{3}q}\exp\left(-\frac{q\varepsilon}{V_0-\zeta}\right)$$

and

$$j=\frac{4\pi e m}{h^3}\mathrm{e}^{-\frac{2}{3}q}\int_0^{\zeta} d\varepsilon\,\varepsilon\exp\left(-\frac{q\varepsilon}{V_0-\zeta}\right).$$

Again, the integrand falls off rapidly with increasing ε so that we may extend the integration to infinity, without noticeable error, and finally arrive at

$$j=\frac{4\pi e m}{h^3}\frac{(V_0-\zeta)^2}{q^2}\mathrm{e}^{-\frac{2}{3}q}. \tag{169.11}$$

Numerical values. According to Eqs. (169.10) and (169.11) the electric current falls rapidly off with decreasing field strength $\mathscr{E}$, and with increasing work function $V_0-\zeta$. If we measure the field strength in volts/cm, the work function in eV and the current density in amp/cm^2, then we get the following numerical relations:

$$\left.\begin{aligned} q&=1.047\times10^{8}(V_0-\zeta)^{\frac{3}{2}}/\mathscr{E}\,;\\ j&=1.59\times10^{10}\frac{(V_0-\zeta)^2}{q^2}\mathrm{e}^{-\frac{2}{3}q}. \end{aligned}\right\} \tag{169.12}$$

For q of the order 1, a current density of the order 10^{10} amp/cm^2 might be expected, i.e. almost every electron hitting the surface would leave the metal. Of course, model and approximations are equally untenable under these extreme conditions. For larger values of q the current rapidly falls off. Thus it may not be unreasonable to ask at which field strength we may expect a current density of 1 amp/cm^2, with different values of the work function. One finds the following pairs of values:

$$\begin{array}{lll} \mathscr{E}=10^6 \text{ volt/cm} & \text{and} & V_0-\zeta=0.083 \text{ eV} \\ 10^7 & & 0.43 \\ 10^8 & & 2.19 \end{array}$$

As $V_0-\zeta$ is always of the order of several eV's in metals, one should not observe any appreciable field emission below 10^8 volts/cm. In fact, experiment shows a threshold field strength of only about 10^6 volts/cm. This wide discrepancy can certainly not be explained by temperature excitation of the Fermi gas, which would only help to lower the work function by an amount between $\frac{1}{10}$ and $\frac{1}{100}$ eV ($kT \sim \frac{1}{30}$ eV at normal temperatures). It can, however, be explained by taking account of the image force, as is shown in the next problem.

Problem 170. Field emission, corrected for image force

The potential threshold for field emission of electrons from a metal surface is essentially lowered by the image force. Its effect upon the electron current emitted shall be investigated.

Solution. The image force is originated by the distortion of the surface charge in the neighbourhood of any electron at a distance $z>0$ outside the metal. It can be calculated from classical electrostatics, neglecting effects of the metal structure if z is appreciably larger than the lattice constant, so that the metal may be treated as a continuum. It then turns out to be

$$V_{\text{image}} = -\frac{e^2}{4z}. \tag{170.1}$$

For smaller values of z the expression is rather bad, as is seen by its unphysical singularity at $z=0$. This error, however, does not affect the following considerations which depend entirely upon the height and breadth of the potential wall *above* electron energy.

This potential wall then becomes

$$V(z) = V_0 - \frac{e^2}{4z} - e\mathscr{E}z \tag{170.2}$$

with the notations of the preceding problem (Fig. 68). We are interested in that part of it that extends between z_1 and z_2, the two solutions of the quadratic equation $V(z)=E_z$. This yields

$$z_{1,2}=\frac{V_0-E_z}{2e\mathscr{E}}\pm\sqrt{\frac{(V_0-E_z)^2}{4e^2\mathscr{E}^2}-\frac{e}{4\mathscr{E}}}. \tag{170.3}$$

Both solutions are real if

$$e\mathscr{E}<\left(\frac{V_0-E_z}{e}\right)^2,$$

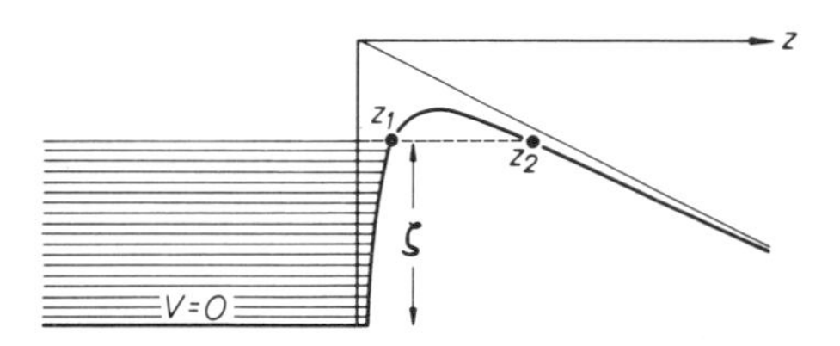

Fig. 68. Same as Fig. 67 but corrected for image force

a condition which is satisfied even for field strengths of about 10^9 volts/cm; for higher values the threshold would be submerged under the Fermi level of the electron sea. For the field strengths used in experiments, which are even below 10^7 volts/cm, we may safely assume

$$e\mathscr{E}\ll\left(\frac{V_0-E_z}{e}\right)^2 \tag{170.4}$$

and accordingly expand the radical in (170.3). The results are

$$z_1=\frac{e^2}{4(V_0-E_z)} \quad \text{and} \quad z_2=z_0-z_1 \tag{170.5}$$

with

$$z_0=\frac{V_0-E_z}{e\mathscr{E}} \tag{170.6}$$

and $z_1\ll z_2$. The summit of the barrier, according to Eq. (170.2), is now, due to the image force, shifted from $z=0$ to $z=\sqrt{e/4\mathscr{E}}$ and instead of V_0 its height is only

$$V_0-e\sqrt{e\mathscr{E}}.$$

According to (170.4) this does not involve much lowering of the threshold but rather a flattening of the summit, so that we may expect a much larger transmission coefficient. Only the neighbourhood of $E_z=\zeta$ will contribute appreciably to the emission current (as in the preceding problem); for this energy $z_1=e^2/[4(V_0-\zeta)]$ is at least of the order of the lattice constant, so that the singularity of the image force potential at $z=0$ will become a matter of indifference.

Again using the WKB approximation, we find the transmission coefficient

$$T=\exp\left\{-2\frac{\sqrt{2m}}{\hbar}\int\limits_{z_1}^{z_2}dz\sqrt{V(z)-E_z}\right\}$$

or, writing $V(z)$ in the form

$$V(z)=\frac{e\mathscr{E}}{z}(z-z_1)(z_2-z),$$

we have

$$-\frac{\hbar}{2\sqrt{2me\mathscr{E}}}\log T=\int\limits_{z_1}^{z_2}dz\sqrt{\frac{(z-z_1)(z_2-z)}{z}}. \tag{170.7}$$

This integral is of elliptic type and may be reduced to standard integrals as follows: use, instead of z, the variable $x=(z-z_1)/(z_2-z_1)$, thus transforming the integral into

$$(z_2-z_1)^{\frac{3}{2}}\int\limits_0^1 dx\sqrt{\frac{x(1-x)}{x+c}}\quad\text{with } c=\frac{z_1}{z_2-z_1}.$$

Next put

$$\frac{z_2-z_1}{z_2}=k^2 \tag{170.8}$$

and

$$x=\frac{(1-k^2)\sin^2\varphi}{1-k^2\sin^2\varphi};$$

then the right-hand side of (170.7) becomes

$$\frac{2k^4}{\sqrt{1-k^2}}z_1^{\frac{3}{2}}\int\limits_0^{\frac{\pi}{2}}\frac{d\varphi\sin^2\varphi\cos^2\varphi}{(1-k^2\sin^2\varphi)^{\frac{5}{2}}}. \tag{170.9}$$

The last integral may be reduced to the two complete elliptic integrals $E(k)$ and $K(k)$:

$$\int_0^{\frac{\pi}{2}} \frac{d\varphi \sin^2\varphi \cos^2\varphi}{(1-k^2\sin^2\varphi)^{\frac{5}{2}}} = \frac{1}{3k^4}\left[\frac{2-k^2}{1-k^2} E(k) - 2K(k)\right]. \qquad (170.10)$$

Eq. (170.10) may be proved as follows. Putting $1-k^2\sin^2\varphi = \Delta^2$, the two complete integrals in standard form are defined by

$$K(k) = \int_0^{\frac{\pi}{2}} \frac{d\varphi}{\Delta}; \qquad E(k) = \int_0^{\frac{\pi}{2}} d\varphi\, \Delta.$$

Now, there can be proved the following identity by simple differentiation,

$$3k^2 \frac{\sin^2\varphi\cos^2\varphi}{\Delta^5} = \frac{2-k^2}{k^2(1-k^2)}\Delta - \frac{2}{k^2\Delta} + \frac{d}{d\varphi}\left\{\sin\varphi\cos\varphi\left(\frac{1}{\Delta^3} - \frac{2-k^2}{(1-k^2)\Delta}\right)\right\}.$$

Integration of this identity yields directly

$$3k^2 \int_0^{\frac{\pi}{2}} d\varphi \frac{\sin^2\varphi\cos^2\varphi}{\Delta^5} = \frac{2-k^2}{k^2(1-k^2)} E(k) - \frac{2}{k^2} K(k),$$

in agreement with (170.10).

Eqs. (170.7) and (170.10) may then be unified into

$$-\frac{\hbar}{2\sqrt{2me\mathscr{E}}} \log T = \frac{2}{3} z_2^{\frac{3}{2}} [(2-k^2)E(k) - 2(1-k^2)K(k)]. \qquad (170.11)$$

This formula may be brought into a much simpler shape. Remembering that $z_1 \ll z_2$ so that $k^2 \simeq 1$, we may replace k^2 by the parameter

$$k'^2 = 1-k^2 = z_1/z_2 \ll 1 \qquad (170.12)$$

and expand (170.11) into an extremely well converging series in this parameter[6] according to

$$K(k) = \Lambda + \tfrac{1}{4}(\Lambda - 1)k'^2 + \cdots; \qquad E(k) = 1 + \tfrac{1}{2}(\Lambda - \tfrac{1}{2})k'^2 + \cdots$$

---

[6] Cf. Jahnke-Emde, 2nd edition (1933), p. 145.

with $\Lambda = \log(4/k')$. If these expansions are put into (170.11), the right-hand side becomes

$$\frac{2}{3} z_2^{\frac{3}{2}} \left[1 + k'^2 \left(\frac{3}{4} - \frac{3}{2} \log \frac{4}{k'} \right) \right] = \frac{2}{3} z_0^{\frac{3}{2}} \left[1 - \frac{3}{2} k'^2 \left(\frac{1}{2} + \log \frac{4}{k'} \right) \right].$$

Were $k' = 0$, we would fall back on the expression (170.9) for T without image force. Let us denote this by T_0; then the result is

$$T = T_0^{1-\lambda} \tag{170.13}$$

with

$$\lambda = \frac{3}{2} k'^2 \left(\frac{1}{2} + \log \frac{4}{k'} \right). \tag{170.14}$$

It remains to evaluate the current integral (169.8) with the new expression for T. Again, as in the preceding problem, it is essentially the vicinity of $E_z = \zeta$ that contributes to the current, so that we may expand λ at $E_z = \zeta$ or $\varepsilon = 0$ and confine ourselves to the linear term in ε. This is practically the same as putting

$$k'^2 \simeq (z_1/z_0)_{E_z=\zeta} = \frac{e^3 \mathscr{E}}{4(V_0 - \zeta)^2} \tag{170.15}$$

and then performing the integration as in Problem 169. Instead of (169.11), which we will denote by j_0, we then find

$$j = j_0 \, \mathrm{e}^{\frac{2}{3} \lambda q} \tag{170.16}$$

with $\lambda \ll 1$; the next better approximation would add a factor

$$[1 - \tfrac{7}{3} \lambda + k'^2]^2$$

in the denominator. The essential thing, of course, is the exponential in (170.16).

Let us finally discuss a few numerical consequences. Besides the relations (169.12) we now get

$$k'^2 = 3.58 \times 10^{-8} \mathscr{E}/(V_0 - \zeta)^2$$

in the same units of volts/cm for $\mathscr{E}$ and eV's for $V_0 - \zeta$. With a reasonable value of the work function, $V_0 - \zeta = 3$ eV, and a field strength $\mathscr{E} = 10^7$ volts/cm, we then arrive at

$$q = 54.5, \quad k'^2 = 0.0397, \quad \lambda = 0.208, \quad \mathrm{e}^{\frac{2}{3}\lambda q} = 1860,$$
$$j_0 = 0.9 \times 10^{-8} \text{ amp/cm}^2,$$
$$j = 1.7 \times 10^{-5} \text{ amp/cm}^2.$$

Problem 171. White dwarf

Let the temperature of a white dwarf be high enough to ionize its atoms practically completely, and low enough to neglect gas pressure and radiation pressure compared to zero-point pressure of the degenerate electron gas. (The latter assumption is rather bad.) The distribution of density through the star shall be calculated for a given total mass of the star, from the equilibrium of zero-point pressure and gravitational pressure.

Solution. In a spherical mass of gas the radial pressure gradient must be in equilibrium with the gravitational force density (barometric formula):

$$\frac{dp}{dr} = -\frac{GM_r}{r^2}\rho. \tag{171.1}$$

Here G is the gravitational constant, M_r the mass inside a sphere of radius r,

$$M_r = 4\pi \int_0^r dr' r'^2 \rho(r'), \tag{171.2}$$

and $\rho(r)$ is the mass density, i.e. the mass of all ions and free electrons inside $1\,\mathrm{cm}^3$ of star matter. For complete ionization with $\mathcal{N}$ electrons per c.c., there are $\mathcal{N}/Z$ ions (nuclei) so that

$$\rho = \frac{\mathcal{N}}{Z} m_H A,$$

with $m_H A$ the mass of one neutral atom. If there are different elements, A and Z represent average values. It should be noted, however, that the ratio

$$\frac{A}{Z} = 2\alpha \tag{171.3}$$

is almost independent of the chemical composition, α varying from 1.0 to 1.3 from light to heavy elements with the one exception of hydrogen where $\alpha = \frac{1}{2}$. Therefore,

$$\rho = 2\alpha m_H \mathcal{N}, \tag{171.4}$$

depends essentially only upon the density of electrons.

The pressure of the electron gas is, according to Problem 167, its zero-point pressure,

$$p_e = \frac{2}{5} \mathcal{N} \frac{\hbar^2}{2m} (3\pi^2 \mathcal{N})^{\frac{2}{3}}. \tag{171.5}$$

The zero-point pressure of the ions, p_i, according to the proportionality with $\mathcal{N}^{\frac{5}{3}}/m$, would be much smaller:

$$p_i/p_e = (\mathcal{N}_i/\mathcal{N})^{\frac{5}{3}}\, m/m_i = Z^{-\frac{5}{3}}\,\frac{m}{m_H A}\,.$$

Even in hydrogen ($Z=1$, $A=1$), we would have $p_i/p_e = 1/1838$, the ratio being much smaller for all other elements. We shall, therefore, neglect p_i and identify p_e, Eq. (171.5), with the total pressure, p.

It is not quite so easy to dispense with temperature effects. Only if $\zeta \gg kT$ can we assume the gas to be extremely degenerate so that its pressure is mainly zero-point pressure. For the electrons, the Fermi energy is

$$\zeta = \frac{\hbar^2}{2m}\left(\frac{3\pi^2\rho}{2\alpha m_H}\right)^{\frac{2}{3}} = 1.64\,\mathrm{eV}\left(\frac{\rho}{\alpha}\right)^{\frac{2}{3}}.$$

This is to be compared with $kT \simeq 100\,\mathrm{eV}$ at 10^6 degrees. Even with $\rho = 10^3$, both quantities would be of the same order of magnitude. The ion gas would not at all be degenerate, then, and contribute in the same order, too.—The radiation pressure is

$$p_R = 2.52 \times 10^{-15}\, T^4\,\mathrm{dyn/cm^2},$$

whereas from (171.5) and (171.4) there follows

$$p_e = 3.16 \times 10^{12} (\rho/\alpha)^{\frac{5}{3}}\ \mathrm{dyn/cm^2}.$$

With $T = 10^6$ degrees, therefore, the radiation pressure is of the order of 10^9 dyn/cm² which may, indeed, be neglected under the density conditions of a white dwarf.

From (171.5) and (171.4) we get the equation of state,

$$p = f\rho^{\frac{5}{3}}; \qquad f = \frac{\hbar^2}{10m}\left(\frac{3\pi^2}{2}\right)^{\frac{2}{3}} (\alpha m_H)^{-\frac{5}{3}}$$
$$= 3.17 \times 10^{12}\,\alpha^{-\frac{5}{3}}\,\mathrm{gm}^{-\frac{2}{3}}\,\mathrm{cm}^4\,\mathrm{sec}^{-2}. \qquad (171.6)$$

Any connection between pressure and density of the form

$$p = f\rho^{1+\frac{1}{n}}$$

is called a *polytrope* of index n. The white dwarf, therefore, is built according to a polytrope of index $n = \frac{3}{2}$.

Putting (171.6) in the equilibrium condition (171.1) we get

$$\frac{5}{3} f \rho^{\frac{2}{3}} \frac{d\rho}{dr} = -\frac{G\rho}{r^2} M_r$$

or, by differentiation,

$$\frac{5f}{3G}\,\frac{d}{dr}\left(r^2 \rho^{-\frac{1}{3}} \frac{d\rho}{dr}\right) = -4\pi r^2 \rho\,. \qquad (171.7)$$

Using, instead of ρ, the dimensionless function

$$\varphi = (\rho/\rho_0)^{\frac{2}{3}} \tag{171.8}$$

with a constant ρ_0 and, instead of the radius r, the dimensionless variable

$$x = r/r_1 \tag{171.9}$$

choosing the unit of length, r_1, according to

$$r_1^2 = \frac{5f}{8\pi G}\rho_0^{-\frac{1}{3}}, \tag{171.10}$$

the differential equation (171.7) is reshaped into

$$\frac{d^2\varphi}{dx^2} + \frac{2}{x}\frac{d\varphi}{dx} + \varphi^{\frac{2}{3}} = 0, \tag{171.11}$$

independent of all physical constants. If we choose ρ_0 to be the density at the centre of the star, we have to solve Eq. (171.11) with the boundary conditions

$$\varphi(0) = 1; \qquad \varphi'(0) = 0. \tag{171.12}$$

The solution of the non-linear equation (171.11) with boundary conditions (171.12) is uniquely to be obtained by numerical integration. This solution decreases monotonously, reaching $\varphi = 0$ at

$$x = X = 3.6537 \tag{171.13a}$$

where its derivative is

$$\left(\frac{d\varphi}{dx}\right)_{x=X} = -D = -0.206. \tag{171.13b}$$

According to (171.8) this zero of φ corresponds to the surface of the star, say, $r = R$. The total mass of the star therefore becomes

$$M = 4\pi\int_0^R dr\, r^2 \rho(r) = 4\pi\rho_0 r_1^3 \int_0^X dx\, x^2 \varphi^{\frac{2}{3}}.$$

This integral can be evaluated without detailed numerical knowledge of the function $\varphi(x)$ by replacing its integrand according to (171.11):

$$x^2\varphi^{\frac{2}{3}} = -\frac{d}{dx}\left(x^2\frac{d\varphi}{dx}\right).$$

The result is

$$M = 4\pi\rho_0 r_1^3 X^2 D = 34.5\,\rho_0 r_1^3. \tag{171.14}$$

If the mass of the star is known by observation, there exist two relations between ρ_0 and r_1, viz. Eqs. (171.10) and (171.14):

$$\left.\begin{aligned} &\rho_0^{\frac{1}{3}} r_1^2 - a \quad \text{with } a = \frac{5f}{8\pi G} = 9.46 \times 10^{18}\, \alpha^{-\frac{5}{3}}\, \mathrm{gm}^{\frac{1}{3}}\, \mathrm{cm}, \\ &\rho_0 r_1^3 = b \quad \text{with } b = 0.0290\, M. \end{aligned}\right\} \tag{171.15}$$

This leads to

$$r_1 = a b^{-\frac{1}{3}}; \qquad \rho_0 = b^2 a^{-3} \tag{171.16}$$

and for the radius R of the star to

$$R = r_1 X = 3.6537\, r_1. \tag{171.17}$$

Finally, the average density is

$$\bar{\rho} = \frac{3D}{X} \rho_0 = 0.169\, \rho_0, \tag{171.18}$$

i.e. about $\frac{1}{6}$ of the central density.

Numerical example. The companion of Sirius (α Can maj), Sirius B, has a mass determined from the motions of Sirius about the centre-of-mass of this binary system. It is about the same as the mass of the sun, viz. $M = 1.94 \times 10^{33}$ gm. This leads to the following numerical values:

$$\begin{aligned} r_1 &= 2.47 \times 10^8\, \alpha^{-\frac{5}{3}}\ \mathrm{cm}; \\ R &= 8.98 \times 10^8\, \alpha^{-\frac{5}{3}}\ \mathrm{cm}; \\ \rho_0 &= 3.73 \times 10^6\, \alpha^5\ \mathrm{gm\ cm^{-3}}; \\ \bar{\rho} &= 6.15 \times 10^5\, \alpha^5\ \mathrm{gm\ cm^{-3}}. \end{aligned}$$

Since observations lead to a radius about $\frac{1}{20}$ of that of the sun, $R_0 = 6.95 \times 10^{10}$ cm, our model leads to $\alpha = 0.445$. This is not very far from $\alpha = 0.5$ for a hydrogen star, but on the wrong side. Our radius, however, is certainly too small. The neglect of a rather important part of the pressure accounts for that: the star will be inflated to a larger radius if temperature effects are accounted for.

Problem 172. Thomas-Fermi approximation

To calculate the electron density of an atom or a positive ion. To obtain a suitable approximation, it shall be supposed that regions within which the electrostatic potential varies but little contain enough electrons to justify treating them statistically.

Solution. There are two basic ideas underlying the model, one electrostatic and the other quantum statistical. Let us start with the electrostatic side of the problem. If $n(r)$ electrons are contained in a unit volume at a distance r from the atomic nucleus, the electrostatic potential originated by both nucleus and electrons satisfies the Poisson equation, the first fundamental equation of our problem:

$$\nabla^2 \Phi = 4\pi e \cdot n(r) \tag{172.1}$$

with $\rho(r) = -e n(r)$ the charge density of the electron cloud. The solution of this equation is subject to the boundary conditions

$$\Phi = \frac{Ze}{r} \quad \text{for } r \to 0 \tag{172.2}$$

in the vicinity of the nuclear charge Ze, and

$$\Phi = \frac{ze}{r} \quad \text{for } r \geqq R \tag{172.3}$$

if R is the radius of the positive ion of charge ze. This radius has still to be determined.

There can be no singularity of charge density at $r = R$, so that not only the potential but the field strength as well must be continuous there. This, instead of (172.3), permits the boundary condition to be written in the form

$$\Phi(R) = \frac{ze}{R} \quad \text{and} \quad \left(\frac{d\Phi}{dr}\right)_R = -\frac{ze}{R^2}. \tag{172.4}$$

We now come to the second principle underlying the calculation, viz. the quantum statistical part of the problem. Considering any volume element inside the atom or ion, we find that the momentum p of an electron found there will be connected with its energy by the relation

$$E = \frac{p^2}{2m} - e\Phi(r).$$

In order to bind the electron, this energy must apparently be smaller inside the atom than the potential energy $-e\Phi(R)$ at its surface. Hence, at a distance r from the nucleus, no electron can have a momentum larger than $p_{\max}$ given by

$$\frac{p_{\max}^2}{2m} = e[\Phi(r) - \Phi(R)]. \tag{172.5}$$

Quantum statistics now couples p_{max} with the electron density $n(r)$ by the relation (cf. Problem 167)

$$n = 2 \cdot \frac{4\pi}{3} p_{max}^3 /(2\pi\hbar)^3 . \tag{172.6}$$

By comparing (172.5) and (172.6), we then arrive at the other fundamental equation

$$n(r) = \frac{1}{3\pi^2 \hbar^3} \{2me[\Phi(r) - \Phi(R)]\}^{\frac{3}{2}} . \tag{172.7}$$

From Eqs. (172.1) and (172.7), the two functions $n(r)$ and $\Phi(r)$ may, in principle, both be determined. Elimination of $n(r)$ and use of the central symmetry of the system lead to

$$\nabla^2 \Phi \equiv \frac{1}{r} \frac{d^2}{dr^2}(r\Phi) = \frac{4e}{3\pi\hbar^3} \{2me[\Phi(r) - \Phi(R)]\}^{\frac{3}{2}} .$$

Using instead of $\Phi(r)$ the dimensionless function

$$\varphi(r) = \frac{r}{Ze} [\Phi(r) - \Phi(R)], \tag{172.8}$$

and instead of r the dimensionless variable

$$x = r/a \quad \text{with} \quad a = \left(\frac{9\pi^2}{128\, Z}\right)^{\frac{1}{3}} \frac{\hbar^2}{me^2} = 0.88534\, Z^{-\frac{1}{3}} \frac{\hbar^2}{me^2} \tag{172.9}$$

we obtain the universal differential equation

$$\frac{d^2\varphi}{dx^2} = \frac{\varphi^{\frac{3}{2}}}{\sqrt{x}}, \tag{172.10}$$

the boundary conditions (172.2) and (172.4) passing over into

$$\varphi(0) = 1 \tag{172.11}$$

and with $X = R/a$,

$$\varphi(X) = 0; \qquad X\varphi'(X) = -\frac{z}{Z} . \tag{172.12}$$

It should be noted that with these boundary conditions all $Z - z$ electrons are indeed enclosed within the sphere of radius R. This can easily be shown by first deriving from (172.7) and (172.9),

$$4\pi \int_0^R n(r) r^2 dr = Z \int_0^X dx \sqrt{x}\, \varphi(x)^{\frac{3}{2}} ,$$

and then using the differential equation (172.10) to replace $\varphi^{\frac{3}{2}}$ by φ'':

$$= Z\int_0^X dx\, x\,\varphi'' = Z[x\varphi' - \varphi]_0^X = Z\{\varphi(0) + X\varphi'(X)\}$$

or with the boundary conditions (172.11) and (172.12),

$$= Z\left(1 - \frac{z}{Z}\right) = Z - z$$

which is indeed the number of electrons.

The problem is thus reduced to the integration of (172.10) with the boundary conditions (172.11) and (172.12). To obtain a general survey of the diversity of solutions, the differential equation may be integrated with $\varphi(0)=1$ and different initial tangent inclinations $\varphi'(0)<0$. In Fig. 69 four such solutions have been drawn. The lines (1) and (2) lead to finite radii X_1, X_2. Since $\varphi'(X)<0$ for these solutions, according to Eq. (172.12), they belong to positive ions. For a neutral atom, (172.12) gives $\varphi'(X)=0$, which is impossible for a finite X, thus leading to line (3) of Fig. 69 with an infinitely large atomic radius. Solutions of type (4)

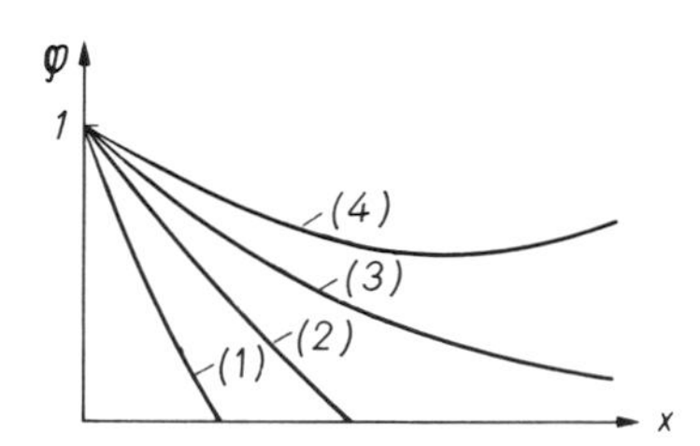

Fig. 69. Solutions of Thomas-Fermi equation (172.10) with different initial tangents

have no direct physical significance for free atoms or ions but may well serve for the description of atoms bound in a crystal lattice under changed boundary conditions.

Our main interest will be concentrated on line (3) of neutral atoms. We shall call this the *standard solution* $\varphi_0(x)$. It is numerically given in the accompanying table. It belongs to the initial tangent inclination $\varphi_0'(0) = -1.58807$. Its asymptotic behaviour is given by $\varphi_0(x) \rightarrow 144/x^3$ (which, by the way, exactly satisfies the differential equation but has a singularity at $x=0$); for practical purposes, however, this expression is rather useless since, even at so large a value as $x=100$, is still differs about 40% from $\varphi_0(x)$. On the other hand, φ_0 should show a much steeper decrease at large x, something certainly of an exponential type. The error of the Thomas-Fermi approximation, as of any statistical model, rapidly increases with small particle numbers, and since the latter decrease below every limit at large distances the method cannot be expected to hold there any longer, whatever good results it may give for the inner parts of the atom.

To obtain other solutions, not too far removed from the standard one, we write

$$\varphi(x) = \varphi_0(x) + k\eta_0(x) \tag{172.13}$$

and linearize in the small deviation $k\eta_0$. From (172.10) we then find

$$\frac{d^2\eta_0}{dx^2} = \frac{3}{2}\left(\frac{\varphi_0}{x}\right)^{\frac{1}{2}} \eta_0 \,. \tag{172.14}$$

In order to satisfy the boundary condition (172.11), we have $\eta_0(0)=0$. We further standardize η_0 by choosing $\eta_0'(0)=1$, satisfying the boundary conditions (172.12) by a suitable choice of the parameter k to be gathered from any one of the following relations:

$$k = -\frac{\varphi_0(X)}{\eta_0(X)}; \tag{172.15a}$$

$$k = -\frac{1}{\eta_0'(X)}\left(\frac{1}{ZX} + \varphi_0'(X)\right); \tag{172.15b}$$

$$k = \varphi'(0) - \varphi_0'(0). \tag{172.15c}$$

Eq. (172.15a) permits a simple relation to be found between k and the ionic radius X. The function $\eta_0(x)$ and its derivative $\eta_0'(x)$ are shown on the table.

x	$\varphi_0(x)$	$-\varphi_0'(x)$	$\eta_0(x)$	$\eta_0'(x)$
0.00	1.0000	1.5881	0.0000	1.0000
0.02	0.9720	1.3093	0.0200	1.0028
0.04	0.9470	1.1991	0.0401	1.0079
0.06	0.9238	1.1177	0.0604	1.0144
0.08	0.9022	1.0516	0.0807	1.0220
0.10	0.8817	0.9954	0.1012	1.0306
0.2	0.7931	0.7942	0.2069	1.0846
0.3	0.7206	0.6618	0.3186	1.1528
0.4	0.6595	0.5646	0.4378	1.2321
0.5	0.6070	0.4894	0.5654	1.3210
0.6	0.5612	0.4292	0.7023	1.4187
0.7	0.5208	0.3798	0.8494	1.5246
0.8	0.4849	0.3386	1.0075	1.6384
0.9	0.4529	0.3038	1.1773	1.7599
1.0	0.4240	0.2740	1.3597	1.8890

x	$\varphi_0(x)$	$-\varphi_0'(x)$	$\eta_0(x)$	$\eta_0'(x)$
1.2	0.3742	0.2259	1.7650	2.1696
1.4	0.3329	0.1890	2.2296	2.4805
1.6	0.2981	0.1601	2.7593	2.8222
1.8	0.2685	0.1370	3.3605	3.1954
2.0	0.2430	0.1182	4.0396	3.6012
2.2	0.2210	0.1028	4.8032	4.0406
2.4	0.2017	0.0900	5.6582	4.5149
2.6	0.1848	0.0793	6.6116	5.0253
2.8	0.1699	0.0702	7.6708	5.5730
3.0	0.1566	0.0625	8.8434	6.1594
3.2	0.1448	0.0558	10.137	6.7858
3.4	0.1343	0.0501	11.561	7.4538
3.6	0.1247	0.0451	13.122	8.1646
3.8	0.1162	0.0408	14.829	8.9198
4.0	0.1084	0.0369	16.693	9.7208
4.5	0.0919	0.0293	22.09	11.93
5.0	0.0788	0.0236	28.68	14.47
5.5	0.0682	0.0192	36.62	17.34
6.0	0.0594	0.0159	46.08	20.59
6.5	0.0522	0.0132	57.27	24.23
7.0	0.0461	0.0111	70.39	28.30
7.5	0.0410	0.0095	85.64	32.81
8.0	0.0366	0.0081	103.27	37.80
8.5	0.0328	0.0070	123.52	43.29
9.0	0.0296	0.0060	146.66	49.32
9.5	0.0268	0.0053	172.94	55.92
10.0	0.0243	0.0046	202.67	63.11

Problem 173. Amaldi correction for a neutral atom

In the Poisson equation underlying the Thomas-Fermi model, it would be more correct to introduce the charge density of all but one of the electrons on the right-hand side, because the equation serves to determine the potential field in which one of the electrons moves. This correction leads to an alteration of the Thomas-Fermi model to be investigated for a neutral atom.

Solution. Instead of (172.1) we now write

$$\nabla^2 \Phi = 4\pi e \cdot \frac{Z-1}{Z} n(r) \tag{173.1}$$

where Φ is the potential field originated by $Z-1$ electrons, acting on the Z'th one. This simple correction makes no difference as to which of the electrons is to be taken to be the probe, a neglect corresponding to a nicety to the statistical picture. The boundary condition for small r is determined by the nucleus and therefore remains unchanged:

$$\Phi(r) = \frac{Ze}{r} \quad \text{for } r \to 0, \tag{173.2}$$

whereas at the atomic surface we now have

$$\Phi(R) = \frac{e}{R}; \quad \left(\frac{d\Phi}{dr}\right)_R = -\frac{e}{R^2} \tag{173.3}$$

for a neutral atom, because there remains a surplus charge e of the nucleus not screened away by the other $Z-1$ electrons, thus still acting on the one considered.

The other fundamental equation originating from quantum statistics remains unaltered, so that we still have

$$n(r) = \frac{1}{3\pi^2 \hbar^3} \{2me[\Phi(r) - \Phi(R)]\}^{\frac{3}{2}}. \tag{173.4}$$

Eliminating $n(r)$ from (173.1) and (173.4) and using again

$$\varphi = \frac{r}{Ze}[\Phi(r) - \Phi(R)], \tag{173.5}$$

the same universal differential equation

$$\frac{d^2\varphi}{dx^2} = \frac{\varphi^{\frac{3}{2}}}{\sqrt{x}} \tag{173.6}$$

will be obtained if only the variable $x = r/\tilde{a}$ is now defined by

$$\tilde{a} = a\left(1 - \frac{1}{Z}\right)^{-\frac{2}{3}} \tag{173.7}$$

with a the characteristic length of Eq. (172.9).

Since $\tilde{a} > a$ this at first sight gives the impression that the atom has been made larger by the correction, in contrast to the physical meaning of the correction in the Poisson equation that lowers the electron-electron repulsive interaction, thus leading to a stronger bond and smaller atom. This discrepancy is, however, resolved by the alteration in the boundary condition. Eq. (173.3), expressed in terms of the function $\varphi(x)$, makes the boundary conditions of the atomic surface become

$$\varphi(X) = 0 \quad \text{and} \quad X\varphi'(X) = -\frac{1}{Z}. \tag{173.8}$$

These conditions can no longer be satisfied except by a finite radius, thus more than compensating for the stretching effect of $\tilde{a}$.

Problem 174. Energy of a Thomas-Fermi atom

The total energy content of a neutral Thomas-Fermi atom shall be calculated. A differential equation for the electron density $n(r)$, resp. for the electrostatic potential $\Phi(r)$ is to be derived by a variational procedure minimizing the energy.

Solution. There are three contributions to the total energy of the atom, viz. the kinetic energy of the electrons, the potential energy $E_{\text{pot}}^{(1)}$ of their interaction with the nucleus of charge Ze, and the potential energy $E_{\text{pot}}^{(2)}$ of their mutual interaction.

The kinetic energy follows from the basic considerations of Problem 167. If $n(r)$ is the density of electrons at some place at a distance r from the nucleus, then the average kinetic energy of an electron at this place is

$$\bar{E}=\tfrac{3}{5}\zeta=\varkappa n^{\frac{2}{3}} \quad \text{with } \varkappa=\frac{3\hbar^2}{10m}(3\pi^2)^{\frac{2}{3}}. \tag{174.1}$$

The total kinetic energy of the electrons then becomes

$$E_{\text{kin}}=\int d\tau\, n(r)\bar{E}(r)$$

or

$$E_{\text{kin}}=\varkappa\int d\tau\, n^{\frac{5}{3}}. \tag{174.2}$$

The two parts of potential energy follow from electrostatics,

$$E_{\text{pot}}^{(1)}=-Ze^2\int\frac{d\tau}{r}n(r) \tag{174.3}$$

and

$$E_{\text{pot}}^{(2)}=\frac{1}{2}e^2\iint d\tau\, d\tau'\frac{n(r)n(r')}{|\boldsymbol{r}-\boldsymbol{r}'|}. \tag{174.4}$$

The total energy, i.e. the sum of (174.2), (174.3), and (174.4),

$$E=\int d\tau\left\{\varkappa n^{\frac{5}{3}}-\frac{Ze^2}{r}n+\frac{1}{2}e^2 n\int d\tau'\frac{n(r')}{|\boldsymbol{r}-\boldsymbol{r}'|}\right\}\equiv\int d\tau\,\eta \tag{174.5}$$

must be minimized by a suitable choice of $n(r)$ with the constraint

$$\int d\tau\, n(r)=Z, \tag{174.6}$$

the latter integral being the total number of electrons. This is a variational problem to be solved by

$$\delta\int d\tau(\eta+\lambda n)=0 \tag{174.7}$$

with a multiplicator λ. Putting Eq. (174.5) into (174.7) we find

$$\int d\tau\, \delta n(r) \left\{ \frac{5}{3} \varkappa n^{\frac{5}{3}} - \frac{Ze^2}{r} + e^2 \int d\tau' \frac{n(r')}{|\boldsymbol{r}-\boldsymbol{r}'|} + \lambda \right\} = 0. \quad (174.8)$$

Here, in the last term, use has been made of the fact that variation of $n(r)$ as well as of $n(r')$ in the double integral twice leads to the same result. The extremal value of E will be obtained if the curly bracket in (174.8) vanishes.

Since for an atom n only depends upon r, not upon direction, the third term in the curly bracket can by expansion into spherical harmonics be written

$$\int d\tau' \frac{n(r')}{|\boldsymbol{r}-\boldsymbol{r}'|} = \frac{4\pi}{r} \int_0^r dr'\, r'^2 n(r') + 4\pi \int_r^\infty dr'\, r' n(r').$$

The variational result from (174.8) therefore becomes

$$\frac{5}{3} \varkappa n^{\frac{2}{3}} - \frac{Ze^2}{r} + \frac{4\pi e^2}{r} \int_0^r dr'\, r'^2 n(r') + 4\pi e^2 \int_r^\infty dr'\, r' n(r') + \lambda = 0.$$

By differentiation with respect to r, λ can be eliminated,

$$\frac{10}{9} \kappa n^{-\frac{1}{3}} \frac{dn}{dr} + \frac{Ze^2}{r^2} - \frac{4\pi e^2}{r^2} \int_0^r dr'\, r'^2 n(r') = 0,$$

the contributions from differentiating the integrals with respect to their limits cancelling each other out. Multiplying by r^2 and again differentiating removes the integral and leaves us with the differential equation

$$\frac{10}{9} \varkappa \frac{d}{dr} \left(r^2 n^{-\frac{1}{3}} \frac{dn}{dr} \right) = 4\pi e^2 r^2 n. \quad (174.9)$$

It is advantageous to pass over from $n(r)$ to the electrostatic potential $\Phi(r)$, as introduced in Problem 172, according to

$$n = \frac{1}{3\pi^2} \left(\frac{2me\Phi}{\hbar^2} \right)^{\frac{3}{2}} \quad (174.10)$$

which satisfies the Poisson equation $\nabla^2 \Phi = 4\pi e n$. With the further abbreviation

$$C = \frac{8}{3\pi} \frac{me^2}{\hbar^3} \sqrt{2me} \quad (174.11)$$

we then find

$$\nabla^2 \Phi = C \Phi^{\frac{3}{2}}, \tag{174.12}$$

an equation which, with

$$\nabla^2 \Phi = \frac{1}{r} \frac{d^2}{dr^2} (r \Phi)$$

has already been derived in Problem 172.

The relations (174.10) and (174.11) enable the fractional powers of n to be eliminated in the energy expressions (174.2), (174.3), (174.4) by putting

$$\text{either } n = \frac{1}{4\pi e} \nabla^2 \Phi \quad \text{or } n = \frac{C}{4\pi e} \Phi^{\frac{3}{2}}. \tag{174.13}$$

We thus obtain

$$E_{\text{kin}} = \frac{3}{20\pi} \int d\tau \, \Phi \nabla^2 \Phi; \tag{174.14a}$$

$$E_{\text{pot}}^{(1)} = -\frac{Ze}{4\pi} \int d\tau \frac{\nabla^2 \Phi}{r}; \tag{174.14b}$$

$$E_{\text{pot}}^{(2)} = \frac{1}{32\pi^2} \iint d\tau \, d\tau_1 \frac{\nabla^2 \Phi \nabla_1^2 \Phi}{|\boldsymbol{r} - \boldsymbol{r}_1|}. \tag{174.14c}$$

These integrals can be considerably simplified by making use of the central symmetry and by introducing instead of $\Phi(r)$ the function

$$\varphi(r) = \frac{r}{Ze} \Phi(r). \tag{174.15}$$

We then have,

$$\nabla^2 \Phi = \frac{Ze}{r} \varphi''(r)$$

and from (174.14a),

$$E_{\text{kin}} = \frac{3}{5} \int_0^\infty dr \, r^2 \cdot \frac{Ze}{r} \varphi \cdot \frac{Ze}{r} \varphi'' = \frac{3}{5} Z^2 e^2 \left\{ -(\varphi \varphi')_{r=0} - \int_0^\infty dr \, \varphi'^2 \right\}. \tag{174.16a}$$

In the same way we find from (174.14b),

$$E_{\text{pot}}^{(1)} = -\frac{Ze}{4\pi}\int \frac{d\tau}{r}\,\frac{Ze}{r}\,\varphi'' = Z^2 e^2 \varphi'(0). \tag{174.16b}$$

In order to evaluate the integral (174.14c), we begin with the inner integral,

$$I(r) = \frac{1}{4\pi}\int d\tau_1 \frac{\nabla_1^2 \Phi(r_1)}{|\boldsymbol{r}-\boldsymbol{r}_1|} = \frac{1}{r}\int_0^r dr_1\, r_1^2 \nabla_1^2 \Phi + \int_r^\infty dr_1\, r_1 \nabla_1^2 \Phi$$

$$= \frac{Ze}{r}\int_0^r dr_1\, r_1\, \varphi''(r_1) + Ze\int_r^\infty dr_1\, \varphi''(r_1) = \frac{Ze}{r}\{(\varphi - r\varphi')_{r=0} - \varphi(r)\}.$$

Then, (174.14c) leads to

$$E_{\text{pot}}^{(2)} = \frac{1}{8\pi}\int d\tau\, I(r)\nabla^2 \Phi = \frac{1}{2} Ze \int_0^\infty dr\, r^2 I(r) \frac{\varphi''}{r}$$

$$= \frac{1}{2} Z^2 e^2 \left\{ \int_0^\infty dr\, \varphi'^2 + [r\varphi'^2]_{r=0} \right\}. \tag{174.16c}$$

It is convenient to use instead of r the dimensionless variable

$$x = \frac{r}{a}; \qquad a = \left(\frac{9\pi^2}{128 Z}\right)^{\frac{1}{3}} \frac{\hbar^2}{me^2} \tag{174.17}$$

defined in Problem 172. Since, for small r or x, we have

$$\varphi(x) = 1 - \mu x + \cdots,$$

we may, in Eqs. (174.16a–c), put

$$-(\varphi\varphi')_{r=0} = \mu/a; \qquad \left(\frac{d\varphi}{dr}\right)_{r=0} = -\mu/a; \qquad \left[r\left(\frac{d\varphi}{dr}\right)^2\right]_{r=0} = 0$$

so that finally there remain the expressions

$$\left.\begin{aligned} E_{\text{kin}} &= \frac{3}{5}\,\frac{Z^2 e^2}{a}(\mu - J); \\ E_{\text{pot}}^{(1)} &= -\frac{Z^2 e^2}{a}\,\mu; \\ E_{\text{pot}}^{(2)} &= \frac{1}{2}\,\frac{Z^2 e^2}{a}\,J \end{aligned}\right\} \tag{174.18}$$

with

$$J = \int_0^\infty dx \left(\frac{d\varphi}{dx}\right)^2 . \tag{174.19}$$

Here, the integral J and the derivative μ are independent of Z since they depend only upon properties of the universal function $\varphi(x)$. Their numerical values can be determined from $\varphi(x)$, Problem 172:

$$\mu = -\varphi'(0) = 1{,}588 \qquad J = 0{,}454 . \tag{174.20}$$

The total energy of the atom, i.e. the sum of the three expressions (174.18),

$$E = -\frac{Z^2 e^2}{a}\left(\frac{2}{5}\mu + \frac{1}{10}J\right) = -0{,}680\frac{Z^2 e^2}{a}, \tag{174.21}$$

then becomes proportional to $Z^{\frac{7}{3}}$ because $a \propto Z^{-\frac{1}{3}}$. Numerically,

$$E = -0{,}7687\, Z^{\frac{7}{3}}\, \mathrm{Ry} = -20{,}93\, Z^{\frac{7}{3}}\, \mathrm{eV} . \tag{174.22}$$

Problem 175. Virial theorem for the Thomas-Fermi atom

To prove the virial theorem for a neutral Thomas-Fermi atom, following the procedure of Problem 151. What relation follows between μ and J of Eq. (174.20) from the virial theorem, and what can be concluded on the ratios of the three parts of energy?

Solution. By scale transformation we replace $n(r)$ by the set of functions

$$n_\lambda(r) = \lambda^3 n(\lambda r)$$

all satisfying the normalization condition

$$\int d\tau\, n_\lambda(r) = Z .$$

The energy expressions (174.2), (174.3), and (174.4) then transform as

$$E_{\mathrm{kin}}(\lambda) = \lambda^2 E_{\mathrm{kin}}; \qquad E^{(1)}_{\mathrm{pot}}(\lambda) = \lambda E^{(1)}_{\mathrm{pot}}; \qquad E^{(2)}_{\mathrm{pot}}(\lambda) = \lambda E^{(2)}_{\mathrm{pot}},$$

thus yielding

$$E(\lambda) = \lambda^2 E_{\mathrm{kin}} + \lambda E_{\mathrm{pot}}$$

which leads through $\partial E(\lambda)/\partial\lambda = 0$ for $\lambda = 1$ as in Problem 151 to the virial theorem

$$2E_{\mathrm{kin}} + E_{\mathrm{pot}} = 0 . \tag{175.1}$$

If the three energy expressions (174.18) are put into this relation, we arrive at

$$\mu = \tfrac{7}{2} J \tag{175.2}$$

which is corroborated by the numerical values (174.20). The energy expressions (174.18) then may all be written in terms of J only, with the result

$$E_{\text{kin}} = \tfrac{3}{2} U; \quad E_{\text{pot}}^{(1)} = -\tfrac{7}{2} U; \quad E_{\text{pot}}^{(2)} = +\tfrac{1}{2} U \tag{175.3}$$

and

$$U = \frac{Z^2 e^2}{a} J. \tag{175.4}$$

The total energy thus becomes the sum

$$E = -\tfrac{3}{2} U \tag{175.5}$$

which again leads to the numerical results given at the end of the preceding problem. A comparison of (175.5) with the kinetic energy (175.3) corroborates the virial theorem.

Problem 176. Tietz approximation of a Thomas-Fermi field

The function

$$\tilde{\varphi}(x) = \frac{1}{(1+\alpha x)^2} \tag{176.1}$$

with a suitable value of α, independent of Z, may be used as a fair approximation to the Thomas-Fermi function $\varphi_0(x)$ for a neutral atom. The constant α shall be determined in such a way as to permit exact normalization of $\tilde{\varphi}$, and a numerical comparison shall be made of $\tilde{\varphi}$ and φ_0.

Solution. In Problem 172 it has been shown that the electron density $n(r)$ and the atomic potential

$$V(r) = -\frac{Z}{r} \varphi_0(r) \tag{176.2}$$

(in atomic units) are coupled by the relation

$$n(r) = \frac{1}{3\pi^2} (-2V)^{\frac{3}{2}}. \tag{176.3}$$

The normalization condition,

$$4\pi \int_0^\infty dr\, r^2 n(r) = Z, \tag{176.4}$$

therefore may be written

$$\frac{4}{3\pi}(2Z)^{\frac{3}{2}}\int_0^\infty dr\, r^{\frac{1}{2}}\,\varphi_0^{\frac{3}{2}} = Z. \tag{176.5}$$

This equation is satisfied exactly by the Fermi function $\varphi_0(x)$ with

$$x = r/a; \qquad a = 0.88534\, Z^{-\frac{1}{3}}. \tag{176.6}$$

We now replace φ_0 by the approximate function $\tilde{\varphi}$, Eq. (176.1), but still keep this normalization. Introduction of the integration variable $y = \alpha x = \dfrac{\alpha}{a} r$ then leads to

$$\frac{8\sqrt{2}}{3\pi}\sqrt{Z}\left(\frac{a}{\alpha}\right)^{\frac{3}{2}}\int_0^\infty \frac{dy\sqrt{y}}{(1+y)^3} = 1. \tag{176.7}$$

The integral can be solved by the substitution of $u = y^2$; one easily verifies

$$\int \frac{dy\sqrt{y}}{(1+y)^3} = 2\int du \frac{u^2}{(u^2+1)^3} = \frac{1}{4}\left[\frac{u(u^2-1)}{(u^2+1)^2} + \tan^{-1} u\right];$$

in the limits $0 \leq y < \infty$ therefore, the integral becomes $\pi/8$ and Eq. (176.7) yields

$$\alpha = \left(\frac{2Z}{9}\right)^{\frac{1}{3}} a = 0.60570\, Z^{\frac{1}{3}}\, a, \tag{176.8}$$

or with Eq. (176.6),

$$\alpha = 0.53625. \tag{176.9}$$

In the accompanying table, the functions φ_0 and $\tilde{\varphi}$ have been compared, using this value of α.

x	$\tilde{\varphi}$	φ_0	$\tilde{\varphi} - \varphi_0$
0	1	1	0
0.1	0.9008	0.8817	+0.0191
0.2	0.8156	0.7931	+0.0225
0.5	0.6219	0.6070	+0.0149
1.0	0.4237	0.4240	−0.0003
2.0	0.2328	0.2430	−0.0102
5.0	0.0738	0.0788	−0.0050
10.0	0.0247	0.0243	+0.0004

Literature. Tietz, T.: J. Chem. Physics **25,** 787 (1956); Z. Naturforsch. **23a,** 191 (1968).—In Tietz's original papers a factor 0.64309 has been used instead of our normalizing factor 0.60570 in Eq. (176.8). Tietz's approximation therefore does not satisfy normalization. His deviations $\tilde{\varphi}-\varphi_0$, however, are somewhat smaller in the most significant region $0<x<0.5$, but are much bigger for $x>1$.

Problem 177. Variational approximation of Thomas-Fermi field

To use a set of Tietz functions

$$\tilde{\varphi}(x)=\frac{1}{(1+\alpha x)^2} \tag{177.1}$$

as trial functions with the Ritz parameter α for the approximate solution of a variational problem equivalent to the Thomas-Fermi differential equation.

Solution. The differential equation

$$\varphi''=x^{-\frac{1}{2}}\varphi^{\frac{3}{2}} \tag{177.2}$$

may be replaced by the variational problem, to make the integral

$$J=\int_0^\infty dx\left(\tfrac{1}{2}\varphi'^2+\tfrac{2}{5}x^{-\frac{1}{2}}\varphi^{\frac{5}{2}}\right) \tag{177.3}$$

an extremum with fixed boundary conditions $\varphi(0)=1$ and $\varphi(\infty)=0$. Putting the trial function (177.1) satisfying the boundary conditions in the integral (177.3) we have

$$J=\int_0^\infty dx\left\{\frac{2\alpha^2}{(1+\alpha x)^6}+\frac{2}{5}x^{-\frac{1}{2}}\frac{1}{(1+\alpha x)^5}\right\}.$$

For the evaluation of the integral we set $\alpha x=t^2$ in the second term; then we may use the formula

$$\int\frac{dt}{(1+t^2)^5}$$
$$=\frac{1}{8}\left\{\frac{t}{(1+t^2)^4}+\frac{7}{6}\frac{t}{(1+t^2)^3}+\frac{35}{24}\frac{t}{(1+t^2)^2}+\frac{35}{16}\frac{t}{1+t^2}+\frac{35}{16}\tan^{-1}t\right\}$$

which can easily be verified. We find

$$J=\tfrac{2}{5}\left(\alpha+\tfrac{35}{128}\alpha^{-\frac{1}{2}}\right). \tag{177.4}$$

The extremum condition $dJ/d\alpha=0$ leads to

$$\alpha=\left(\tfrac{35}{256}\right)^{\frac{2}{3}}=0.570. \tag{177.5}$$

This value of α deviates only slightly from $\alpha = 0.536$ which is found to satisfy the normalization condition

$$\int_0^\infty dx \sqrt{x}\, \varphi^{\frac{3}{2}} = 1 \tag{177.6}$$

of the preceding problem; with the present value of α it yields

$$\alpha^{-\frac{3}{2}} \cdot \frac{\pi}{8} = \frac{32}{35},$$

i.e. the approximate field minimizing the integral J corresponds to $\frac{32}{35}Z$ electrons instead of to Z in the atom.

Problem 178. Screening of K electrons

To determine the screening correction to the binding energy of a K electron by using the Tietz approximation to the Thomas-Fermi model.

Solution. Suppose one of the two K electrons and one unit of the nuclear charge to be removed from the atom of charge Z. The result will be a neutral atom of charge $Z-1$. Then, add again the removed nuclear charge but neglect its influence upon the $Z-1$ remaining electrons. The result (in atomic units) is a charge distribution with the electrostatic potential

$$\Phi(r) = \frac{1}{r} + \frac{Z-1}{r}\varphi(x) \tag{178.1}$$

with $\varphi(x)$ the Thomas-Fermi function to the variable

$$x = \frac{r}{a}; \qquad a = 0.88534(Z-1)^{-\frac{1}{3}}. \tag{178.2}$$

The potential energy of an electron (charge -1) moving in this potential would be

$$V(r) = -\Phi(r); \tag{178.3}$$

this therefore is the potential energy field in which the removed K electron would move if replaced into the atom.

Let us now apply perturbation theory. Without screening we should have

$$V_0(r) = -\frac{Z}{r};$$

with screening we have

$$V(r) = -\frac{1}{r} + \frac{Z-1}{r}\varphi(x) = V_0(r) + \frac{Z-1}{r}[1-\varphi(x)]. \qquad (178.4)$$

The eigenvalue E_0 and eigenfunction $u_0(r)$ of the K electron without screening are

$$E_0 = -\tfrac{1}{2}Z^2; \qquad u_0 = \frac{Z^{\frac{3}{2}}}{\sqrt{\pi}} e^{-Zr}. \qquad (178.5)$$

The first-order energy shift due to screening then becomes

$$\Delta E_s = \int d\tau\, u_0^* \frac{Z-1}{r}(1-\varphi)u_0$$

or, after inserting u_0,

$$\Delta E_s = 4Z^3(Z-1)\int_0^\infty dr\, r\, e^{-2Zr}[1-\varphi(x)]. \qquad (178.6)$$

We now are prepared to introduce the Tietz approximation (Problem 176),

$$\varphi(x) = \frac{1}{(1+\alpha x)^2}; \qquad \alpha = 0.53625 \qquad (178.7)$$

To evaluate the integral (178.6) we then use the auxiliary variable

$$t = \beta(1+\alpha x) \qquad (178.8)$$

with

$$\beta = 2Z\frac{a}{\alpha} = 3.302\, Z(Z-1)^{-\frac{1}{3}} \qquad (178.9)$$

thus obtaining

$$\Delta E_s = Z(Z-1)e^{\beta}\int_\beta^\infty dt\, e^{-t}\left(t-\beta-\frac{\beta^2}{t}+\frac{\beta^3}{t^2}\right). \qquad (178.10)$$

The exponential integral occurring here

$$E_1(\beta) = \int_\beta^\infty \frac{dt}{t} e^{-t}, \qquad (178.11)$$

is a well-known function whose asymptotic behaviour at large values of β is described by the semiconvergent series

$$E_1(\beta) = \frac{e^{-\beta}}{\beta}\left\{1 - \frac{1!}{\beta} + \frac{2!}{\beta^2} - \frac{3!}{\beta^3} \pm \cdots\right\}. \tag{178.12}$$

The integral over the last term in (178.10) may be reduced to this function,

$$E_2(\beta) = \int\limits_\beta^\infty \frac{dt}{t^2} e^{-t} = \frac{e^{-\beta}}{\beta} - E_1(\beta). \tag{178.13}$$

Thus we find

$$\Delta E_s = Z(Z-1)\{1+\beta^2 - \beta^2(1+\beta)e^\beta E_1(\beta)\}, \tag{178.14}$$

or by expansion, according to (178.12), for $\beta \gg 1$, the semiconvergent series

$$\Delta E_s = Z(Z-1)\frac{4}{\beta}\left(1 - \frac{9/2}{\beta} + \frac{24}{\beta^2} - \frac{150}{\beta^3} \pm \cdots\right). \tag{178.15}$$

Turning now to the numerical evaluation of the energy shift ΔE_s, we find β from (178.9) indeed to be large (cf. table next page) so that (178.15) is a reasonable approximation. In the column marked $(1-\cdots)$ in the table the values of the bracket in (178.15) are reproduced; the series converges rapidly. The energy shifts, ΔE_s, lie between 26% and 12% of $|E_0|$ if Z varies from 20 to 80. They are therefore corrections only which, however, are not so very small that a second-order perturbation calculation would not change them by several per cent. On the other hand, the Thomas-Fermi model used is too rough to make such a change in the values physically significant.

In x-ray spectroscopy it is customary to describe the energy shift by a screening constant s defined by

$$E = -\tfrac{1}{2}Z^2 + \Delta E_s = -\tfrac{1}{2}(Z-s)^2; \tag{178.16}$$

this definition renders

$$s = Z\left(1 - \sqrt{1 - \frac{\Delta E_s}{|E_0|}}\right). \tag{178.17}$$

The screening constants given in the table are computed according to this formula. Since $\Delta E_s \lesssim \frac{1}{4}|E_0|$, we may roughly expand the radical in (178.17) and write

$$s \simeq Z\frac{\Delta E_s}{2|E_0|} = \frac{\Delta E_s}{Z},$$

which is about proportional to $Z^{\frac{1}{3}}$. If we try a rough formula of this type,

$$s = \sigma Z^{\frac{1}{3}},$$

we obtain the following pairs of values:

$Z = 20$	50	80
$\sigma = 1.03$	1.12	1.15

Z	β	$(1-\cdots)$	ΔE_s	$\lvert E_0 \rvert$	s	ΔE_r	s corr.
20	24.75	0.847	52.2	200	2.81		
30	32.25	0.880	95.2	450	3.36		
40	38.96	0.897	144.3	800	3.79	− 17.1	3.35
50	45.12	0.910	198.0	1250	4.13	− 41.7	3.26
60	50.90	0.921	257.3	1800	4.42	− 86.3	2.94
70	56.35	0.927	318.6	2450	4.71	−160	2.34
80	61.56	0.932	384.0	3200	4.96	−273	1.62

It is, however, useless to go into such detail. One glance at the experimental values shows that our s values are rather good up to about $Z = 50$ but that, instead of the predicted slow rise of s with increasing Z beyond this value, the s values do not exceed a maximum of 3.7 and then, first slowly and above $Z = 70$ rapidly, begin to fall again. Such a discrepancy at high values of Z clearly must be explained as a relativistic effect. This is, in essence, if only in a rough way, shown in the last two columns of the table. According to relativistic quantum mechanics (cf. Problem 203) the unperturbed K electron energy is lowered by the amount

$$\Delta E_r = -\frac{1}{8} Z^2 \left(\frac{Z}{137}\right)^2;$$

these shifts have to be added to the ΔE_s screening shifts before calculating screening constants as given in the last column. Since

$$\Delta E = \Delta E_s + \Delta E_r$$

becomes, for large values of Z, increasingly smaller than the original ΔE_s, the deviations from the unscreened nuclear field, i.e. the corrected s values, will also become increasingly smaller. This is corroborated by experiment. In a strict sense, of course, relativistic corrections should not be applied only to E_0 but to ΔE_s as well. The results may therefore still be a little rough, but scarcely more so than corresponds to the general application of the Thomas-Fermi field which neglects all special shell structure effects.

V. Non-Stationary Problems

Problem 179. Two-level system with time-independent perturbation

Given an atomic system with only two stationary states $|1\rangle$ and $|2\rangle$ and energies $\hbar\omega_1 < \hbar\omega_2$. At the time $t=0$, the system being in its ground state, a perturbation W not depending upon time is switched on. The probability shall be calculated of finding the system in either state at the time t.

Solution. Let H be the hamiltonian of the unperturbed system with

$$H|1\rangle = \hbar\omega_1|1\rangle; \qquad H|2\rangle = \hbar\omega_2|2\rangle \tag{179.1}$$

defining its two stationary states. Then, the Schrödinger equation with perturbation,

$$-\frac{\hbar}{i}\dot{\psi} = (H+W)\psi, \tag{179.2}$$

is to be solved in terms of the stationary functions:

$$\psi(t) = c_1(t)\,\mathrm{e}^{-i\omega_1 t}|1\rangle + c_2(t)\,\mathrm{e}^{-i\omega_2 t}|2\rangle. \tag{179.3}$$

It must be possible thus to construct the *exact* solution, because $|1\rangle$ and $|2\rangle$ form a complete orthonormal set so that (179.3) is just the expansion of ψ with respect to this set with time-dependent coefficients. The latter have to be determined with initial conditions

$$c_1(0) = 1; \qquad c_2(0) = 0. \tag{179.4}$$

If we put (179.3) into (179.2) and multiply[1] by either $\langle 1|$ or $\langle 2|$, we find two differential equations of the first order for the coefficients:

---

[1] "Multiplication" here means the formation of scalar products in Hilbert space, i.e. matrix elements between a pair of states.

$$\begin{aligned} -\frac{\hbar}{i}\dot{c}_1\, e^{-i\omega_1 t} &= \langle 1|W|1\rangle c_1\, e^{-i\omega_1 t} + \langle 1|W|2\rangle c_2\, e^{-i\omega_2 t}, \\ -\frac{\hbar}{i}\dot{c}_2\, e^{-i\omega_2 t} &= \langle 2|W|1\rangle c_1\, e^{-i\omega_1 t} + \langle 2|W|2\rangle c_2\, e^{-i\omega_2 t}. \end{aligned} \tag{179.5}$$

Let us briefly write

$$\langle\mu|W|\nu\rangle = W_{\nu\mu},$$

then, in consequence of the hermiticity of the operator W, the diagonal matrix elements W_{11} and W_{22} are real, whereas the complex off-diagonal elements are conjugate:

$$W_{12} = W_{21}^{*}.$$

Using the abbreviation

$$\omega_0 = \omega_2 - \omega_1, \tag{179.6}$$

so that $\hbar\omega_0$ is the energy difference between the two levels, Eqs. (179.5) may be written

$$\left.\begin{aligned} \hbar i \dot{c}_1 &= W_{11}\, c_1 + W_{21}\, e^{-i\omega_0 t}\, c_2, \\ \hbar i \dot{c}_2 &= W_{12}\, e^{i\omega_0 t}\, c_1 + W_{22}\, c_2. \end{aligned}\right\} \tag{179.7}$$

Apparently it is possible to solve these equations by

$$c_1 = A\, e^{-i\omega t}; \qquad c_2 = B\, e^{-i(\omega-\omega_0)t}. \tag{179.8}$$

This can immediately be seen when (179.8) is put into (179.7) leading to the linear equations

$$\begin{aligned} (W_{11} - \hbar\omega)A + W_{21}B &= 0, \\ W_{12}A + (W_{22} - \hbar\omega + \hbar\omega_0)B &= 0. \end{aligned}$$

The determinant of these two equations vanishes for the two frequencies

$$\omega_{\mathrm{I,II}} = \frac{W_{11}}{\hbar} + \frac{1}{2}\gamma \pm \sigma \tag{179.9}$$

with

$$\begin{aligned} \hbar\gamma &= W_{22} - W_{11} + \hbar\omega_0, \\ \hbar\sigma &= \sqrt{\tfrac{1}{4}\gamma^2 \hbar^2 + |W_{12}|^2}\,. \end{aligned} \tag{179.10}$$

Further, we obtain

$$B_{\mathrm{I,II}} = \frac{\hbar\omega_{\mathrm{I,II}} - W_{11}}{W_{21}} A_{\mathrm{I,II}}. \tag{179.11}$$

Thence,

$$c_1(t) = A_{\mathrm{I}} \mathrm{e}^{-i\omega_{\mathrm{I}} t} + A_{\mathrm{II}} \mathrm{e}^{-i\omega_{\mathrm{II}} t};$$

$$c_2(t) = \frac{1}{W_{21}} \mathrm{e}^{i\omega_0 t} \{(\hbar\omega_{\mathrm{I}} - W_{11}) A_{\mathrm{I}} \mathrm{e}^{-i\omega_{\mathrm{I}} t} + (\hbar\omega_{\mathrm{II}} - W_{11}) A_{\mathrm{II}} \mathrm{e}^{-i\omega_{\mathrm{II}} t}\}.$$

The initial conditions (179.4) permit evaluation of the constants A_{I} and A_{II}. After a straightforward computation, we then arrive at the result:

$$c_1(t) = \exp\left[-i\left(\frac{W_{11}}{\hbar} + \frac{1}{2}\gamma\right)t\right]\left\{\cos\sigma t + i\frac{\gamma}{2\sigma}\sin\sigma t\right\}; \qquad (179.12\mathrm{a})$$

$$c_2(t) = -i\frac{W_{12}}{\hbar\sigma}\exp\left[-i\left(\frac{W_{11}}{\hbar} + \frac{1}{2}\gamma - \omega_0\right)t\right]\sin\sigma t. \qquad (179.12\mathrm{b})$$

The probability, then, of finding the system in the excited state will be

$$|c_2(t)|^2 = \frac{|W_{12}|^2}{\hbar^2\sigma^2}\sin^2\sigma t$$

or, using (179.10),

$$|c_2(t)|^2 = \frac{4|W_{12}|^2}{(\hbar\gamma)^2 + 4|W_{12}|^2}\sin^2\sigma t. \qquad (179.13)$$

The probability of finding it in the original ground state again, on the other hand, becomes

$$|c_1(t)|^2 = \cos^2\sigma t + \left(\frac{\gamma}{2\sigma}\right)^2 \sin^2\sigma t$$

or

$$|c_1(t)|^2 = 1 - \frac{4|W_{12}|^2}{(\hbar\gamma)^2 + 4|W_{12}|^2}\sin^2\sigma t. \qquad (179.14)$$

Note that the sum of (179.13) and (179.14) will be 1. The system is oscillating between the two levels with a time period π/σ.

Problem 180. Periodic perturbation of a two-level system

Given the same two-level system as in the preceding problem. At the time $t=0$, however, let a *periodic* perturbation $W\cos\omega t$ be switched on (e.g. a light wave) with its frequency ω *almost* corresponding to the energy difference $\hbar\omega_0 = \hbar(\omega_2 - \omega_1)$ of the two levels. Again the probability of finding the system in either state after switching off the perturbation at the time t shall be found.

Solution. The Schrödinger equation

$$-\frac{\hbar}{i}\dot{\psi} = [H + W\cos\omega t]\psi \tag{180.1}$$

is to be solved by the function

$$\psi(t) = c_1(t)\,\mathrm{e}^{-i\omega_1 t}|1\rangle + c_2(t)\,\mathrm{e}^{-i\omega_2 t}|2\rangle \tag{180.2}$$

with the initial condition

$$\psi(0) = 1, \quad \text{or} \quad c_1(0) = 1; \qquad c_2(0) = 0. \tag{180.3}$$

Here, $|1\rangle$ and $|2\rangle$ are solutions of the stationary state equations

$$H|1\rangle = \hbar\omega_1|1\rangle; \qquad H|2\rangle = \hbar\omega_2|2\rangle, \tag{180.4}$$

and $|1\rangle$ and $|2\rangle$ may be supposed to be orthonormal. By putting (180.2) into (180.1) and by scalar multiplication of the result with either $\langle 1|$ or $\langle 2|$, we arrive at two differential equations for $c_1(t)$ and $c_2(t)$, viz.

$$\begin{aligned} -\frac{\hbar}{i}\dot{c}_1\,\mathrm{e}^{-i\omega_1 t} &= \cos\omega t\{\langle 1|W|1\rangle c_1\,\mathrm{e}^{-i\omega_1 t} + \langle 1|W|2\rangle c_2\,\mathrm{e}^{-i\omega_2 t}\}; \\ -\frac{\hbar}{i}\dot{c}_2\,\mathrm{e}^{-i\omega_2 t} &= \cos\omega t\{\langle 2|W|1\rangle c_1\,\mathrm{e}^{-i\omega_1 t} + \langle 2|W|2\rangle c_2\,\mathrm{e}^{-i\omega_2 t}\}. \end{aligned} \tag{180.5}$$

Let us here introduce the above abbreviation

$$\omega_0 = \omega_2 - \omega_1$$

and write

$$\omega - \omega_0 = \Delta\omega; \tag{180.6}$$

then it is to be supposed that

$$|\Delta\omega| \ll \omega_0. \tag{180.7}$$

Eq. (180.5) thus becomes

$$i\dot{c}_1 = \frac{1}{2\hbar}\{\langle 1|W|1\rangle(\mathrm{e}^{i\omega t} + \mathrm{e}^{-i\omega t})c_1 + \langle 1|W|2\rangle(\mathrm{e}^{i\Delta\omega t} + \mathrm{e}^{-i(\omega+\omega_0)t})c_2\};$$

$$i\dot{c}_2 = \frac{1}{2\hbar}\{\langle 2|W|1\rangle(\mathrm{e}^{i(\omega+\omega_0)t} + \mathrm{e}^{-i\Delta\omega t})c_1 + \langle 2|W|2\rangle(\mathrm{e}^{i\omega t} + \mathrm{e}^{-i\omega t})c_2\}.$$

Here we meet a very pronounced distinction between high-frequency terms with frequencies of the orders ω and 2ω, and low-frequency terms with $\Delta\omega$. Averaging over a time interval $2\pi/\omega$, all high-frequency contributions will cancel. So, if we replace c_1 and c_2 by

$$C_\mu(t) = \frac{1}{2\tau}\int\limits_{t-\tau}^{t+\tau} dt'\,c_\mu(t') \quad \text{with} \quad \tau = \pi/\omega,$$

these averages satisfy the much simpler equations

$$\left.\begin{aligned} i\dot{C}_1 &= \frac{1}{2\hbar}\langle 1|W|2\rangle\, e^{i\Delta\omega t}\, C_2; \\ i\dot{C}_2 &= \frac{1}{2\hbar}\langle 2|W|1\rangle\, e^{-i\Delta\omega t}\, C_1 \end{aligned}\right\} \tag{180.8}$$

where the slowly varying factors $\exp(\pm i\Delta\omega t)$ on the right-hand side are treated as constants in averaging.

The differential equations (180.8) permit exact solution by differentiation of either, and elimination of the other variable and its first derivative by using the original equations (180.8). We thus arrive at

$$\left.\begin{aligned} \ddot{C}_1 - i\Delta\omega\,\dot{C}_1 + \tfrac{1}{4}\Omega^2 C_1 = 0; \\ \ddot{C}_2 + i\Delta\omega\,\dot{C}_2 + \tfrac{1}{4}\Omega^2 C_2 = 0 \end{aligned}\right\} \tag{180.9}$$

with

$$\Omega^2 = \frac{1}{\hbar^2}\langle 1|W|2\rangle\langle 2|W|1\rangle = \frac{1}{\hbar^2}|\langle 2|W|1\rangle|^2. \tag{180.10}$$

Use of the abbreviation

$$R = \sqrt{\Omega^2 + (\Delta\omega)^2} \tag{180.11}$$

leads to the following solutions satisfying the initial conditions (180.3):

$$\left.\begin{aligned} C_1 &= e^{i\frac{\Delta\omega}{2}t}\left\{\cos\frac{Rt}{2} + A\sin\frac{Rt}{2}\right\}; \\ C_2 &= e^{-i\frac{\Delta\omega}{2}t}\, B\sin\frac{Rt}{2}. \end{aligned}\right. \tag{180.12}$$

The remaining two integration constants A and B can be computed by putting (180.12) into the first-order equations (180.8). The result is

$$A = -i\frac{\Delta\omega}{R}; \qquad B = -i\frac{\langle 2|W|1\rangle}{\hbar R}. \tag{180.13}$$

The probability of finding the system, at the time t, in the excited state is now

$$|C_2|^2 = \frac{\Omega}{\Omega^2 + (\Delta\omega)^2}\sin^2\frac{Rt}{2} \tag{180.14}$$

and that of finding it in its ground state again is

$$|C_1|^2 = \cos^2\frac{Rt}{2} + \frac{(\Delta\omega)^2}{\Omega^2 + (\Delta\omega)^2}\sin^2\frac{Rt}{2}. \tag{180.15}$$

According to (180.14), the excitation is a typical resonance process, its probability rapidly decreasing with increasing values of $|\Delta\omega|$. Needless to say, this holds only as long as condition (180.7) remains satisfied. The process is periodically repeated with the frequency R determined by Eq. (180.11), i.e. mainly by the value of the matrix element, so that after a time interval

$$t_n = \frac{2\pi n}{R}, \qquad n=1,2,3,\ldots \tag{180.16}$$

the system will be found in the ground state again. If the periodic perturbation is performed, e.g. by a light wave switched on at time $t=0$ and switched off again at $t=t_n$, no resultant change will have affected the system.

Application. Let the hamiltonian H describe the Zeeman effect in a one-electron S state produced by a magnetic field $\mathscr{H}_0$ in z direction. Then $\hbar\,(\omega_2-\omega_1)$ is the level splitting between the two spin orientations, i.e. $\omega_0=2\mu\mathscr{H}_0/\hbar$ (with $|2\rangle$ the upper, $|1\rangle$ the lower state). Let now the perturbation consist of a periodic magnetic field, $\mathscr{H}'\cos\omega t$, so that

$$W=-\mu(\boldsymbol{\sigma}\cdot\mathscr{H}')\cos\omega t, \qquad \mu=-\frac{e\hbar}{2mc}.$$

If the field $\mathscr{H}'$ is parallel to $\mathscr{H}_0$, the matrix element $\langle 1|W|2\rangle$ vanishes; the states $|1\rangle$ and $|2\rangle$ are then independently perturbed and no transitions are induced. If, on the other hand, $\mathscr{H}'$ is chosen perpendicular to $\mathscr{H}_0$, say in x direction, the diagonal matrix elements of W vanish and we find exactly the case described above with

$$\langle 1|W|2\rangle=\langle 2|W|1\rangle=-\mu\mathscr{H}'.$$

Resonance will occur in this device if $\omega\simeq\omega_0$, as before, and the system will alternate between its two magnetic states. This is the simplest case of *paramagnetic resonance*.

Problem 181. Dirac perturbation method

Let an atomic system have the non-degenerate stationary states ψ_k. Let it be in its ground state ψ_0 at time $t=0$ and then a perturbation be switched on (depending upon, or independent of, time) inducing transitions to other states ψ_l. The probability shall be determined of finding the system, after switching off the perturbation at time t, in a state ψ_l, supposing the perturbation to be small.

Solution. Let the unperturbed states satisfy the Schrödinger equation

$$-\frac{\hbar}{i}\dot{\psi}_k=H\psi_k \quad \text{with} \quad \psi_k=|k\rangle\,\mathrm{e}^{-i\omega_k t}; \qquad E_k=\hbar\omega_k \tag{181.1}$$

and

$$\langle l|k\rangle=\delta_{kl}. \tag{181.2}$$

After switching on the perturbation W, we have the differential equation

$$-\frac{\hbar}{i}\dot{\psi} = (H + W)\psi \tag{181.3}$$

with a wave function ψ which may be expanded into a series

$$\psi = \sum_k a_k(t)\psi_k . \tag{181.4}$$

In consequence of (181.2), it then follows from (181.4) that[2]

$$\sum_k |a_k(t)|^2 = 1 . \tag{181.5}$$

Each $|a_k|^2$ then is the probability of finding the system in the state ψ_k at the time t.

Introducing the sum (181.4) into the differential equation (181.3) we find

$$-\frac{\hbar}{i}\sum_k (\dot{a}_k - i\omega_k a_k)\psi_k = \sum_k (\hbar\omega_k + W) a_k \psi_k$$

or, forming the scalar product in Hilbert space with $\langle l|$ and making use of (181.2),

$$\dot{a}_l = -\frac{i}{\hbar}\sum_k \mathrm{e}^{-i(\omega_k - \omega_l)t}\langle l|W|k\rangle a_k . \tag{181.6}$$

In this equation, so far, nothing has yet been neglected. It corresponds to the fact that the time rate of any state $|l\rangle$ depends upon *all* states of the system combining with $|l\rangle$ under the action of the perturbation. This, of course, is a consequence of (181.5): If one of the coefficients, a_l, is changed, the other coefficients are bound to change as well in order to keep the sum (181.5) constant. (Cf. Problem 179 for a system with two states only.)

If the perturbation is *small*, we may in first approximation insert on the right-hand side of (181.6) the initial values

$$a_k(0) = \delta_{k0} . \tag{181.7}$$

Then, the set of equations (181.6) becomes, for $l \neq 0$,

$$\dot{a}_l = -\frac{i}{\hbar}\mathrm{e}^{-i(\omega_0 - \omega_l)t}\langle l|W|0\rangle . \tag{181.8}$$

---

[2] If a continuous part of the spectrum exists besides the discontinuous part, it can, by the use of a periodicity volume, be transformed into a formally discontinuous spectrum and thus be included into the sums (181.4) and (181.5) without further mathematical difficulties.

(This is a much more specialized behaviour than that of Problem 179, because it neglects backward transitions from $|l\rangle$ to $|0\rangle$ and the like.) Integration of (181.8) yields

$$a_l(t) = -\frac{i}{\hbar}\int_0^t dt \langle l|W|0\rangle \, e^{-i(\omega_0-\omega_l)t}. \tag{181.9}$$

The integral, of course, depends very much on how the perturbation W, and thus its matrix element, depend upon time.

Our approximation is valid only if

$$|\langle l|W|0\rangle|\, t/\hbar \ll 1, \tag{181.10}$$

so that the coefficient $a_l(t)$ remains small throughout. Since it follows from (181.10) that

$$(\omega_l-\omega_0)\,t \ll \frac{\hbar(\omega_l-\omega_0)}{|\langle l|W|0\rangle|}$$

and the excitation energy in the numerator is usually much larger than the matrix element in the denominator, the exponent in (181.8) or (181.9) may still be quite large so that there occur periodic oscillations of $a_l(t)$ which do not quite agree with the basic idea underlying our first-order perturbation approximation. However, in the following problems we shall show how to eliminate this difficulty.

Problem 182. Periodic perturbation: Resonance

Let the atomic system of the preceding problem be perturbed by a periodic field

$$W(t) = \mathscr{W} e^{-i\omega t} + \mathscr{W}^\dagger e^{i\omega t}. \tag{182.1}$$

Discuss resonance absorption and the effect of a finite frequency width of the irradiated field upon the transitions.

Solution. If (182.1) is put into the general first-order perturbation formula (182.9) and the integration is performed, we get

$$a_l(t) = -\frac{i}{\hbar}\left\{\langle l|\mathscr{W}|0\rangle \frac{e^{i(\omega_l-\omega_0-\omega)t}-1}{i(\omega_l-\omega_0-\omega)} + \langle l|\mathscr{W}^\dagger|0\rangle \frac{e^{i(\omega_l-\omega_0+\omega)t}-1}{i(\omega_l-\omega_0+\omega)}\right\}. \tag{182.2}$$

The excitation energy, $E_{ex}=\hbar(\omega_l-\omega_0)$ being positive, the first term has resonance if $\hbar\omega=E_{ex}$ whereas the second term never shows resonance. Thus, if Bohr's frequency condition

$$\omega=\omega_l-\omega_0 \tag{182.3}$$

holds, the system can absorb energy from the alternating field applied:

$$|a_l(t)|^2=\frac{4|\langle l|\mathscr{W}|0\rangle|^2}{\hbar^2}\cdot\frac{\sin^2\frac{1}{2}(\omega_l-\omega_0-\omega)t}{(\omega_l-\omega_0-\omega)^2}. \tag{182.4}$$

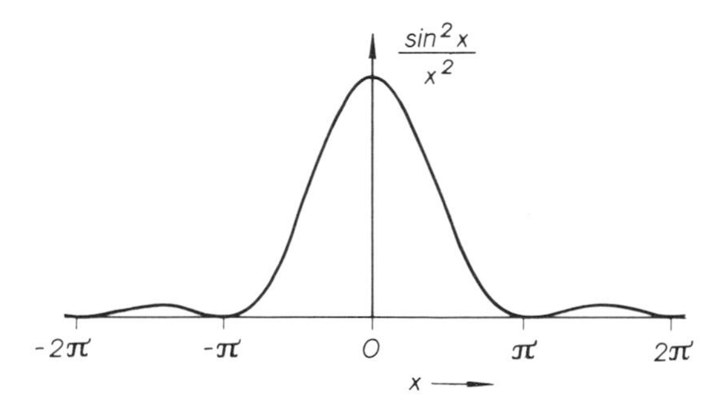

Fig. 70. The natural line shape $\sin^2 x/x^2$

This formula should still be corrected for the finite frequency width of the irradiated field. Let $\rho(\omega)\,d\omega$ be its intensity between ω and $\omega+d\omega$, then we have

$$|a_l(t)|^2=\int d\omega\;\rho(\omega)\cdot 4|\langle l|\mathscr{W}|0\rangle|^2\,\frac{\sin^2\frac{1}{2}(\omega_l-\omega_0-\omega)t}{\hbar^2(\omega_l-\omega_0-\omega)^2} \tag{182.5}$$

or, with

$$\tfrac{1}{2}[\omega-(\omega_l-\omega_0)]\,t=x$$

as integration variable,

$$|a_l(t)|^2=2t\int dx\,\rho\,(\omega_l-\omega_0+2x/t)\left|\langle l|\frac{\mathscr{W}}{\hbar}|0\rangle\right|^2\cdot\frac{\sin^2 x}{x^2}.$$

Here, the last factor, $\sin^2 x/x^2$, has a pronounced maximum at $x=0$ whence it decreases rapidly on both sides (Fig. 70) so that $|x|<\pi$ brings the main contribution to the integral

$$\int dx\frac{\sin^2 x}{x^2}=\pi.$$

Within this range of x values we have $|2x/t| < 2\pi/t$ or, since condition (181.10) of the preceding problem must hold, i.e. since

$$|\langle l|\mathscr{W}|0\rangle| \ll \frac{\hbar}{t},$$

and since this matrix element will usually be very small compared with the excitation energy, we find that the argument of ρ may be replaced simply by $\omega_l - \omega_0$. A similar argument obtains for the matrix element which too may be treated as a constant, independent of x, so that we arrive at

$$|a_l(t)|^2 = 2\pi t \left|\langle l|\frac{\mathscr{W}}{\hbar}|0\rangle\right|^2 \rho(\omega_l - \omega_0). \tag{182.6}$$

The probability of finding the system in any state $|l\rangle$ thus increases in proportion to the time. Therefore we may reasonably define a *transition probability*

$$P_l = \frac{1}{t}|a_l(t)|^2 \tag{182.7}$$

independent of time which becomes

$$P_l = 2\pi \left|\langle l|\frac{\mathscr{W}}{\hbar}|0\rangle\right|^2 \rho(\omega_l - \omega_0). \tag{182.8}$$

NB. The last result shows a close similarity to the Golden Rule to be discussed in Problem 183. It should, however, be borne in mind that the Golden Rule describes summation over close-lying *final* states, whereas we have introduced a summation over a continuum of *initial* field properties. It has not been shown here that this summation must necessarily be performed in the probabilities, Eq. (182.5), and not in the amplitude formula (182.2).

Problem 183. Golden Rule for scattering

Let a beam of particles with initial momentum $\boldsymbol{p}_i = \hbar \boldsymbol{k}_i$ be elastically scattered by a potential $W(r)$ into states of final momentum $\boldsymbol{p}_f = \hbar \boldsymbol{k}_f$ within the solid angle element $d\Omega_f$. The differential cross section $d\sigma/d\Omega_f$ shall be derived by the Dirac perturbation method.

Solution. We may gather from (181.9) that, in the first Dirac approximation,

$$a_f(t) = -\frac{i}{\hbar}\int_0^t dt \langle f|W|i\rangle e^{-i(\omega_i - \omega_f)t}. \tag{183.1}$$

This may be integrated, if the matrix element is supposed not to depend upon time, so that we arrive at the basic formula

$$|a_f(t)|^2 = \frac{4|\langle f|W|i\rangle|^2}{\hbar^2(\omega_i-\omega_f)^2}\sin^2\frac{(\omega_i-\omega_f)t}{2}. \tag{183.2}$$

Now both the initial and final states are lying in the continuous spectrum. Using a normalization volume V, the respective wave functions are

$$|i\rangle = V^{-\frac{1}{2}}\,e^{i\boldsymbol{k}_i\cdot\boldsymbol{r}}; \qquad \langle f| = V^{-\frac{1}{2}}\,e^{-i\boldsymbol{k}_f\cdot\boldsymbol{r}}. \tag{183.3}$$

There are, even for finite volume V, a great many final states in the vicinity of $\langle f|$ and, in the limit $V\to\infty$, there will be an infinite number of them in an infinitesimal surrounding. To ask what is the probability of *one* final state $\langle f|$ with sharp $\boldsymbol{k}_f$ thus becomes meaningless; we may only ask with what probability a certain interval will be reached.

Let $\rho_f(E_f)\,dE_f$ levels be lying in the interval dE_f at the final energy E_f with their momenta within the solid angle element $d\Omega_f$, then the transition probability per unit time to this angular interval becomes

$$dT = \frac{1}{t}\int dE_f\,\rho_f(E_f)\,|a_f(t)|^2. \tag{183.4}$$

This definition is reasonable only because this expression does not depend on time and the integral goes over a very narrow energy region.

If now we introduce the variable

$$x = \tfrac{1}{2}(\omega_f-\omega_i)\,t$$

and put $E_f = \hbar\omega_f$ so that we get

$$dE_f = \frac{2\hbar}{t}\,dx$$

we find, according to (183.2) and (183.4),

$$dT = \frac{2}{\hbar}\int dx\,\rho_f(E_f)\,\frac{\sin^2 x}{x^2}\,|\langle f|W|i\rangle|^2$$

where the considerations of the preceding problem will again hold for the integral, so that

$$dT = \frac{2\pi}{\hbar}\,\rho_f(E)\,|\langle f|W|i\rangle|^2 \tag{183.5}$$

since integration over an extremely large interval in x about $x=0$ corresponds to a very narrow one only in energy about the resonance energy $E_f=E_i$. The differential notation, dT, is appropriate because of the infinitesimal interval $d\Omega_f$ still contained in ρ_f (which itself might perhaps better be written $d\rho_f$). Eq. (183.5) is called the *Golden Rule*.

This transition probability still depends in an obvious way upon the normalization volume V and the initial velocity $v_0=\hbar k_i/m$ of the particles hitting the obstacle,

$$dT=\frac{v_0}{V}\,d\sigma, \tag{183.6}$$

where $d\sigma$ is independent of V and therefore a quantity of physical significance. It has the dimension of cm^2 and is identical with the differential cross section thus to be found from the Golden Rule expression (183.5):

$$d\sigma=\frac{2\pi}{\hbar}\,\rho_f(E)\,\frac{V}{v_0}\,|\langle f|W|i\rangle|^2. \tag{183.7}$$

Here we still have to evaluate the final state density ρ_f and the matrix element.

The final state density may be derived from the fact that one state (if the particles have no spin) falls in an element d^3p of momentum space of the amount $(2\pi\hbar)^3/V$. Therefore, in an arbitrary element d^3p there will lie

$$\frac{d^3p\,V}{8\pi^3\hbar^3}$$

states. With

$$d^3p=p^2\,dp\,d\Omega=mp\,dE\,d\Omega$$

we have

$$\rho_f\,dE_f=\frac{mp_f\,dE_f\,d\Omega_f\,V}{8\pi^3\hbar^3}, \tag{183.8}$$

so that

$$\frac{d\sigma}{d\Omega_f}=\left(\frac{mV}{2\pi\hbar^2}\right)^2\frac{k_f}{k_i}|\langle f|W|i\rangle|^2\,d\Omega_f \tag{183.9}$$

where we may omit the factor $k_f/k_i=1$ for elastic scattering.

The matrix element, for a potential $W(r)$ of central symmetry, formed with the plane waves (183.3) runs

$$\langle f|W|i\rangle=\frac{1}{V}\int d^3x\,e^{-i\boldsymbol{K}\cdot\boldsymbol{r}}\,W(r) \tag{183.10}$$

with $\boldsymbol{K}=\boldsymbol{k}_f-\boldsymbol{k}_i$ the momentum transfer vector (in units of $\hbar$). The integration over the polar angles leads to

$$\langle f|W|i\rangle = \frac{4\pi}{V}\int_0^\infty dr\,r^2\,W(r)\frac{\sin Kr}{Kr} \tag{183.11}$$

so that finally we arrive at the cross section formula

$$\frac{d\sigma}{d\Omega_f} = \left|\frac{2m}{\hbar^2}\int_0^\infty dr\,r^2\,W(r)\frac{\sin Kr}{Kr}\right|^2, \tag{183.12}$$

i.e. the result of the first Born approximation (cf. Problem 105). This of course, was only to be expected since we treated the scattering potential as a perturbation already in our starting equation (183.1) and consequently used plane waves to describe the initial and final states.

Problem 184. Born scattering in momentum space

To derive the differential scattering cross section in momentum space by a time-dependent perturbation method in the first approximation. Let the perturbation be switched on at $t=0$ and be constant thereafter.

Solution. According to Problem 14 the time-dependent Schrödinger equation

$$-\frac{\hbar}{i}\dot{\psi} = -\frac{\hbar^2}{2m}\nabla^2\psi + V(\boldsymbol{r})\psi \tag{184.1}$$

corresponds to the integro-differential equation in momentum space,

$$-\frac{\hbar}{i}\dot{f}(\boldsymbol{k},t) = \frac{\hbar^2}{2m}k^2 f(\boldsymbol{k},t) + \int d^3k'\,W(\boldsymbol{k}-\boldsymbol{k}')\,f(\boldsymbol{k}',t) \tag{184.2}$$

with $f(\boldsymbol{k},t)$ the Fourier transform of $\psi(\boldsymbol{r},t)$ and $W(\boldsymbol{k})$ of $V(\boldsymbol{r})$ in the normalization used in Problem 14. Eq. (184.2) may be somewhat simplified by writing f in the form

$$f(\boldsymbol{k},t)=v(\boldsymbol{k},t)\,\mathrm{e}^{-i\omega t};\qquad \omega = \frac{\hbar}{2m}k^2. \tag{184.3}$$

We then find

$$\frac{\partial v(\boldsymbol{k},t)}{\partial t} = -\frac{i}{\hbar}\int d^3k'\,W(\boldsymbol{k}-\boldsymbol{k}')\,\mathrm{e}^{i(\omega-\omega')t}\,v(\boldsymbol{k}',t). \tag{184.4}$$

To solve (184.4) in first approximation we replace v under the integral by the unperturbed function

$$v_0(\boldsymbol{k},t)=C(2\pi)^{\frac{3}{2}}\,\delta(\boldsymbol{k}-\boldsymbol{k}_0), \tag{184.5}$$

i.e. by the Fourier transform of the plane wave

$$\psi_0(\boldsymbol{r},t)=C\,e^{i\boldsymbol{k}_0\cdot\boldsymbol{r}}. \tag{184.6}$$

Eq. (184.4) then simply becomes

$$\frac{\partial v(\boldsymbol{k},t)}{\partial t} = -\frac{i}{\hbar}\,C(2\pi)^{\frac{3}{2}}\,W(\boldsymbol{k}-\boldsymbol{k}_0)\,e^{i(\omega-\omega_0)t}$$

and yields the integral

$$v(\boldsymbol{k},t)=-\frac{C}{\hbar}(2\pi)^{\frac{3}{2}}\,W(\boldsymbol{k}-\boldsymbol{k}_0)\frac{e^{i(\omega-\omega_0)t}-1}{\omega-\omega_0}. \tag{184.7}$$

The probability of finding a particle of momentum $\boldsymbol{k}$ within d^3k at the time t is then (cf. Problem 15)

$$|v(\boldsymbol{k},t)|^2\,d^3k = \frac{|C|^2(2\pi)^3}{\hbar^2}\,|W(\boldsymbol{k}-\boldsymbol{k}_0)|^2\,t^2\,\frac{\sin^2 x}{x^2}\,k^2\,dk\,d\Omega$$

with the abbreviation

$$x=\tfrac{1}{2}(\omega-\omega_0)\,t.$$

This expression has still to be integrated over the energy resonance (cf. Problem 183); since

$$k\,dk = \frac{2m}{\hbar t}\,dx$$

and

$$\int_{-\infty}^{+\infty} dx\,\frac{\sin^2 x}{x^2} = \pi,$$

we find an expression linear in t, and the transition probability, defined by

$$dT=\frac{1}{t}\,d\Omega\int dk\,k^2\,|v(\boldsymbol{k},t)|^2, \tag{184.8}$$

becomes

$$dT=\frac{|C|^2(2\pi)^4\,m\,k}{\hbar^3}\,|W(\boldsymbol{k}-\boldsymbol{k}_0)|^2\,d\Omega. \tag{184.9}$$

The differential cross section is defined by putting

$$dT = |C|^2 v_0 d\sigma = |C|^2 \frac{\hbar k}{m} d\sigma; \tag{184.10}$$

therefore

$$d\sigma = \frac{(2\pi)^4 m^2}{\hbar^4} |W(\boldsymbol{k}-\boldsymbol{k}_0)|^2 d\Omega. \tag{184.11}$$

This general formula may still be simplified for a central-force potential $V(r)$ which permits integration over the polar angles in its Fourier transform:

$$W(\boldsymbol{k}-\boldsymbol{k}_0) = \frac{1}{(2\pi)^3} \int d^3x \, e^{i\boldsymbol{k}\cdot\boldsymbol{r}} V(r) = \frac{4\pi}{(2\pi)^3} \int_0^\infty dr \, r^2 V(r) \frac{\sin Kr}{Kr} \tag{184.12}$$

with $\boldsymbol{K} = \boldsymbol{k} - \boldsymbol{k}_0$ the momentum transfer (in units of $\hbar$), i.e. with

$$K = 2k_0 \sin\frac{\vartheta}{2} \tag{184.13}$$

where ϑ is the angle of deflection. Putting (184.12) into (184.11) we finally arrive at the well-known result of Born,

$$\frac{d\sigma}{d\Omega} = \left\{ \frac{2m}{\hbar^2} \int_0^\infty dr \, r^2 V(r) \frac{\sin Kr}{Kr} \right\}^2. \tag{184.14}$$

NB. The transition from the momentum function f to v is, generally speaking, the transition from Schrödinger to interaction representation as v would no longer depend upon time if there were no interaction $W(\boldsymbol{k})$.—The cross section formula (184.14) has also been derived in Problem 105.—It should further be noted that the Fourier transform $W(\boldsymbol{k}-\boldsymbol{k}_0)$ is—except for a normalization factor—the matrix element of the potential $V(\boldsymbol{r})$ in ordinary space:

$$(2\pi)^3 W(\boldsymbol{k}-\boldsymbol{k}_0) = \int d^3k' \, e^{-i\boldsymbol{k}\cdot\boldsymbol{r}} V(\boldsymbol{r}) e^{i\boldsymbol{k}_0\cdot\boldsymbol{r}} = \langle \boldsymbol{k}|V|\boldsymbol{k}_0 \rangle.$$

This permits e.g. to write the cross section formula (184.11) as well in the form

$$d\sigma = \left| \frac{m}{2\pi\hbar^2} \langle \boldsymbol{k}|V|\boldsymbol{k}_0 \rangle \right|^2 d\Omega. \tag{184.15}$$

Problem 185. Coulomb excitation of an atom

Let an electron of velocity v pass an alkali atom at a distance b ("collision parameter"); let its velocity be large compared to the velocity of the valence electron in the atom. Its Coulomb interaction with the latter may originate an excitation, the cross section of which shall be determined by a perturbation procedure.

Solution. The position of the atomic electron is shown in Fig. 71, its position $\boldsymbol{r}$ relative to the atomic core having coordinates x, y, z, as indicated. Its interaction with the electron passing at the point P is

$$V(t) = \frac{e^2}{R(t)} \tag{185.1}$$

with

$$R^2 = (v\,t - x)^2 + y^2 + (b - z)^2 . \tag{185.2}$$

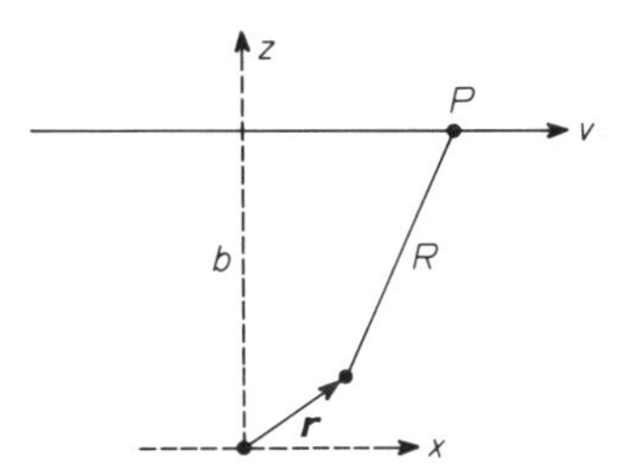

Fig. 71. Notations indicating position of the two interacting particles

Here any deflection or deceleration of the passing electron has been neglected. Its classical treatment, of course, precludes effects of overlap such as exchange phenomena. It is a reasonable approximation for a wavelength small compared to the distance b and to the atomic dimensions. Let us further assume the collision parameter b to be very large compared to the atomic dimensions so that we may expand

$$\frac{1}{R} = \frac{1}{\sqrt{(v\,t)^2 + b^2}} \left\{ 1 + \frac{v\,t\,x + b\,z}{(v\,t)^2 + b^2} \right\} . \tag{185.2'}$$

The matrix element between the ground state (subscript: 0) and an excited state (k) then becomes, in this approximation,

$$\langle k|V(t)|0\rangle = \frac{e^2}{[(vt)^2+b^2]^{\frac{3}{2}}}\{vt\langle k|x|0\rangle + b\langle k|z|0\rangle\}, \qquad (185.3)$$

the first (static) term in (185.2′), only contributing to the diagonal elements $\langle 0|V|0\rangle$ of elastic scattering.

If the unperturbed wave functions, $|k\rangle$, satisfy a Schrödinger equation

$$H|k\rangle = \hbar\omega_k|k\rangle$$

the perturbed wave function may be written

$$\psi = e^{-i\omega_0 t}|0\rangle + \sum_k{}' a_k(t)e^{-i\omega_k t}|k\rangle \qquad (185.4)$$

with (cf. Problem 181)

$$a_k(t) = -\frac{i}{\hbar}\int_{-\infty}^{t} dt\, e^{i(\omega_k-\omega_0)t}\langle k|V(t)|0\rangle. \qquad (185.5)$$

The probability of finding the atom in the excited state $|k\rangle$ after passage of the colliding electron is then

$$p_k = |a_k(\infty)|^2, \qquad (185.6)$$

and the excitation cross section of the state $|k\rangle$ is obtained by integration over the collision parameter:

$$\sigma_k = 2\pi\int_0^\infty db\, b\,|a_k(\infty)|^2. \qquad (185.7)$$

Thus the problem is reduced mainly to the calculation of $a_k(\infty)$ with the matrix element (185.3):

$$a_k(\infty) = -\frac{i}{\hbar}e^2\int_{-\infty}^{+\infty} dt\,\frac{e^{i(\omega_k-\omega_0)t}}{[v^2t^2+b^2]^{\frac{3}{2}}}\{vt\langle k|x|0\rangle + b\langle k|z|0\rangle\}.$$

With the abbreviations

$$\beta = \frac{(\omega_k-\omega_0)b}{v}; \qquad s = \frac{v}{b}t \qquad (185.8)$$

this may be written

$$a_k(\infty) = -\frac{ie^2}{\hbar v b}\left\{\langle k|x|0\rangle \int_{-\infty}^{+\infty} ds \frac{s e^{i\beta s}}{(1+s^2)^{\frac{3}{2}}} + \langle k|z|0\rangle \int_{-\infty}^{+\infty} ds \frac{e^{i\beta s}}{(1+s^2)^{\frac{3}{2}}}\right\}. \quad (185.9)$$

The two integrals can be determined using the integral representation of the modified Hankel function

$$\int_0^\infty ds \frac{\cos\beta s}{(1+s^2)^{\frac{3}{2}}} = \beta K_1(\beta). \quad (185.10)$$

This integral may equally well be written

$$\frac{1}{2}\int_0^\infty \frac{ds\, e^{i\beta s}}{(1+s^2)^{\frac{3}{2}}} + \frac{1}{2}\int_0^\infty \frac{ds\, e^{-i\beta s}}{(1+s^2)^{\frac{3}{2}}} = \frac{1}{2}\int_{-\infty}^{+\infty} \frac{ds\, e^{i\beta s}}{(1+s^2)^{\frac{3}{2}}}$$

so that

$$\int_{-\infty}^{+\infty} ds \frac{e^{i\beta s}}{(1+s^2)^{\frac{3}{2}}} = 2\beta K_1(\beta). \quad (185.11)$$

The derivative with respect to β is

$$i\int_{-\infty}^{+\infty} ds \frac{s\, e^{i\beta s}}{(1+s^2)^{\frac{3}{2}}} = 2\frac{d}{d\beta}[\beta K_1(\beta)] = -2\beta K_0(\beta). \quad (185.12)$$

We therefore obtain

$$a_k(\infty) = -\frac{e^2}{\hbar v}\cdot\frac{2}{b}\{\langle k|x|0\rangle(-\beta K_0(\beta)) + i\langle k|z|0\rangle \beta K_1(\beta)\}. \quad (185.13)$$

If the atoms are not oriented, in the statistical average

$$\overline{\langle k|x|0\rangle} = \overline{\langle k|z|0\rangle} = 0; \qquad \overline{|\langle k|x|0\rangle|^2} = \overline{|\langle k|z|0\rangle|^2} \quad (185.14)$$

so that

$$\overline{|a_k(\infty)|^2} = \left(\frac{e^2}{\hbar v}\right)^2 \cdot \frac{4}{b^2}\, \overline{|\langle k|x|0\rangle|^2}\, \beta^2 [K_0^2(\beta) + K_1^2(\beta)]. \quad (185.15)$$

According to (185.7) and (185.8) this leads to the excitation cross section of the state $|k\rangle$:

$$\sigma_k = 8\pi\left(\frac{e^2}{\hbar v}\right)^2 \overline{|\langle k|x|0\rangle|^2} \int_0^\infty d\beta\, \beta[K_0^2(\beta) + K_1^2(\beta)]. \quad (185.16)$$

The last integral can be evaluated for a finite lower bound

$$\int_{\beta}^{\infty} d\beta\,\beta[K_0^2(\beta)+K_1^2(\beta)]=\beta K_0(\beta)K_1(\beta). \tag{185.17}$$

For small values of β, the limiting values to the functions are

$$\beta K_1(\beta)\to 1; \qquad K_0(\beta)\to C+\log\frac{2}{\beta} \tag{185.18}$$

with $C=0.5772\ldots$ the Euler constant. The integral therefore diverges at small values of β or, according to (185.8), of the collision parameter b. This divergence is caused by the expansion (185.2′) of the interaction which only holds if the distance r of the atomic electron from the atomic nucleus is small compared to the collision parameter b. The divergence can to some extent be remedied by a cut-off at $b=r_0$, with r_0 something like the atomic radius or, using (185.8), at

$$\beta_{\min}=\frac{r_0(\omega_k-\omega_0)}{v}. \tag{185.19}$$

Since v shall be large compared to the atomic electron velocity, $\beta_{\min}\ll 1$ so that the formulae (185.18) obtain. We then arrive at the cross section formula

$$\sigma_k=8\pi\left(\frac{e^2}{\hbar v}\right)^2\overline{|\langle k|x|0\rangle|^2}\left\{\log\frac{2v}{r_0(\omega_k-\omega_0)}-C\right\}. \tag{185.20}$$

No very exact knowledge of the cut-off radius is needed, since the logarithm varies rather slowly with its argument.

Literature. The method is an abridged form of that used in the theory of nuclear Coulomb excitation, cf. Alder, K., Winther, W.: Dan. Mat.-Fys. Medd. **29,** 19 (1955).

Problem 186. Photoeffect

A light wave of linear polarization ($\mathscr{E}\parallel x$, $\mathscr{H}\parallel y$) propagating in z direction falls on a hydrogen atom with the electron in its ground state. What is the angular distribution of photoelectrons emitted? What is the differential cross section of photo-emission? Retardation effects shall be neglected and the final state be approximated by a plane wave.

Solution. The light wave may be described by a vector potential $\boldsymbol{A}$ with

$$A_x=\frac{c}{\omega}\mathscr{E}_0\cos\left[\omega\left(t-\frac{z}{c}\right)+\delta\right]; \quad A_y=0; \quad A_z=0, \tag{186.1}$$

then the following field strength components can be derived:

$$\mathscr{E}_x = -\frac{1}{c}\dot{A}_x = \mathscr{E}_0 \sin\left[\omega\left(t-\frac{z}{c}\right)+\delta\right];$$

$$\mathscr{H}_y = \frac{\partial A_x}{\partial z} = \mathscr{E}_0 \sin\left[\omega\left(t-\frac{z}{c}\right)+\delta\right],$$

all other components vanishing. The average Poynting vector in z direction is

$$\overline{S} = \frac{c}{4\pi}\overline{\mathscr{E}_x \mathscr{H}_y} = \frac{c}{8\pi}\mathscr{E}_0^2$$

so that there are

$$n = \frac{c\mathscr{E}_0^2}{8\pi\hbar\omega} \tag{186.2}$$

incident photons per cm^2 and sec.

The interaction energy between light and electron is, according to Problem 125,

$$W = -\frac{e\hbar}{mc} i(\boldsymbol{A}\cdot\nabla) = \mathsf{W}\,\mathrm{e}^{-i\omega t} + \mathsf{W}^\dagger \mathrm{e}^{i\omega t} \tag{186.3}$$

where

$$\mathsf{W} = -\frac{e\hbar}{2m\omega}\mathscr{E}_0\, i\,\mathrm{e}^{-i\delta}\frac{\partial}{\partial x}. \tag{186.4}$$

Here the retardation factor, $\exp\left(i\frac{\omega z}{c}\right)$ has been taken $=1$.

We then may apply the method developped in Problem 182. Only the interaction term W leads to a resonance denominator $\omega_f-\omega_i-\omega$, thus satisfying energy conservation. Putting

$$x = \tfrac{1}{2}(\omega_f-\omega_i-\omega)\,t$$

we get

$$|a_f(t)|^2 = \frac{4}{\hbar^2}|\langle f|\mathsf{W}|i\rangle|^2 \frac{\sin^2 x}{x^2}$$

and thence the transition probability P_f, from the initial state $|i\rangle$ to the final state $|f\rangle$,

$$P_f = \frac{2\pi}{\hbar}\rho_f |\langle f|\mathsf{W}|i\rangle|^2. \tag{186.5}$$

Here ρ_f, the final state density in the energy scale refers to the outgoing electron and is, according to (183.8),

$$\rho_f = \frac{mV}{8\pi^3 \hbar^2} k_f \, d\Omega_f \tag{186.6}$$

if V is the normalization volume and $\hbar k_f$ the momentum of the photoelectron. The differential cross section for photo-emission into the solid angle element $d\Omega_f$ is P_f/n so that, gathering expressions from (186.2) and (186.4–6), we arrive at the general formula

$$d\sigma = \frac{e^2 V}{2\pi m c \omega} k_f \, d\Omega_f \left| \langle f | \frac{\partial}{\partial x} | i \rangle \right|^2 . \tag{186.7}$$

We now are dealing with wave functions in a potential field of central symmetry, with the ground state $|i\rangle$ independent of polar angles. Then,

$$\frac{\partial}{\partial x} |i\rangle = \frac{d|i\rangle}{dr} \sin\vartheta \cos\varphi \tag{186.8}$$

is proportional to a spherical harmonic of the first order so that the matrix element must necessarily vanish if the electron is not emitted in a P state.

Let the final state be approximately described by a plane wave,

$$|f\rangle = V^{-\frac{1}{2}} e^{i k_f r \cos\gamma} = \frac{1}{kr} \sum_{l=0}^{\infty} (2l+1) i^l j_l(k_f r) P_l(\cos\gamma) \tag{186.9}$$

where γ is the angle between the directions of $\boldsymbol{k}_f$ with polar angles Θ, Φ and $\boldsymbol{r}$ with angles ϑ and φ. Then only the term $l=1$ (P term) can contribute to the matrix element so that

$$\langle f | \frac{\partial}{\partial x} | i \rangle = \frac{3i}{\sqrt{V}} \int_0^\infty dr \, r^2 \frac{j_1(kr)}{kr} \frac{d|i\rangle}{dr} \oint d\Omega \cos\gamma \sin\vartheta \cos\varphi .$$

Because of

$$\cos\gamma = \cos\vartheta \cos\Theta + \sin\vartheta \sin\Theta (\cos\varphi \cos\Phi + \sin\varphi \sin\Phi)$$

the angular integral yields

$$\sin\Theta \cos\Phi \oint d\Omega \sin^2\vartheta \cos^2\varphi = \frac{4\pi}{3} \sin\Theta \cos\Phi$$

so that we arrive at

$$\langle f|\frac{\partial}{\partial r}|i\rangle = \frac{4\pi i}{\sqrt{V}}\sin\Theta\cos\Phi\int_0^\infty dr\, r^2\frac{j_1(k_f r)}{k_f r}\frac{d|i\rangle}{dr}$$

and, according to (186.7),

$$\frac{d\sigma}{d\Omega_f} = \frac{8\pi e^2}{m c\omega k_f}\left\{\int_0^\infty dr\, r\, j_1(k_f r)\frac{d|i\rangle}{dr}\right\}^2 \sin^2\Theta\cos^2\Phi. \tag{186.10}$$

Here the radial function j_1 stemming from the plane-wave approximation should be replaced by a more correct expression for quantitative calculations. The angular distribution of the photo-electrons, however, is correct and in complete agreement with the classical expectation, since $\sin^2\Theta\cos^2\Phi$ has a maximum in the direction x of the electrical field strength.

NB. With

$$|i\rangle = \pi^{-\frac{1}{2}}a^{-\frac{3}{2}}e^{-\frac{r}{a}}; \qquad a = \frac{\hbar^2}{(Z-s)me^2}$$

for a K shell electron (screening constant s, cf. Problem 178) we get for the integral in (186.10)

$$J = \int_0^\infty dr\, r\, j_1(k_f r)\frac{d|i\rangle}{dr} = -\frac{1}{k_f^2 a^2\sqrt{\pi a}}\int_0^\infty dx\, x\left(\frac{\sin x}{x} - \cos x\right)e^{-\frac{x}{k_f a}}$$

where $x = k_f r$. This integral can be evaluated in an elementary way and yields

$$J = -\frac{2}{\sqrt{\pi a}}\cdot\frac{k_f^2 a^2}{(1+k_f^2 a^2)^2}.$$

This formula can only hold for $k_f a \gg 1$, because otherwise the plane-wave approximation would be quite insufficient; so we may write

$$J \simeq -\frac{2}{\sqrt{\pi a}}(k_f a)^{-2}$$

and

$$\frac{d\sigma}{d\Omega_f} = \frac{32 e^2}{m c a^5}\cdot\frac{1}{\omega k_f^5}$$

or, with

$$\frac{\hbar^2 k_f^2}{2m} = \hbar\omega_f,$$

finally,

$$\frac{d\sigma}{d\Omega_f} \simeq 8(Z-s)^5\sin^2\Theta\cos^2\Phi\cdot\frac{e^2}{\hbar c}\cdot\frac{m^2 e^{10}}{\hbar^7}\cdot\frac{1}{\omega\omega_f^2 k_f}.$$

This formula shows the main features of more exact calculations: rapid increase with $Z-s$, rapid decrease with growing quantum energy $\hbar\omega$, roughly proportional to $\omega^{-3.5}$, order of magnitude in atomic units $e^2/\hbar c$, and correct angular distribution.

Literature. Stobbe, M.: Ann. Physik **7**, 661 (1930). — *Retardation effects in hydrogen:* Sommerfeld, A., Schur, G., Ann. Physik **4**, 409 (1930). — *Relativistic treatment:* Sauter, F., Ann. Physik **11**, 454 (1931).

187. Dispersion of light. Oscillator strengths

A light wave as defined in the preceding problem (but take $\delta=0$) interacts with an atom. The induced polarization shall be calculated from which the oscillator strengths may be derived. Only one electron shall be taken into account, and all matrix elements, neglecting retardation, reduced to matrix elements of electrical dipole moments.

Solution. In the Dirac notation of Problem 181 the atomic wave function under the action of the light wave may be written

$$\psi=\sum_l a_l(t)|l\rangle\,\mathrm{e}^{-i\omega_l t}$$

if $|l\rangle$ is a state of the unperturbed atom. Using (182.2) for the coefficient $a_l(t)$ and omitting all switch-on effects of the light wave, we arrive at

$$\psi=|0\rangle\,\mathrm{e}^{-i\omega_0 t}-\frac{1}{\hbar}\sum_l\left\{\langle l|\mathrm{W}|0\rangle\frac{\mathrm{e}^{i(\omega_l-\omega_0-\omega)t}}{\omega_l-\omega_0-\omega}+\langle l|\mathrm{W}^\dagger|0\rangle\frac{\mathrm{e}^{i(\omega_l-\omega_0+\omega)t}}{\omega_l-\omega_0+\omega}\right\}|l\rangle\,\mathrm{e}^{-i\omega_l t}.$$

Here $|0\rangle$ denotes the atomic ground state and $|l\rangle$ any excited state so that $\omega_l-\omega_0>0$ and only the first term in the sum shows resonance. Neglecting the other term we then may start from the wave function

$$\psi=\left\{|0\rangle-\frac{1}{\hbar}\mathrm{e}^{-i\omega t}\sum_l\frac{\langle l|\mathrm{W}|0\rangle}{\omega_l-\omega_0-\omega}|l\rangle\right\}\mathrm{e}^{-i\omega_0 t}. \tag{187.1}$$

Now we know that for the optical properties the induced dipole moment $\boldsymbol{p}_{\mathrm{ind}}$ plays the essential role. It follows from (187.1) according to

$$\boldsymbol{p}_{\mathrm{ind}}=-e\left\{\int d^3x\,\psi^*\boldsymbol{r}\,\psi-\langle 0|\boldsymbol{r}|0\rangle\right\}. \tag{187.2}$$

Neglecting second-order corrections we thus get

$$\boldsymbol{p}_{\mathrm{ind}}=\frac{e}{\hbar}\sum_l\frac{\langle 0|\boldsymbol{r}|l\rangle\langle l|\mathrm{W}|0\rangle\,\mathrm{e}^{-i\omega t}+\langle l|\boldsymbol{r}|0\rangle\langle l|\mathrm{W}|0\rangle^*\,\mathrm{e}^{i\omega t}}{\omega_l-\omega_0-\omega}. \tag{187.3}$$

This expression can be much simplified if we replace the matrix elements $\langle l|\mathsf{W}|0\rangle$ of the interaction

$$\mathsf{W} = -\frac{e\hbar}{2m\omega}\mathscr{E}_0\, i\frac{\partial}{\partial x} \tag{187.4}$$

by matrix elements of the atomic moment $\boldsymbol{p}$, in field direction,

$$\langle l|p_x|0\rangle = -e\langle l|x|0\rangle . \tag{187.5}$$

This can be accomplished by using the relation

$$\langle l|\frac{\partial}{\partial x}|k\rangle = \frac{m}{\hbar}(\omega_k - \omega_l)\langle l|x|k\rangle \tag{187.6}$$

holding between any pair of states $|l\rangle$ and $|k\rangle$.

Eq. (187.6) may be derived from the two Schrödinger equations

$$\left\{-\frac{\hbar^2}{2m}\nabla^2 + V\right\}|k\rangle = \hbar\omega_k|k\rangle ; \qquad \left\{-\frac{\hbar^2}{2m}\nabla^2 + V\right\}\langle l| = \hbar\omega_l\langle l|$$

from which there follows, by multiplying by x, forming matrix elements and subtracting both equations in order to eliminate the term with V:

$$-\frac{\hbar^2}{2m}\{\langle l|x\nabla^2|k\rangle - \langle k|x\nabla^2|l\rangle^*\} = \hbar(\omega_k - \omega_l)\langle l|x|k\rangle . \tag{187.6'}$$

Now, according to

$$\int d^3x(vx)\nabla^2 u = -\int d^3x\nabla(vx)\cdot\nabla u = -\int d^3x\left(x\nabla v\cdot\nabla u + v\frac{\partial u}{\partial x}\right),$$

the curly bracket on the left-hand side may be reshaped into

$$-\langle l|\frac{\partial}{\partial x}|k\rangle + \langle k|\frac{\partial}{\partial x}|l\rangle^* = -2\langle l|\frac{\partial}{\partial x}|k\rangle ,$$

proportional to the left-hand side of (187.6) whereas on the right-hand side of (187.6′) we already have the dipole matrix element wanted.

Using (187.6), we may write the induced dipole moment (187.3) in the simpler form

$$\boldsymbol{p}_{\mathrm{ind}} = -\frac{\mathscr{E}_0}{2i\hbar}\sum_l \frac{\langle 0|\boldsymbol{p}|l\rangle\langle l|p_x|0\rangle\,\mathrm{e}^{-i\omega t} - \langle l|\boldsymbol{p}|0\rangle\langle l|p_x|0\rangle^*\,\mathrm{e}^{i\omega t}}{\omega_l - \omega_0 - \omega} . \tag{187.7}$$

In a statistical distribution of atoms with dipole moments $\boldsymbol{p}$ oriented in all directions, the y and z components of $\boldsymbol{p}_{\mathrm{ind}}$ will cancel and an induced dipole moment remain only in the x direction, i.e. parallel to the electrical field applied. In the first term of (187.7) we further use the hermiticity of $\langle 0|\boldsymbol{p}|l\rangle = \langle l|\boldsymbol{p}|0\rangle^*$ and then the statistical average relation

$$\overline{|\langle l|p_x|0\rangle|^2} = \tfrac{1}{3}|\langle l|\boldsymbol{p}|0\rangle|^2$$

where the expression on the right-hand side needs no further averaging, being independent of the atom's orientation. Thus we arrive at the result

$$\boldsymbol{p}_{\text{ind}} = \frac{\mathscr{E}_0}{3\hbar} \sum_l \frac{|\langle l|\boldsymbol{p}|0\rangle|^2}{\omega_l - \omega_0 - \omega} \sin\omega t .$$

Since $\mathscr{E}_0 \sin\omega t = \mathscr{E}$ is the instantaneous value of the field strength we may define atomic polarizability α in the usual way by $\boldsymbol{p}_{\text{ind}} = \alpha\mathscr{E}$ so that

$$\alpha = \frac{1}{3} \sum_l \frac{|\langle l|\boldsymbol{p}|0\rangle|^2}{\hbar(\omega_l - \omega_0 - \omega)} . \tag{187.8}$$

In classical optics, the index of refraction, n, is derived from the formula

$$\frac{n^2-1}{n^2+2} = \frac{4\pi}{3} N\alpha \tag{187.9}$$

with N the number of atoms per unit volume. (This expression is called the *refraction*.) The classical polarizability is then evaluated as a sum over all the electrons contributing (subscript λ) and is written in the form

$$\frac{n^2-1}{n^2+2} = \frac{4\pi}{3} N \frac{e^2}{m} \sum_\lambda \frac{f_\lambda}{\omega_\lambda^2 - \omega^2} \tag{187.10}$$

where ω_λ is the eigenfrequency of the λ'th electron and f_λ, the so-called *oscillator strength*, gives the number of electrons per atom in a state of eigenfrequency ω_λ, in this classical picture. That the oscillator strengths turned out experimentally not to be integers raised the first doubts in this classical picture.

The quantum theoretical formula (187.8) gives formally a very similar result. We may write,

$$\alpha = \frac{1}{3} \sum_l \frac{(\omega_l - \omega_0 + \omega)|\langle l|\boldsymbol{p}|0\rangle|^2}{\hbar[(\omega_l - \omega_0)^2 - \omega^2]} \simeq \frac{2\omega}{3\hbar} \sum_l \frac{|\langle l|\boldsymbol{p}|0\rangle|^2}{(\omega_l - \omega_0)^2 - \omega^2} ;$$

then (187.9) leads to the refraction

$$\frac{n^2-1}{n^2+2} = \frac{4\pi}{3} N \frac{e^2}{m} \sum_l \frac{f_l}{(\omega_l - \omega_0)^2 - \omega^2} \tag{187.11}$$

with the oscillator strengths

$$f_l = \frac{2m\omega}{3\hbar e^2} |\langle l|\boldsymbol{p}|0\rangle|^2 . \tag{187.12}$$

The formal similarity of the quantum theoretical result (187.11) with the classical one, (187.10), however, is deceptive. In Eq. (187.11) the sum does not run over electrons but over excited states so that summation over a multitude of terms is necessary even for our one-electron problem. The eigenfrequencies ω_λ are replaced by frequency differences, $\omega_l - \omega_0$. Finally, the oscillator strengths f_l no longer mean numbers of electrons but are rather involved intensity constants to be computed from dipole transition matrix elements according to (187.12). It is therefore no longer surprising that these numbers turn out not to be integers.

Problem 188. Spin flip in a magnetic resonance device

Let a particle of spin $\frac{1}{2}\hbar$ and of magnetic moment μ pass in y direction through an homogeneous magnetic field $\mathscr{H}_0$ parallel to the z axis. The particle spin in this field will be oriented in either $+z$ or $-z$ direction. Let us assume it to point in positive z direction. When passing the point $y=0$ at the time $t=0$, the particle enters an additional homogeneous field $\mathscr{H}'$ parallel to the x axis. It leaves this auxiliary field at $y=l$ at $t=t_0$. What is the probability of a spin flip during this time interval?

Solution. In the Schrödinger equation

$$-\frac{\hbar}{i}\dot{\psi} = -\frac{\hbar^2}{2m}\nabla^2\psi - \boldsymbol{\mu}\cdot\mathscr{H}\psi \tag{188.1}$$

the last term is the interaction energy of the magnetic field $\mathscr{H}$ with the magnetic moment $\boldsymbol{\mu}$ of the particle. The latter is defined as the vector operator

$$\boldsymbol{\mu} = \mu\boldsymbol{\sigma} \tag{188.2}$$

with $\boldsymbol{\sigma}$ the spin vector whose three components are the Pauli matrices (cf. Problem 129).

For $t<0$, only the field $\mathscr{H}_0 \| z$ is acting on the particle; the solution of (188.1) then is

$$\psi = e^{i(ky-\omega t)}\begin{pmatrix}1\\0\end{pmatrix} \tag{188.3}$$

and the energy of the particle

$$\hbar\omega = \frac{\hbar^2 k^2}{2m} - \mu\mathscr{H}_0. \tag{188.4}$$

If now, at $t=0$, the field $\mathscr{H}'\|x$ is switched on, the state of the particle undergoes a change so that its wave function may be written

$$\psi = \mathrm{e}^{i(ky-\omega_0 t)}\left\{a(t)\begin{pmatrix}1\\0\end{pmatrix}+b(t)\begin{pmatrix}0\\1\end{pmatrix}\right\} = \mathrm{e}^{i(ky-\omega_0 t)}\begin{pmatrix}a(t)\\b(t)\end{pmatrix} \tag{188.5}$$

with the abbreviation

$$\hbar\omega = \frac{\hbar^2 k^2}{2m}. \tag{188.6}$$

If path curvature by the Lorentz force acting perpendicular to the y direction may be neglected, it is safe to assume that the momentum $\hbar k$ in y direction is unchanged throughout.

Putting (188.5) in (188.1) we have to be a little cautious about the magnetic interaction:

$$(\boldsymbol{\mu}\cdot\mathscr{H}) = \mu(\mathscr{H}_0\sigma_z + \mathscr{H}'\sigma_x) = \mu\begin{pmatrix}\mathscr{H}_0 & \mathscr{H}'\\ \mathscr{H}' & -\mathscr{H}_0\end{pmatrix}$$

and therefore

$$(\boldsymbol{\mu}\cdot\mathscr{H})\begin{pmatrix}a\\b\end{pmatrix} = \mu\begin{pmatrix}\mathscr{H}_0 a+\mathscr{H}' b\\ \mathscr{H}' a-\mathscr{H}_0 b\end{pmatrix}.$$

Separating the two components of the Schrödinger equation, we thus arrive at

$$\left.\begin{aligned} -\frac{\hbar}{i}\dot{a} &= -\mu(\mathscr{H}_0 a+\mathscr{H}' b);\\ -\frac{\hbar}{i}\dot{b} &= -\mu(\mathscr{H}' a-\mathscr{H}_0 b).\end{aligned}\right\} \tag{188.7}$$

To solve these equations we set

$$a(t) = A\,\mathrm{e}^{i\omega' t};\qquad b(t) = B\,\mathrm{e}^{i\omega' t};$$

then (188.7) becomes the algebraic set of linear equations

$$\begin{aligned}&(\mu\mathscr{H}_0-\hbar\omega')A+\mu\mathscr{H}' B=0;\\ &\mu\mathscr{H}' A-(\mu\mathscr{H}_0+\hbar\omega')B=0.\end{aligned}$$

The determinant vanishes for two different values of ω',

$$\omega' = \pm\frac{\mu}{\hbar}\sqrt{\mathscr{H}_0^2+\mathscr{H}'^2}; \tag{188.8}$$

in what follows we shall write ω' and $-\omega'$ for these two roots. Then the set (188.7) will be solved by

$$a(t) = A_{\mathrm{I}}\,\mathrm{e}^{i\omega' t} + A_{\mathrm{II}}\,\mathrm{e}^{-i\omega' t};$$
$$b(t) = B_{\mathrm{I}}\,\mathrm{e}^{i\omega' t} + B_{\mathrm{II}}\,\mathrm{e}^{-i\omega' t}$$

with

$$B_{\mathrm{I,II}} = \frac{\pm\sqrt{\mathscr{H}_0^2 + \mathscr{H}'^2} - \mathscr{H}_0}{\mathscr{H}'} A_{\mathrm{I,II}}\,.$$

There still remain two integration constants, A_{I} and A_{II}, to be determined from the initial conditions,

$$a(0) = 1; \qquad b(0) = 0. \tag{188.9}$$

This leads to

$$A_{\mathrm{I}} = \frac{\sqrt{\mathscr{H}_0^2 + \mathscr{H}'^2} + \mathscr{H}_0}{2\sqrt{\mathscr{H}_0^2 + \mathscr{H}'^2}}; \qquad A_{\mathrm{II}} = \frac{\sqrt{\mathscr{H}_0^2 + \mathscr{H}'^2} - \mathscr{H}_0}{2\sqrt{\mathscr{H}_0^2 + \mathscr{H}'^2}}$$

and yields after some straightforward reshaping the formulae

$$\left.\begin{aligned} a(t) &= \cos\omega' t + i\frac{\mathscr{H}_0}{\sqrt{\mathscr{H}_0^2 + \mathscr{H}'^2}} \sin\omega' t; \\ b(t) &= i\frac{\mathscr{H}'}{\sqrt{\mathscr{H}_0^2 + \mathscr{H}'^2}} \sin\omega' t. \end{aligned}\right\} \tag{188.10}$$

It can easily be checked that

$$|a(t)|^2 + |b(t)|^2 = 1\,.$$

The probability of spin flip, i.e. of finding the particle with spin downward, in the negative z direction, after its leaving the auxiliary field $\mathscr{H}'$ at the time $t = t_0 = l/v$ will be, according to (188.10) and (188.8),

$$|b(t_0)|^2 = \frac{\mathscr{H}'^2}{\mathscr{H}_0^2 + \mathscr{H}'^2} \sin^2\left\{\frac{\mu}{\hbar}\sqrt{\mathscr{H}_0^2 + \mathscr{H}'^2}\, t_0\right\}. \tag{188.11}$$

Eq. (188.11) shows that the experimental device may be used to determine the magnetic moment of an atom of spin $\frac{1}{2}\hbar$ (as an alkali atom in its ground state). If atoms are focussed if they do not undergo spin flip, but defocussed if they do, and the magnetic fields are varied during the course of the experiment, the beam intensity will become a minimum,

$$|a(t_0)|^2_{\min} = \frac{\mathscr{H}_0^2}{\mathscr{H}_0^2 + \mathscr{H}'^2}$$

if

$$\frac{\mu}{\hbar}\sqrt{\mathscr{H}_0^2 + \mathscr{H}'^2}\,\frac{l}{v} = \frac{\pi}{2}. \tag{188.12}$$

Such a determination is, of course, possible only if a velocity selection is applied and v is well known. Our representation of the method has necessarily been simplified by neglecting detail like the deflection by the Lorentz force, field inhomogeneities used for focussing, stray fields and —most important—the non-static changes of the magnetic moment by the Zeeman effect. Our problem corresponds rather to the Paschen-Back effect of decoupled moments which, however, may contradict the idea of fields weak enough to allow perpendicular momentum transfer to be neglected.

NB. A particle "at the point $y=0$ at the time $t=0$, etc." should, of course, be described by a wave packet, cf. Problem 17. For the present purpose, however, this is of little avail and has therefore been omitted.

VI. The Relativistic Dirac Equation

Remark. In this chapter we use the fourth coordinate $x_4 = ict$ and Euclidian metric. Greek subscripts (e.g. x_μ) run over $\mu = 1, 2, 3, 4$, Latin subscripts (x_k) over $k = 1, 2, 3$, only.

Problem 189. Iteration of the Dirac equation

To derive commutation relations of the γ's from the relativistic dispersion law of plane Dirac waves. An irreducible representation of the γ's shall then be given that makes γ_4 diagonal.

Solution. If the Dirac equation for the force-free case,

$$\sum_\mu \gamma_\mu \partial_\mu \psi + \varkappa \psi = 0 \tag{189.1}$$

is to be solved by a plane wave

$$\psi = C\,\mathrm{e}^{i(\boldsymbol{k}\boldsymbol{r} - \omega t)} \tag{189.2}$$

the γ's must satisfy the algebraic relation

$$i \sum_\mu k_\mu \gamma_\mu + \varkappa = 0; \qquad k_4 = \frac{\omega}{c}. \tag{189.3}$$

Now the γ's must be independent of the special choice of the k_μ's. The latter can only be eliminated from (189.3) by using the relativistic dispersion law,

$$\sum_\mu k_\mu^2 \equiv k^2 - \frac{\omega^2}{c^2} = -\varkappa^2, \tag{189.4}$$

which can be constructed from (189.3) by iteration:

$$\varkappa^2 = \left(i \sum_\mu k_\mu \gamma_\mu \right)^2 = -\sum_\mu \sum_\nu k_\mu k_\nu \gamma_\mu \gamma_\nu. \tag{189.5}$$

The latter equation becomes identical with (189.4) if, and only if, the double sum is reduced to diagonal terms of appropriate normalization, in fact, if

$$\gamma_\mu \gamma_\nu + \gamma_\nu \gamma_\mu = 2\delta_{\mu\nu}. \tag{189.6}$$

If an analogous procedure is applied directly to the Dirac equation (189.1) without using plane waves, we have

$$\varkappa^2 \psi = \Big(\sum_\mu \gamma_\mu \partial_\mu\Big)^2 \psi = \sum_\mu \sum_\nu \gamma_\mu \gamma_\nu \partial_\mu \partial_\nu .$$

With the anticommutators (189.6) this simply leads to

$$\Box^2 \psi - \varkappa^2 \psi = 0, \tag{189.7}$$

i.e. to the Klein-Gordon equation.

There exist irreducible representations of the γ_μ's by 4×4 matrices. If γ_μ is one such representation, any unitary transformation $U^\dagger \gamma_\mu U$ will produce another set. Therefore, one of the matrices, say γ_4, may always be supposed to be diagonal. As $\gamma_4^2 = 1$, its eigenvalues must be $+1$ and -1. Thus we may start the construction of the matrix set by writing

$$\gamma_k = \begin{pmatrix} \boldsymbol{A}_k & \boldsymbol{B}_k \\ \boldsymbol{C}_k & \boldsymbol{D}_k \end{pmatrix} \quad (k=1,2,3); \qquad \gamma_4 = \begin{pmatrix} \boldsymbol{1} & \boldsymbol{0} \\ \boldsymbol{0} & -\boldsymbol{1} \end{pmatrix} \tag{189.8}$$

with all bold face letters meaning 2×2 matrices. From (189.6) we then find

$$\gamma_i \gamma_k + \gamma_k \gamma_i = \begin{pmatrix} \boldsymbol{B}_i \boldsymbol{C}_k + \boldsymbol{B}_k \boldsymbol{C}_i; & \boldsymbol{0} \\ \boldsymbol{0} & \boldsymbol{C}_i \boldsymbol{B}_k + \boldsymbol{C}_k \boldsymbol{B}_i \end{pmatrix} = 2\delta_{ik}; \tag{189.9a}$$

$$\gamma_k \gamma_4 + \gamma_4 \gamma_k = \begin{pmatrix} 2\boldsymbol{A}_k; & \boldsymbol{0} \\ \boldsymbol{0}; & -2\boldsymbol{D}_k \end{pmatrix} = 0; \tag{189.9b}$$

From (189.9b) we get $\boldsymbol{A}_k = 0$, $\boldsymbol{D}_k = 0$ for the three first matrices, and from (189.9a) it follows that

$$\boldsymbol{B}_k \boldsymbol{C}_k = 1; \qquad \boldsymbol{B}_i \boldsymbol{C}_k + \boldsymbol{C}_k \boldsymbol{B}_i = 0. \tag{189.10}$$

These are three anticommutation rules for 2×2 matrices allowing reduction to the Pauli matrices, σ_k (cf. Problem 129). If a and b are two numbers, Eqs. (189.10) are satisfied by

$$\boldsymbol{B}_k = a\sigma_k; \qquad \boldsymbol{C}_k = b\sigma_k; \qquad ab = 1, \tag{189.11}$$

so that any representation

$$\gamma_k = \begin{pmatrix} \boldsymbol{0}; & b^{-1}\sigma_k \\ b\sigma_k; & \boldsymbol{0} \end{pmatrix}; \qquad \gamma_4 = \begin{pmatrix} \boldsymbol{1} & \boldsymbol{0} \\ \boldsymbol{0} & -\boldsymbol{1} \end{pmatrix} \tag{189.12}$$

satisfies the commutation rules (189.6). The *standard representation* (often used in the following problems) is obtained by the choice $b=i$; the four matrices then become

$$\gamma_1=\begin{pmatrix}0&0&0&-i\\0&0&-i&0\\0&i&0&0\\i&0&0&0\end{pmatrix};\quad \gamma_2=\begin{pmatrix}0&0&0&-1\\0&0&1&0\\0&1&0&0\\-1&0&0&0\end{pmatrix};$$
$$\gamma_3=\begin{pmatrix}0&0&-i&0\\0&0&0&i\\i&0&0&0\\0&-i&0&0\end{pmatrix};\quad \gamma_4=\begin{pmatrix}1&0&0&0\\0&1&0&0\\0&0&-1&0\\0&0&0&-1\end{pmatrix}. \tag{189.13}$$

NB. With $b=a=1$ one obtains the set of matrices

$$\alpha_k=\begin{pmatrix}0&\sigma_k\\\sigma_k&0\end{pmatrix} \tag{189.14}$$

instead of the three γ_k's. Together with $\gamma_4\equiv\beta$ as above, they obey the same commutation laws. They are connected with the γ's by the relations

$$\gamma_k=-i\beta\alpha_k;\qquad \gamma_4=\beta. \tag{189.15}$$

The α's are used to advantage in the Dirac hamiltonian, cf. Problem 200.

Problem 190. Plane Dirac waves of positive energy

To determine in standard representation the spinor amplitudes of plane Dirac waves of positive and negative helicity, but of positive energy only.

Solution. With

$$\psi=C\,e^{i(\boldsymbol{k}\boldsymbol{r}-\omega t)} \tag{190.1}$$

we obtain [cf. Eq. (189.3)] in standard representation the four component equations

$$\begin{aligned}
(k_x-ik_y)C_4+k_zC_3+\left(-\frac{\omega}{c}+\varkappa\right)C_1&=0,\\
(k_x+ik_y)C_3-k_zC_4+\left(-\frac{\omega}{c}+\varkappa\right)C_2&=0,\\
-(k_x-ik_y)C_2-k_zC_1+\left(\frac{\omega}{c}+\varkappa\right)C_3&=0,\\
-(k_x+ik_y)C_1+k_zC_2+\left(\frac{\omega}{c}+\varkappa\right)C_4&=0.
\end{aligned} \tag{190.2}$$

Using the abbreviations

$$k\eta = \frac{\omega}{c} - \varkappa; \qquad \frac{k}{\eta} = \frac{\omega}{c} + \varkappa, \tag{190.3}$$

we have in η a suitable parameter in which to express the most important particle quantities, viz. its momentum

$$p = \hbar k = mc\frac{2\eta}{1-\eta^2}, \tag{190.4a}$$

its kinetic energy

$$E = \hbar\omega = mc^2\frac{1+\eta^2}{1-\eta^2} \tag{190.4b}$$

and its velocity

$$v = c\frac{2\eta}{1+\eta^2}. \tag{190.4c}$$

It is also useful to introduce two polar angles ϑ and φ, determining the direction of the vector $\boldsymbol{k}$,

$$k_x \pm i k_y = k \sin\vartheta \, e^{\pm i\varphi}; \qquad k_z = k\cos\vartheta. \tag{190.5}$$

The equations (190.2) may then be written in the form

$$\begin{aligned}
\sin\vartheta \, e^{-i\varphi} C_4 + \cos\vartheta \, C_3 - \eta \, C_1 &= 0,\\
\sin\vartheta \, e^{+i\varphi} C_3 - \cos\vartheta \, C_4 - \eta \, C_2 &= 0,\\
-\sin\vartheta \, e^{-i\varphi} C_2 - \cos\vartheta \, C_1 + \frac{1}{\eta} C_3 &= 0,\\
-\sin\vartheta \, e^{+i\varphi} C_1 + \cos\vartheta \, C_2 + \frac{1}{\eta} C_4 &= 0.
\end{aligned} \tag{190.6}$$

This is a homogeneous system of linear equations for the four C_μ's. Its determinant vanishes, as may be easily checked, but it can also be factorized in two factors each of which is zero. It is not therefore possible to express all the C_μ's as multiples of one of them, and there remain two C_μ's to be chosen arbitrarily. Therefore we shall proceed in another way.

We first look for eigenfunctions of the helicity operator

$$\mathsf{h} = \frac{1}{k} \sum_j \sigma_j k_j, \tag{190.7}$$

i.e. of the operator "spin component in the direction of $\boldsymbol{k}$". Since, in standard representation, the spin matrices consist merely of doubled Pauli 2×2 matrices s_j,

$$\sigma_j = \begin{pmatrix} s_j & 0 \\ 0 & s_j \end{pmatrix},$$

we find the definition (190.7):

$$\mathsf{h} = \begin{pmatrix} \cos\vartheta; & \sin\vartheta\, e^{-i\varphi}; & 0; & 0 \\ \sin\vartheta\, e^{i\varphi}; & -\cos\vartheta; & 0; & 0 \\ 0; & 0; & \cos\vartheta; & \sin\vartheta\, e^{-i\varphi} \\ 0; & 0; & \sin\vartheta\, e^{i\varphi}; & -\cos\vartheta \end{pmatrix}. \tag{190.8}$$

Let the eigenvalue be h; then the eigenvalue problem $\mathsf{h}\,C = h\cdot C$ can be decomposed into one pair of equations for C_1 and C_2,

$$\begin{aligned} C_1\cos\vartheta + C_2 \sin\vartheta\, e^{-i\varphi} &= h\,C_1; \\ C_1 \sin\vartheta\, e^{i\varphi} - C_2\cos\vartheta &= h\,C_2 \end{aligned} \tag{190.9}$$

and the same pair for C_3 and C_4. The determinant of (190.9) vanishes if $h = \pm 1$; thus we arrive at two solutions:

$h = +1$ (spin parallel to $\boldsymbol{k}$):

$$C_2 = \tan\frac{\vartheta}{2}\, e^{i\varphi} C_1; \qquad C_4 = \tan\frac{\vartheta}{2}\, e^{i\varphi} C_3 \tag{190.10}$$

and

$h = -1$ (spin antiparallel to $\boldsymbol{k}$):

$$C_2 = -\cot\frac{\vartheta}{2}\, e^{i\varphi} C_1; \qquad C_4 = -\cot\frac{\vartheta}{2}\, e^{i\varphi} C_3. \tag{190.11}$$

C_1 and C_3 may still be chosen arbitrarily.

Let us now put these results into the set of Eqs. (190.6). With the elementary identities

$$\sin\vartheta\tan\frac{\vartheta}{2} + \cos\vartheta = 1; \qquad \sin\vartheta\cot\frac{\vartheta}{2} - \cos\vartheta = 1;$$

$$\sin\vartheta - \cos\vartheta\tan\frac{\vartheta}{2} = \tan\frac{\vartheta}{2}; \qquad \sin\vartheta + \cos\vartheta\cot\frac{\vartheta}{2} = \cot\frac{\vartheta}{2}$$

all four equations (190.6) reduce to

$$C_3 = \eta C_1 \qquad \text{for } h = +1 \tag{190.12}$$

and to

$$C_3 = -\eta C_1 \quad \text{for } h = -1, \tag{190.13}$$

respectively. If, therefore, we normalize so that[1]

$$\int_V d^3x\, \psi^\dagger \psi = \int_V d^3x\, C^\dagger C = 1 \tag{190.14}$$

we obtain the following spinor amplitudes:

$$\text{for } h = +1 \qquad C_+ = \frac{1}{\sqrt{V(1+\eta^2)}} \begin{pmatrix} \cos\frac{\vartheta}{2}\, e^{-\frac{i}{2}\varphi} \\ \sin\frac{\vartheta}{2}\, e^{+\frac{i}{2}\varphi} \\ \eta\cos\frac{\vartheta}{2}\, e^{-\frac{i}{2}\varphi} \\ \eta\sin\frac{\vartheta}{2}\, e^{+\frac{i}{2}\varphi} \end{pmatrix} \tag{190.15}$$

and

$$\text{for } h = -1 \qquad C_- = \frac{1}{\sqrt{V(1+\eta^2)}} \begin{pmatrix} \sin\frac{\vartheta}{2}\, e^{-\frac{i}{2}\varphi} \\ -\cos\frac{\vartheta}{2}\, e^{+\frac{i}{2}\varphi} \\ -\eta\sin\frac{\vartheta}{2}\, e^{-\frac{i}{2}\varphi} \\ \eta\cos\frac{\vartheta}{2}\, e^{+\frac{i}{2}\varphi} \end{pmatrix}. \tag{190.16}$$

NB. The unrelativistic case, according to (190.4c), leads to $\eta \ll 1$. The components ψ_3 and ψ_4 of the spinor then may be neglected, and we fall back upon the two-component Pauli spin theory.

---

[1] It should be noted that this normalization is Lorentz-invariant, the integral being proportional to the electric charge total inside the volume V. Another Lorentz-invariant normalization often used is $\bar{\psi}\psi = 1$.

Problem 191. Transformation properties of a spinor

How does a spinor ψ transform under infinitesimal Lorentz transformation?

Solution. An infinitesimal Lorentz transformation is defined by

$$x'_\mu = x_\mu + \sum_\rho \varepsilon_{\mu\rho} x_\rho; \qquad \varepsilon_{\mu\rho} = -\varepsilon_{\rho\mu}; \qquad |\varepsilon_{\mu\rho}| \ll 1. \tag{191.1}$$

Here all ε_{kl} are real, and the three ε_{k4} purely imaginary. The Dirac equation

$$\sum_\mu \gamma_\mu D_\mu \psi + \varkappa \psi = 0 \tag{191.2}$$

shall be transformed into

$$\sum_\mu \gamma_\mu D'_\mu \psi' + \varkappa \psi' = 0 \tag{191.2'}$$

with unchanged coefficients γ_μ and $\varkappa$. The operators D_μ are four-vectors transforming under the same law as the coordinates (191.1):

$$D'_\mu = D_\mu + \sum_\rho \varepsilon_{\mu\rho} D_\rho; \tag{191.3}$$

the transformation formula of ψ may be

$$\psi' = (1+\xi)\psi \tag{191.4}$$

with an infinitesimal ξ linear in the $\varepsilon_{\mu\rho}$'s which is a Clifford number.

We start with Eq. (191.2′) in which we put D'_μ from (191.3) and ψ' from (191.4):

$$\sum_\mu \gamma_\mu \Big(D_\mu + \sum_\rho \varepsilon_{\mu\rho} D_\rho\Big)(1+\xi)\psi + \varkappa(1+\xi)\psi = 0.$$

If we multiply from the left side by $(1-\xi)$, the last term goes over into $\varkappa\psi$, i.e. into the last term of (191.2). The operator ξ therefore must be chosen in such a way as to make

$$\sum_\mu (1-\xi)\gamma_\mu \Big(D_\mu + \sum_\rho \varepsilon_{\mu\rho} D_\rho\Big)(1+\xi)\psi = \sum_\mu \gamma_\mu D_\mu \psi. \tag{191.5}$$

From this equation ξ shall be determined. Neglecting all second-order contributions, we find

$$\sum_\mu (\gamma_\mu \xi - \xi \gamma_\mu) D_\mu \psi + \sum_\mu \sum_\rho \varepsilon_{\mu\rho} \gamma_\mu D_\rho \psi = 0$$

or, exchanging the dummies μ and ρ in the double sum (and writing ν instead of ρ),

$$\sum_\mu \Big\{(\gamma_\mu \xi - \xi \gamma_\mu) - \sum_\nu \varepsilon_{\mu\nu} \gamma_\nu\Big\} D_\mu \psi = 0.$$

Since this relation is supposed to hold for any ψ, each sum term is bound to vanish separately:

$$\gamma_\mu \xi - \xi \gamma_\mu = \sum_\nu \varepsilon_{\mu\nu} \gamma_\nu . \tag{191.6}$$

The only Clifford number linear in the $\varepsilon_{\mu\nu}$'s satisfying these four commutation relations is

$$\xi = \tfrac{1}{4} \sum_\rho \sum_\sigma \varepsilon_{\rho\sigma} \gamma_\rho \gamma_\sigma . \tag{191.7}$$

This can easily be proved by direct evaluation of the commutators (191.6): We have

$$\gamma_\mu \xi - \xi \gamma_\mu = \tfrac{1}{4} \sum_\rho \sum_\sigma \varepsilon_{\rho\sigma} (\gamma_\mu \gamma_\rho \gamma_\sigma - \gamma_\rho \gamma_\sigma \gamma_\mu);$$

here we find

$$\gamma_\rho \gamma_\sigma \gamma_\mu = \gamma_\rho (-\gamma_\mu \gamma_\sigma + 2\delta_{\mu\sigma}) = (\gamma_\mu \gamma_\rho - 2\delta_{\mu\rho}) \gamma_\sigma + 2\gamma_\rho \delta_{\mu\sigma}$$

so that

$$\gamma_\mu \gamma_\rho \gamma_\sigma - \gamma_\rho \gamma_\sigma \gamma_\mu = 2(\gamma_\sigma \delta_{\mu\rho} - \gamma_\rho \delta_{\mu\sigma});$$

therefore

$$\begin{aligned} \gamma_\mu \xi - \xi \gamma_\mu &= \tfrac{1}{2} \sum_\rho \sum_\sigma \varepsilon_{\rho\sigma} (\gamma_\sigma \delta_{\mu\rho} - \gamma_\rho \delta_{\mu\sigma}) \\ &= \tfrac{1}{2} \Big(\sum_\sigma \varepsilon_{\mu\sigma} \gamma_\sigma - \sum_\rho \varepsilon_{\rho\mu} \gamma_\rho \Big) = \sum_\nu \varepsilon_{\mu\nu} \gamma_\nu , \end{aligned}$$

and that was to be proved.

Hence the spinor ψ transforms according to

$$\psi' = \psi + \tfrac{1}{4} \sum_\rho \sum_\sigma \varepsilon_{\rho\sigma} \gamma_\rho \gamma_\sigma \psi . \tag{191.8}$$

Problem 192. Lorentz covariants

Which Lorentz covariants of the form

$$G = \bar{\psi} \Gamma \psi ; \qquad \bar{\psi} = \psi^\dagger \gamma_4 \tag{192.1}$$

can be constructed, Γ being one of the 16 basis elements of the Clifford algebra?

Solution. The 16 basis elements of the algebra can be grouped into five sets as follows:

(1) $1,$

(2) $\gamma_1, \gamma_2, \gamma_3, \gamma_4,$

(3) $\gamma_1 \gamma_2, \gamma_1, \gamma_3, \gamma_1 \gamma_4, \gamma_2 \gamma_3, \gamma_2 \gamma_4, \gamma_3 \gamma_4,$

(4) $\gamma_2 \gamma_3 \gamma_4, \gamma_3 \gamma_4 \gamma_1, \gamma_4 \gamma_1 \gamma_2, \gamma_1 \gamma_2 \gamma_3,$

(5) $\gamma_1 \gamma_2 \gamma_3 \gamma_4 .$ (192.2)

Expressions of the form (192.1) will be formed with each of these five sets separately.

Before performing this programme in detail, however, let us investigate the transformation properties of any of the 16 quantities (192.1) under the infinitesimal Lorentz transformation

$$x'_\mu = x_\mu + \sum_\rho \varepsilon_{\mu\rho} x_\rho . \tag{192.3}$$

In the preceding problem it has been shown that ψ under this transformation becomes

$$\psi' = (1+\xi)\psi \tag{192.4a}$$

with

$$\xi = \tfrac{1}{4}\sum_\rho \sum_\sigma \varepsilon_{\rho\sigma}\gamma_\rho\gamma_\sigma ; \qquad \gamma_\mu \xi - \xi\gamma_\mu = \sum_\nu \varepsilon_{\mu\nu}\gamma_\nu . \tag{192.4b}$$

The transform of G is

$$G' = \psi'^\dagger \gamma_4 \Gamma \psi' = \psi^\dagger (1+\xi^\dagger)\gamma_4 \Gamma (1+\xi)\psi .$$

Formally we may write

$$G' = \bar{\psi}\Gamma'\psi ; \qquad \Gamma' = \gamma_4 (1+\xi^\dagger)\gamma_4 \Gamma (1+\xi) . \tag{192.5}$$

This can be further simplified when we know a little more about the hermitian conjugate $\xi^\dagger$. According to Eq. (192.4b), it should obey the commutation relation

$$\xi^\dagger \gamma_4 - \gamma_4 \xi^\dagger = \sum_k \varepsilon^*_{4k}\gamma_k$$

since $\gamma_\nu^\dagger = \gamma_\nu$. Now, the rotation angles ε_{4k} are purely imaginary so that $\varepsilon^*_{4k} = -\varepsilon_{4k}$, i. e.

$$\gamma_4 \xi^\dagger - \xi^\dagger \gamma_4 = \sum_k \varepsilon_{4k}\gamma_k$$

and

$$\gamma_4 \xi^\dagger \gamma_4 = \gamma_4 \Big\{ \gamma_4 \xi^\dagger - \sum_k \varepsilon_{4k}\gamma_k \Big\} = \xi^\dagger - \sum_k \varepsilon_{4k}\gamma_4\gamma_k . \tag{192.6}$$

It further follows from (192.4b) with $\varepsilon^*_{kl} = \varepsilon_{kl}$ (real rotations in 3-space) that

$$\xi = \tfrac{1}{4}\sum_k \sum_l \varepsilon_{kl}\gamma_k\gamma_l + \tfrac{1}{2}\sum_k \varepsilon_{k4}\gamma_k\gamma_4 \tag{192.7a}$$

has the hermitian conjugate

$$\xi^\dagger = -\tfrac{1}{4}\sum_k \sum_l \varepsilon_{kl}\gamma_k\gamma_l + \tfrac{1}{2}\sum_k \varepsilon_{k4}\gamma_k\gamma_4 \tag{192.7b}$$

so that

$$\xi + \xi^\dagger = \sum_k \varepsilon_{k4}\gamma_k\gamma_4 . \tag{192.8}$$

Putting (192.8) into (192.6) we obtain

$$\gamma_4 \xi^\dagger \gamma_4 = -\xi. \tag{192.9}$$

With this last result, the "transformed operator" Γ', Eq. (192.5), may finally be written

$$\Gamma' = \Gamma + (\Gamma \xi - \xi \Gamma). \tag{192.10}$$

We now may easily apply this simple result to the five sets of quantities defined by (192.2), consecutively.

1. With $\Gamma = 1$, Eq. (192.10) gives immediately $\Gamma' = 1$ so that we find

$$G = \bar{\psi}\psi \to G' = \bar{\psi}\psi; \qquad G' = G. \tag{192.11}$$

The quantity G therefore behaves as a *scalar.*

2. With $\Gamma = \gamma_\mu$, Eqs. (192.10) and (192.4b) yield

$$\Gamma' = \gamma_\mu + \sum_\nu \varepsilon_{\mu\nu} \gamma_\nu$$

so that we arrive at the transformation formulae

$$G_\mu = \bar{\psi} \gamma_\mu \psi \to G'_\mu = G_\mu + \sum_\nu \varepsilon_{\mu\nu} G_\nu, \tag{192.12}$$

i. e. the G_μ's transform as the components of a *vector.*

3. It is suitable first to decompose the products $\gamma_\mu \gamma_\nu$ into a symmetrical and an antisymmetrical part,

$$\gamma_\mu \gamma_\nu = \tfrac{1}{2}(\gamma_\mu \gamma_\nu + \gamma_\nu \gamma_\mu) + \tfrac{1}{2}(\gamma_\mu \gamma_\nu - \gamma_\nu \gamma_\mu).$$

The first part reduces to $\delta_{\mu\nu}$, i. e. to the scalar (192.11) multiplied by the unit tensor. New evidence apparently comes only from the antisymmetrical part; so we confine our discussion to the combination

$$\sigma_{\mu\nu} = \tfrac{1}{2}(\gamma_\mu \gamma_\nu - \gamma_\nu \gamma_\mu). \tag{192.13}$$

According to (192.4b) we have

$$\xi \gamma_\mu \gamma_\nu = \Big(\gamma_\mu \xi - \sum_\rho \varepsilon_{\mu\rho} \gamma_\rho\Big)\gamma_\nu = \gamma_\mu \Big(\gamma_\nu \xi - \sum_\rho \varepsilon_{\nu\rho} \gamma_\rho\Big) - \sum_\rho \varepsilon_{\mu\rho} \gamma_\rho \gamma_\nu$$

or

$$\gamma_\mu \gamma_\nu \xi - \xi \gamma_\mu \gamma_\nu = \sum_\rho (\varepsilon_{\nu\rho} \gamma_\mu \gamma_\rho + \varepsilon_{\mu\rho} \gamma_\rho \gamma_\nu)$$

so that the transformation formula runs as follows:

$$G_{\mu\nu} = \bar{\psi} \sigma_{\mu\nu} \psi \to G'_{\mu\nu} = G_{\mu\nu} + \sum_\rho (\varepsilon_{\mu\rho} G_{\rho\nu} + \varepsilon_{\nu\rho} G_{\mu\rho}). \tag{192.14}$$

This is the behaviour of a *tensor* (of rank 2) under infinitesimal rotation.

4. The products of three γ's can be written in a simpler way by introducing the Clifford number

$$\gamma_5 = \gamma_1 \gamma_2 \gamma_3 \gamma_4. \tag{192.15}$$

Then the four products become

$$\gamma_2\gamma_3\gamma_4=\gamma_1\gamma_5; \quad -\gamma_3\gamma_4\gamma_1=\gamma_2\gamma_5; \quad \gamma_4\gamma_1\gamma_2=\gamma_3\gamma_5; \quad -\gamma_1\gamma_2\gamma_3=\gamma_4\gamma_5.$$

Since γ_5 anticommutes with all four γ_μ,

$$\gamma_\mu\gamma_5+\gamma_5\gamma_\mu=0, \tag{192.16}$$

it will commute with ξ so that Eq. (192.10) applied to $\Gamma=\gamma_\mu\gamma_5$ leads to

$$\Gamma'=\gamma_\mu\gamma_5+(\gamma_\mu\gamma_5\xi-\xi\gamma_\mu\gamma_5)=[\gamma_\mu+(\gamma_\mu\xi-\xi\gamma_\mu)]\gamma_5.$$

Thus we essentially fall back upon case 2 and find the transformation laws of the components of a vector:

$$G_\mu=\bar{\psi}\gamma_\mu\gamma_5\psi\rightarrow G'_\mu=G_\mu+\sum_\nu \varepsilon_{\mu\nu}G_\nu. \tag{192.17}$$

Strictly speaking, this is not a (polar) vector but a *pseudovector*, as will be shown in the following problem.

5. According to (192.15) and (192.16) we obtain

$$G=\bar{\psi}\gamma_5\psi\rightarrow G'=G, \tag{192.18}$$

i. e. for this last combination there holds the transformation law of a scalar. We shall see in the next problem that the quantity is more correctly classified a *pseudoscalar*.

Problem 193. Parity transformation

How do the five Lorentz covariants of the preceding problem transform under reflection of space coordinates (i. e. under parity transformation)?

Solution. We start by investigating the behaviour of the spinor ψ under the reflection process under consideration. It is defined by the postulate that the Dirac equation

$$\sum_\mu \gamma_\mu D_\mu\psi+\varkappa\psi=0 \tag{193.1}$$

shall transform into

$$\sum_\mu \gamma_\mu D'_\mu\psi'+\varkappa\psi'=0 \tag{193.1'}$$

under the parity transformation

$$x'_k=-x_k; \quad x'_4=x_4. \tag{193.2}$$

The same relations as (193.2) will hold for the operators ∂_μ. A more detailed investigation is necessary of the vector potential. The electric field components $\mathscr{E}_k$ are coupled with the components A_μ by the relations

$$\mathscr{E}_k = -\frac{1}{c}\dot{A}_k - \partial_k \Phi = -i(\partial_4 A_k - \partial_k A_4).$$

Since the electric field is a polar 3-vector, it changes sign with the x_k's under transformation. This leads to

$$A'_k = -A_k; \quad A'_4 = A_4. \tag{193.3}$$

Thus A_μ undergoes the same transformation as ∂_μ and, in consequence thereof, D_μ transforms in the same way. Therefore, instead of Eq. (1') we may write

$$-\sum_k \gamma_k D_k \psi' + \gamma_4 D_4 \psi' + \varkappa \psi' = 0.$$

It is immediately seen that with

$$\psi' = \gamma_4 \psi \tag{193.4}$$

this leads back to (193.1), thus determining the parity transformation of a spinor.

Any quantity

$$G = \overline{\psi} \Gamma \psi = \psi^\dagger \gamma_4 \Gamma \psi$$

then transforms into

$$G' = \psi'^\dagger \gamma_4 \Gamma \psi' = \psi^\dagger \Gamma \gamma_4 \psi = \overline{\psi} \gamma_4 \Gamma \gamma_4 \psi$$

so that we may write

$$G = \overline{\psi} \Gamma \psi \to G' = \overline{\psi} \Gamma' \psi; \quad \Gamma' = \gamma_4 \Gamma \gamma_4. \tag{193.5}$$

Applied to the five covariants of the preceding problem we have

$$(1) \quad G = \overline{\psi}\psi; \quad \Gamma = 1; \quad \Gamma' = 1; \quad G' = G; \tag{193.6}$$

$$(5) \quad G = \overline{\psi}\gamma_5\psi; \quad \Gamma = \gamma_5; \quad \Gamma' = \gamma_4\gamma_5\gamma_4 = -\gamma_5; \quad G' = -G. \tag{193.7}$$

The behaviour of these two quantities was the same under rotation, but it now is opposite under space reflection, (1) being called a (genuine) *scalar* and (5) a *pseudoscalar*.

$$(2) \quad G_\mu = \overline{\psi}\gamma_\mu\psi; \quad \Gamma = \gamma_\mu; \quad \begin{aligned} \Gamma'_k &= \gamma_4\gamma_k\gamma_4 = -\gamma_k; & G'_k &= -G_k; \\ \Gamma'_4 &= \gamma_4 & G'_4 &= +G_4. \end{aligned} \tag{193.8}$$

$$(4) \quad G_\mu = \overline{\psi}\gamma_\mu\gamma_5\psi; \quad \Gamma = \gamma_\mu\gamma_5;$$

$$\begin{aligned} \Gamma'_k &= \gamma_4\gamma_k\gamma_5\gamma_4 = +\gamma_k\gamma_5; & G'_k &= +G_k; \\ \Gamma'_4 &= \gamma_5\gamma_4 = -\gamma_4\gamma_5; & G'_4 &= -G_4. \end{aligned} \tag{193.9}$$

Both quantities behave as 4-vectors under rotation, but reversely under space reflection, (2) being called a (polar or genuine) *vector* and (4) an axial or *pseudovector*.

(3)
$$G_{\mu\nu} = \bar{\psi}\sigma_{\mu\nu}\psi; \quad \Gamma = \tfrac{1}{2}(\gamma_\mu\gamma_\nu - \gamma_\nu\gamma_\mu);$$
$$\Gamma'_{kl} = \Gamma_{kl}; \quad G'_{kl} = G_{kl}$$
$$\Gamma'_{k4} = -\Gamma_{k4}; \quad G'_{k4} = -G_{k4}. \tag{193.10}$$

Since there is only one tensor, no further classification is necessary.

Problem 194. Charge conjugation

To construct from the spinor ψ solving the Dirac equation for a particle of charge e the charge conjugate spinor ψ_c describing the behaviour of a particle of the opposite charge $-e$.

Solution. Let us use the abbreviation $a_\mu = (e/\hbar c)A_\mu$ with A_μ the electromagnetic 4-potential in this problem. Then the Dirac equation for the particle of charge e runs

$$\sum_\mu \gamma_\mu(\partial_\mu - i a_\mu)\psi + \varkappa\psi = 0. \tag{194.1}$$

The equation to be constructed for the opposite charge then must be

$$\sum_\mu \gamma_\mu(\partial_\mu + i a_\mu)\psi_c + \varkappa\psi_c = 0. \tag{194.2}$$

The charge conjugate ψ_c which solves the latter equation shall be connected with the solution ψ of (194.1).

The operator $\partial_\mu + i a_\mu$ occurring in (194.2) can be introduced by using the adjoint equation of (194.1),

$$\sum_\mu (\partial_\mu + i a_\mu)\bar{\psi}\gamma_\mu - \varkappa\bar{\psi} = 0; \qquad \bar{\psi} = \psi^\dagger\gamma_4. \tag{194.3}$$

Transposing the latter again, we find

$$\sum_\mu \tilde{\gamma}_\mu(\partial_\mu + i a_\mu)\tilde{\bar{\psi}} - \varkappa\tilde{\bar{\psi}} = 0$$

where

$$\tilde{\bar{\psi}} = \tilde{\gamma}_4\psi^*. \tag{194.4}$$

If we multiply the last equation with a Clifford number C,

$$\sum_\mu C\tilde{\gamma}_\mu(\partial_\mu + i a_\mu)\tilde{\gamma}_4\psi^* - \varkappa C\tilde{\gamma}_4\psi^* = 0,$$

the result becomes identical with (194.2) if we choose C so that the two relations

$$-C\tilde{\gamma}_\mu\tilde{\gamma}_4\psi^* = \gamma_\mu\psi_c; \qquad C\tilde{\gamma}_4\psi^* = \psi_c \tag{194.5}$$

are satisfied. Here ψ_c may be eliminated in order first to determine C:

$$-C\tilde{\gamma}_\mu\tilde{\gamma}_4\psi^* = \gamma_\mu C\tilde{\gamma}_4\psi^*$$

where we omit $\tilde{\gamma}_4\psi^*$ and are left with the four relations

$$\gamma_\mu C = -C\tilde{\gamma}_\mu \tag{194.6}$$

to determine C.

Since the problem is homogeneous, there is of course always an arbitrary factor in ψ_c. It is reasonable to fix it as far as possible by postulating that charge conjugation shall not alter the normalization,

$$\psi_c^\dagger\psi_c = \psi^\dagger\psi = (\psi^\dagger\psi)^*. \tag{194.7}$$

Now, from (194.5) we have

$$\psi_c^\dagger = \tilde{\psi}\gamma_4^* C^\dagger,$$

hence

$$\psi_c^\dagger\psi_c = \tilde{\psi}\gamma_4^* C^\dagger C\tilde{\gamma}_4\psi^* = (\psi^*)^\dagger(\gamma_4^* C^\dagger C\tilde{\gamma}_4)(\psi^*),$$

and that becomes identical with (194.7) if

$$\gamma_4^* C^\dagger C\tilde{\gamma}_4 = 1$$

or, since $\gamma_4^\dagger = \gamma_4$, $\tilde{\gamma}_4 = \gamma_4^*$,

$$C^\dagger C = 1. \tag{194.8}$$

It follows that C is an unitary operator.

Specializing to standard representation we note that

$$\tilde{\gamma}_1 = -\gamma_1; \quad \tilde{\gamma}_2 = +\gamma_2; \quad \tilde{\gamma}_3 = -\gamma_3; \quad \tilde{\gamma}_4 = +\gamma_4. \tag{194.9}$$

Thence Eq. (194.6) leads to C commuting with γ_1 and γ_3, but anti-commuting with γ_2 and γ_4. This will be performed by

$$C = \gamma_2\gamma_4 \tag{194.10}$$

which is the only one of the 16 basis elements of the Clifford algebra satisfying the four conditions (194.6). It may be noted that from (194.10) we find

$$C^\dagger = -C; \quad C^2 = -1. \tag{194.11}$$

It then follows from (194.5) that the charge conjugate wave function in standard representation is

$$\psi_c = \gamma_2\psi^*. \tag{194.12}$$

Problem 195. Mixed helicity states

A plane Dirac wave runs in z direction. It shall be shown that it is impossible to construct a spinor amplitude which simultaneously makes ψ an eigenfunction of σ_x.

Solution. a) For a plane wave,

$$\psi = C\,e^{i(kz-\omega t)} \tag{195.1}$$

there follows, from the Dirac equation, the algebraic relation

$$\Omega C \equiv \left(ik\gamma_3 - \frac{\omega}{c}\gamma_4 + \varkappa\right) C = 0 \tag{195.2}$$

for the spinor amplitude C. The operator Ω defined by (195.2), however, does not commute with

$$\sigma_x = -i\gamma_2\gamma_3 \tag{195.3}$$

since

$$\sigma_x \Omega = +k\gamma_2 + i\frac{\omega}{c}\gamma_2\gamma_3\gamma_4 - i\varkappa\gamma_2\gamma_3$$

but

$$\Omega\sigma_x = -k\gamma_2 + i\frac{\omega}{c}\gamma_2\gamma_3\gamma_4 - i\varkappa\gamma_2\gamma_3 .$$

Therefore ψ cannot be eigenfunction to both operators.

b) In standard representation Eq. (195.2) would run as follows:

$$\begin{aligned} kC_3 + \left(-\frac{\omega}{c} + \varkappa\right) C_1 &= 0; \\ -kC_4 + \left(-\frac{\omega}{c} + \varkappa\right) C_2 &= 0; \\ -kC_1 + \left(\frac{\omega}{c} + \varkappa\right) C_3 &= 0; \\ kC_2 + \left(\frac{\omega}{c} + \varkappa\right) C_4 &= 0. \end{aligned} \tag{195.2'}$$

With the abbreviations

$$\frac{\omega}{c} - \varkappa = k\eta; \qquad \frac{\omega}{c} + \varkappa = \frac{k}{\eta} \tag{195.3'}$$

this leads to

$$C_3 = \eta C_1; \qquad C_4 = -\eta C_2 . \tag{195.4}$$

On the other hand, the eigenvalue problem

$$\sigma_x C = \lambda \cdot C \tag{195.5}$$

with λ being an eigenvalue demands

$$\sigma_x C = \begin{pmatrix} 0 & 1 & 0 & 0 \\ 1 & 0 & 0 & 0 \\ 0 & 0 & 0 & 1 \\ 0 & 0 & 1 & 0 \end{pmatrix} \begin{pmatrix} C_1 \\ C_2 \\ C_3 \\ C_4 \end{pmatrix} = \begin{pmatrix} C_2 \\ C_1 \\ C_4 \\ C_3 \end{pmatrix} = \begin{pmatrix} \lambda C_1 \\ \lambda C_2 \\ \lambda C_3 \\ \lambda C_4 \end{pmatrix}$$

so that

$$C_2 = \lambda C_1; \quad C_1 = \lambda C_2$$

and

$$C_4 = \lambda C_3; \quad C_3 = \lambda C_4 . \tag{195.6}$$

Both pairs of equations can only be satisfied for $\lambda = \pm 1$. Using (195.6) we may eliminate C_2 and C_4 from Eq. (195.4) thus arriving at

$$C_3 = \eta C_1 \quad \text{and} \quad \lambda C_3 = -\eta \lambda C_1 ,$$

two relations which contradict each other. Therefore the spinor C satisfying (195.4) cannot simultaneously satisfy (195.6) as had to be proved.

NB. In the unrelativistic limit $\eta \to 0$, the amplitudes C_3 and C_4 and hence the second pair of equations (195.6) drop out so that no contradiction remains.

Problem 196. Spin expectation value

The expectation value of σ_x shall be calculated for a superposition of two plane waves in z direction having opposite helicities.

Solution. With the spinor amplitudes [cf. Eqs. (190.15, 16) for $\vartheta = 0$]

$$C_+ = \frac{1}{\sqrt{V(1+\eta^2)}} \begin{pmatrix} 1 \\ 0 \\ \eta \\ 0 \end{pmatrix} \quad \text{and} \quad C_- = \frac{1}{\sqrt{V(1+\eta^2)}} \begin{pmatrix} 0 \\ -1 \\ 0 \\ \eta \end{pmatrix} \tag{196.1}$$

for positive and negative helicities, respectively, we construct the mixed state amplitude,

$$C = C_+ \cos\alpha \, e^{i\beta} + C_- \sin\alpha \, e^{-i\beta} \tag{196.2}$$

of the same normalization,

$$\int d^3x \, C^\dagger C = 1 \tag{196.3}$$

with arbitrary but real constants α and β. The expectation value of σ_x is then defined by

$$\langle\sigma_x\rangle = \int C^\dagger \sigma_x C d^3x. \tag{196.4}$$

From

$$\sqrt{V(1+\eta^2)}\;\sigma_x C_+ = \begin{pmatrix} 0 \\ 1 \\ 0 \\ \eta \end{pmatrix}; \qquad \sqrt{V(1+\eta^2)}\;\sigma_x C_- = \begin{pmatrix} -1 \\ 0 \\ \eta \\ 0 \end{pmatrix}$$

we obtain the products

$$C_+^\dagger \sigma_x C_+ = 0; \qquad C_-^\dagger \sigma_x C_+ = -\frac{1}{V}\frac{1-\eta^2}{1+\eta^2};$$

$$C_+^\dagger \sigma_x C_- = -\frac{1}{V}\frac{1-\eta^2}{1+\eta^2}; \qquad C_-^\dagger \sigma_x C_- = 0.$$

Then,

$$\langle\sigma_x\rangle = -\cos\alpha\sin\alpha(e^{2i\beta}+e^{-2i\beta})\frac{1-\eta^2}{1+\eta^2}$$

or

$$\langle\sigma_x\rangle = -\sin 2\alpha\cos 2\beta\,\frac{1-\eta^2}{1+\eta^2}. \tag{196.5}$$

The absolute value of the expectation value of σ_x therefore always turns out to be smaller than 1. In the extreme relativistic case where η approches unity, σ_x becomes very small so that almost complete orientation of the spin parallel or antiparallel to the direction of propagation is obtained. In the unrelativistic limit, on the other hand, where η is very small, polarization perpendicular to the direction of propagation becomes possible with $\beta=0$ and $\alpha=\mp\frac{\pi}{4}$ leading to $\langle\sigma_x\rangle=\pm 1$.

Problem 197. Algebraic properties of a Dirac wave spinor

Given a potential $V(z)$. The wave spinor of a state, with spin in either positive or negative z direction, may not depend upon x and y (one-dimensional problem). As far as possible, the Clifford algebra shall be used without recourse to matrix representations. There will then remain four functions of z satisfying a set of coupled differential equations. They shall finally be expressed by the four component wave functions in standard representation.

Solution. The wave spinor may be written

$$\psi(z,t) = e^{-iEt/\hbar} u(z) \tag{197.1}$$

where the spinor $u(z)$ satisfies the one-dimensional Dirac equation

$$\gamma_3 \frac{du}{dz} + \gamma_4 Q(z) u + \varkappa u = 0; \quad Q(z) = \frac{V(z) - E}{\hbar c}. \tag{197.2}$$

Since this equation is entirely built up within a sub-body of which the Clifford numbers $1, \gamma_3, \gamma_4, \gamma_3\gamma_4$ form the basis, it should be solved by a spinor of the form

$$v(z) = A(z) + B(z)\gamma_3 + C(z)\gamma_4 + D(z)\gamma_3\gamma_4. \tag{197.3}$$

Of course, if v solves the Dirac equation (197.2), so does any spinor

$$u = v\Gamma \tag{197.4}$$

with Γ any constant Clifford number, including elements formed with γ_1 and γ_2. On the other hand, v commutes with the spin operator

$$\sigma_z = -i\gamma_1\gamma_2 \tag{197.5}$$

of which it yet is no eigenspinor. The extension (197.4), however, permits the solution of the Dirac equation to be made an eigenspinor of σ_z. We find

$$\sigma_z u = \sigma_z v\Gamma = v\sigma_z \Gamma.$$

So, if Γ is any eigenspinor of σ_z,

$$\sigma_z \Gamma = \pm \Gamma, \tag{197.6}$$

we arrive at

$$\sigma_z u = \pm u. \tag{197.7}$$

The two eigenvalues $+1$ and -1 are called *helicities* (cf. Problem 190). Now it can easily be seen that

$$\Gamma_+ = 1 - i\gamma_1\gamma_2 = 1 + \sigma_z \tag{197.8a}$$

and

$$\Gamma_- = 1 + i\gamma_1\gamma_2 = 1 - \sigma_z \tag{197.8b}$$

are such eigenspinors with eigenvalues ± 1 of σ_z:

$$\sigma_z \Gamma_\pm = \sigma_z(1 \pm \sigma_z) = \sigma_z \pm 1 = \pm(1 \pm \sigma_z) = \pm \Gamma_\pm.$$

Our argument thus leads to

$$u(z) = v(z)(1 \mp i\gamma_1\gamma_2) \tag{197.9}$$

where $v(z)$ still remains to be determined by setting it into the Dirac equation (197.2). A simple calculation along these lines then leads to

$$(B'+QC+\varkappa A)+\gamma_3(A'-QD+\varkappa B)+\gamma_4(D'+QA+\varkappa C)$$
$$+\gamma_3\gamma_4(C'-QB+\varkappa D)=0 \qquad (197.10)$$

with the prime denoting differentiation. This expression is zero if, and only if, each of the four brackets vanishes. Thus we find the four functions A, B, C, D to satisfy a set of coupled differential equations.

$$B'+QC+\varkappa A=0; \qquad A'-QD+\varkappa B=0;$$
$$D'+QA+\varkappa C=0; \qquad C'-QB+\varkappa D=0. \qquad (197.11)$$

These equations become even simpler if combined into two pairs,

$$(B-D)'+(\varkappa-Q)(A-C)=0;$$
$$(A-C)'+(\varkappa+Q)(B-D)=0 \qquad (197.12\text{a})$$

and

$$(B+D)'+(\varkappa+Q)(A+C)=0;$$
$$(A+C)'+(\varkappa-Q)(B+D)=0, \qquad (197.12\text{b})$$

so that the first pair of equations, (197.12a), only connects the two functions

$$w_1=\tfrac{1}{2}(B-D); \qquad w_3=\tfrac{1}{2}(A-C) \qquad (197.13\text{a})$$

with one another and the other pair, (197.12b), the two functions

$$w_2=\tfrac{1}{2}(A+C); \qquad w_4=\tfrac{1}{2}(B+D). \qquad (197.13\text{b})$$

Putting these into $v(z)$, Eq. (197.3), we finally find

$$v(z)=(w_2+w_4\gamma_3)(1+\gamma_4)+(w_3+w_1\gamma_3)(1-\gamma_4). \qquad (197.14)$$

Each of the two terms in (197.14) *separately* satisfies the Dirac equation (197.2) if Eqs. (197.12a, b) are satisfied. Multiplication on the right-hand side of each term with either Γ_+ or Γ_-, Eqs. (197.8a, b), makes it an eigenspinor of σ_z too.

It remains to show how the four functions w_μ are connected with the four component wave functions u_μ of the standard representation. In this matrix description the Dirac equation (197.2) consists of the following component equations:

$$\left.\begin{aligned}-iu_3'+(Q+\varkappa)u_1&=0,\\ iu_4'+(Q+\varkappa)u_2&=0,\\ iu_1'+(\varkappa-Q)u_3&=0,\\ -iu_2'+(\varkappa-Q)u_4&=0.\end{aligned}\right\} \qquad (197.15)$$

Comparing this set with (197.12a, b) we are led to identify

$$u_1 = w_1; \quad u_3 = i w_3; \quad u_2 = i w_2; \quad u_4 = w_4 \tag{197.16}$$

or

$$A = -i(u_2 + u_3); \quad B = u_4 + u_1;$$
$$C = -i(u_2 - u_3); \quad D = u_4 - u_1. \tag{197.17}$$

For $u_2 = u_4 = 0$, the helicity will be $+1$, for $u_1 = u_3 = 0$, it will be -1.

Problem 198. Current in algebraic formulation

To determine the components of the electrical current for the eigenspinor

$$u(z) = (w_3 + w_1 \gamma_3)(1 - \gamma_4)(1 - i \gamma_1 \gamma_2) \tag{198.1}$$

of the preceding problem.

Solution. The components of the electrical four-current are defined by

$$s_\mu = i e c \bar{u} \gamma_\mu u; \quad \bar{u} = u^\dagger \gamma_4 \tag{198.2}$$

where, in our example,

$$u^\dagger = (1 - i \gamma_1 \gamma_2)(1 - \gamma_4)(w_3^* + w_1^* \gamma_3) \tag{198.3}$$

since the Clifford numbers $i\gamma_1\gamma_2, \gamma_4, \gamma_3$ are hermitian operators. Thus we obtain

$$s_\mu = i e c (1 - i \gamma_1 \gamma_2)(1 - \gamma_4)(w_3^* + w_1^* \gamma_3) \gamma_4 \gamma_\mu (w_3 + w_1 \gamma_3)(1 - \gamma_4)(1 - i \gamma_1 \gamma_2). \tag{198.4}$$

For the components s_1 and s_2 the Clifford number γ_μ may be shifted through two places towards the end of the expression, whereas γ_4 may be shifted one place towards its front:

$$s_{1,2} = i e c (1 - i \gamma_1 \gamma_2)(\gamma_4 - 1)(w_3^* - w_1^* \gamma_3)(w_3 - w_1 \gamma_3)(1 + \gamma_4) \gamma_{1,2} (1 - i \gamma_1 \gamma_2).$$

The operator $1 - i\gamma_1\gamma_2$ commutes with γ_3 as well as with γ_4, so that

$$s_{1,2} = i e c (\gamma_4 - 1)\{(|w_3|^2 + |w_1|^2) - (w_1^* w_3 + w_3^* w_1)\gamma_3\}(1 + \gamma_4)(1 - i \gamma_1 \gamma_2)\gamma_{1,2}(1 - i \gamma_1 \gamma_2).$$

The last three factors give either, for $\mu = 1$,

$$(1 - i \gamma_1 \gamma_2)(\gamma_1 - i \gamma_2) = \gamma_1 + i \gamma_2 - i \gamma_2 - \gamma_1 = 0$$

or, for $\mu = 2$,

$$(1 - i \gamma_1 \gamma_2)(\gamma_2 + i \gamma_1) = \gamma_2 - i \gamma_1 + i \gamma_1 - \gamma_2 = 0.$$

Therefore, as was to be expected, no current components exist perpendicular to the z direction.

For s_3 we may write in a similar way,

$$s_3 = iec(\gamma_4 - 1)(w_3^* - w_1^* \gamma_3)(w_3 \gamma_3 + w_1)(1-\gamma_4)(1 - i\gamma_1\gamma_2)^2.$$

The square at the end of this expression gives

$$(1 - i\gamma_1\gamma_2)^2 = 2(1 - i\gamma_1\gamma_2) \tag{198.5}$$

so that we find

$$s_3 = 2iec(\gamma_4 - 1)\{(w_3^* w_1 - w_1^* w_3) + (|w_3|^2 - |w_1|^2)\gamma_3\}(1-\gamma_4)(1 - i\gamma_1\gamma_2).$$

Shifting the front factor $(\gamma_4 - 1)$ one place towards the right, we get

$$\begin{aligned} s_3 = 2iec\{&(w_1^* w_3 - w_3^* w_1)(1-\gamma_4) \\ &+ (|w_1|^2 - |w_3|^2)\gamma_3(1+\gamma_4)\}(1-\gamma_4)(1 - i\gamma_1\gamma_2). \end{aligned}$$

Since

$$(1+\gamma_4)(1-\gamma_4) = 0; \quad (1-\gamma_4)^2 = 2(1-\gamma_4) \tag{198.6}$$

the second term in the curly bracket does not contribute and we arrive at

$$s_3 = 4iec(w_1^* w_3 - w_3^* w_1)(1-\gamma_4)(1 - i\gamma_1\gamma_2). \tag{198.7}$$

Finally, we get in a similar fashion

$$\begin{aligned} s_4 &= iec(1 - i\gamma_1\gamma_2)(1-\gamma_4)(w_3^* + w_1^*\gamma_3)(w_3 + w_1\gamma_3)(1-\gamma_4)(1 - i\gamma_1\gamma_2) \\ &= 2iec(1-\gamma_4)\{(|w_1|^2 + |w_3|^2) + (w_1^* w_3 + w_3^* w_1)\gamma_3\}(1-\gamma_4)(1 - i\gamma_1\gamma_2) \\ &= 4iec(|w_1|^2 + |w_3|^2)(1-\gamma_4)(1 - i\gamma_1\gamma_2). \end{aligned} \tag{198.8}$$

The expressions for s_3 and s_4 are still Clifford numbers but of the same shape. They are to be compared with the normalization expression,

$$\bar{u}u = (1 - i\gamma_1\gamma_2)(1-\gamma_4)(w_3^* + w_1^*\gamma_3)\gamma_4(w_3 + w_1\gamma_3)(1-\gamma_4)(1 - i\gamma_1\gamma_2)$$

which by the same procedure may be brought into the form

$$\bar{u}u = 4(|w_1|^2 - |w_3|^2)(1-\gamma_4)(1 - i\gamma_1\gamma_2). \tag{198.9}$$

Gathering up the results, except for a common factor

$$\Gamma = 4(1-\gamma_4)(1 - i\gamma_1\gamma_2),$$

we just have simple c-number expressions for the current density in z direction,

$$s_3 = iec(w_1^* w_3 - w_3^* w_1)\Gamma, \tag{198.10}$$

for the charge density ρ following from $s_4 = ic\rho$,

$$\rho = e(|w_1|^2 + |w_3|^2)\Gamma \tag{198.11}$$

and for the normalization expression

$$\bar{u}u = (|w_1|^2 - |w_3|^2)\,\Gamma. \tag{198.12}$$

Using the standard components, $u_1 = w_1$ and $u_3 = i w_3$, introduced in the preceding problem, these expression may as well be written

$$s_3 = ec(u_1^* u_3 + u_3^* u_1)\,\Gamma; \quad \rho = e(|u_1|^2 + |u_3|^2)\,\Gamma;$$
$$\bar{u}u = (|u_1|^2 - |u_3|^2)\,\Gamma. \tag{198.13}$$

It should be noted that in standard representation the operator Γ becomes very simple. We have

$$1-\gamma_4 = \begin{pmatrix} 0 & 0 & 0 & 0 \\ 0 & 0 & 0 & 0 \\ 0 & 0 & 2 & 0 \\ 0 & 0 & 0 & 2 \end{pmatrix}; \qquad 1 - i\gamma_1\gamma_2 = 1 + \sigma_3 = \begin{pmatrix} 2 & 0 & 0 & 0 \\ 0 & 0 & 0 & 0 \\ 0 & 0 & 2 & 0 \\ 0 & 0 & 0 & 0 \end{pmatrix};$$

thence we find the product

$$\Gamma = 16 \begin{pmatrix} 0 & 0 & 0 & 0 \\ 0 & 0 & 0 & 0 \\ 0 & 0 & 1 & 0 \\ 0 & 0 & 0 & 0 \end{pmatrix} \tag{198.14}$$

a matrix which, in diagonal form, consists of only one element.

Problem 199. Conduction current and polarization current

a) The electrical current density (particle charge e),

$$s_\nu = iec\bar{\psi}\gamma_\nu\psi; \quad s_k = j_k; \quad s_4 = ic\rho \tag{199.1}$$

shall be shown to satisfy the equation of continuity,

$$\sum_\mu \frac{\partial s_\mu}{\partial x_\mu} = 0 \quad \text{or} \quad \operatorname{div}\boldsymbol{j} + \frac{\partial \rho}{\partial t} = 0. \tag{199.2}$$

b) The vector s_ν shall be decomposed,

$$s_\nu = s_\nu^C + s_\nu^P \tag{199.3}$$

so that the space part of s_ν^C, the *conduction current*, is of the same form as the unrelativistic expression for j_k. The remaining part, s_k^P, is then called the *polarization current*.

Solution. a) In order to prove (199.1) we have to supplement the Dirac equation

$$\sum_\mu \gamma_\mu(\partial_\mu - i a_\mu)\psi + \varkappa\psi = 0; \qquad \partial_\mu = \frac{\partial}{\partial x_\mu}; \qquad a_\mu = \frac{e}{\hbar c} A_\mu \quad (199.4\,\text{a})$$

by an analogous differential equation for $\bar\psi = \psi^\dagger \gamma_4$. The operators

$$D_k = \partial_k - i a_k, \qquad D_4 = \partial_4 - i a_4$$

have the complex conjugates

$$D_k^* = \partial_k + i a_k, \qquad D_4^* = -(\partial_4 + i a_4),$$

since x_k and a_k are real, but x_4 and a_4 imaginary. The conjugate equation of (199.4a),

$$\sum_\mu D_\mu^* \psi^\dagger \gamma_\mu + \varkappa \psi^\dagger = 0$$

or, with $\bar\psi = \psi^\dagger \gamma_4$,

$$-\sum_k D_k^* \bar\psi \gamma_k + D_4^* \bar\psi \gamma_4 + \varkappa \bar\psi = 0$$

may be written

$$\sum_\mu (\partial_\mu + i a_\mu)\bar\psi \gamma_\mu - \varkappa \bar\psi = 0. \qquad (199.4\,\text{b})$$

From (199.4a, b) it then follows by elimination of the mass terms that

$$\sum_\mu \{\bar\psi \gamma_\mu (\partial_\mu - i a_\mu)\psi + (\partial_\mu + i a_\mu)\bar\psi \gamma_\mu \cdot \psi\} = 0.$$

Here the a_μ terms cancel and the rest may be written

$$\sum_\mu \partial_\mu (\bar\psi \gamma_\mu \psi) = 0,$$

in agreement with the equation of continuity (199.2).

b) The unrelativistic expression of the space part of the electrical current is, according to Problem 126,

$$j_k = \frac{i e \hbar}{2m}(\psi \partial_k \psi^* - \psi^* \partial_k \psi + 2 i a_k \psi^* \psi), \qquad (199.5)$$

i. e. mainly a bilinear combination of wave functions and their space derivatives. In order to give s_ν, Eq. (199.1), a similar form, we may there replace either $\bar\psi$ according to (199.4a) or ψ according to (199.4b) by their first derivatives:

$$s_\nu = \frac{iec}{\varkappa} \sum_\mu (\partial_\mu + i a_\mu)\bar\psi \gamma_\mu \cdot \gamma_\nu \psi = -\frac{iec}{\varkappa} \bar\psi \gamma_\nu \sum_\mu \gamma_\mu (\partial_\mu - i a_\mu)\psi.$$

Symmetrizing by taking half the sum of these two expressions and putting $ec/\varkappa = e\hbar/m$, we find

$$s_\nu = \frac{ie\hbar}{2m}\sum_\mu \left\{ \frac{\partial\bar{\psi}}{\partial x_\mu}\gamma_\mu\gamma_\nu\psi - \bar{\psi}\gamma_\nu\gamma_\mu\frac{\partial\psi}{\partial x_\mu} + i a_\mu \bar{\psi}(\gamma_\mu\gamma_\nu+\gamma_\nu\gamma_\mu)\psi \right\}. \qquad (199.6)$$

Using the commutation rules

$$\gamma_\mu\gamma_\nu+\gamma_\nu\gamma_\mu = 2\delta_{\mu\nu}$$

in the second and third terms, we may reshape this into

$$s_\nu = \frac{ie\hbar}{2m}\left\{\sum_\mu\left(\frac{\partial\bar{\psi}}{\partial x_\mu}\gamma_\mu\gamma_\nu\psi + \bar{\psi}\gamma_\mu\gamma_\nu\frac{\partial\psi}{\partial x_\mu}\right) - 2\bar{\psi}\frac{\partial\psi}{\partial x_\nu} + 2ia_\nu\bar{\psi}\psi\right\}$$

or, splitting off the sum its diagonal term $\mu=\nu$,

$$s_\nu = \frac{ie\hbar}{2m}\left\{\frac{\partial\bar{\psi}}{\partial x_\nu}\psi - \bar{\psi}\frac{\partial\psi}{\partial x_\nu} + 2ia_\nu\bar{\psi}\psi\right\} + \frac{ie\hbar}{2m}\sum_\mu{}' \frac{\partial}{\partial x_\mu}(\bar{\psi}\gamma_\mu\gamma_\nu\psi). \qquad (199.7)$$

The first term of this decomposition exactly matches Eq. (199.5) and is therefore the conduction current s_ν^C as defined above. The other term then is the so-called polarization current,

$$s_\nu^P = \frac{ie\hbar}{2m}\sum_\mu{}' \frac{\partial}{\partial x_\mu}(\bar{\psi}\gamma_\mu\gamma_\nu\psi). \qquad (199.8)$$

NB. This decomposition has first been studied by W. Gordon, Z. Physik **50**, 630 (1928). The space part of s_ν^P may be written

$$\mathbf{s}^P = \frac{e}{m}\operatorname{curl}(\bar{\psi}\mathbf{S}\psi) - \frac{e\hbar i}{2mc}\frac{\partial}{\partial t}(\bar{\psi}\boldsymbol{\alpha}\psi)$$

where $S_k = \frac{\hbar}{2}\sigma_k$ are the components of the spin vector (written in 4×4 reducible matrices) and the α_k are the matrices defined at the end of Problem 189. In a plane wave the polarization current vanishes.

Problem 200. Splitting up of Dirac equations into two pairs

Write the Dirac equation in hamiltonian form and split up the resulting four-component equation of standard representation into a pair of two component equations. Pauli matrices will occur in the latter. Show that for rest mass zero (e. g. for a neutrino) two two-component theories are possible.

Solution. The Dirac equation

$$\left.\begin{aligned} &\sum_{\mu} \gamma_\mu D_\mu \psi + \varkappa \psi = 0; \qquad D_\mu = \partial_\mu - \frac{ie}{\hbar c} A_\mu; \\ &A_4 = i\Phi; \qquad e\Phi = V; \qquad \partial_4 = -\frac{i}{c}\partial_t \end{aligned}\right\} \tag{200.1}$$

can be written, distinguishing time and space derivatives,

$$\sum_{n=1}^{3} \gamma_n D_n \psi + \gamma_4 \left(-\frac{i}{c}\partial_t + \frac{V}{\hbar c} \right) \psi + \varkappa \psi = 0 .$$

Multiplication from the left by $c\hbar\gamma_4$ renders

$$\hbar c \sum_n \gamma_4 \gamma_n D_n \psi - i\hbar \partial_t \psi + V\psi + mc^2 \gamma_4 \psi = 0$$

or

$$-\frac{\hbar}{i}\frac{\partial \psi}{\partial t} = H\psi$$

with

$$H = \hbar c \sum_{n=1}^{3} \gamma_4 \gamma_n \left(\partial_n - \frac{ie}{\hbar c} A_n \right) + V + mc^2 \gamma_4 \tag{200.2}$$

the hamiltonian.

In standard representation we have

$$\gamma_n = \begin{pmatrix} \mathit{0} & -i\boldsymbol{s}_n \\ i\boldsymbol{s}_n & \mathit{0} \end{pmatrix}; \qquad \gamma_4 = \begin{pmatrix} \mathit{1} & \mathit{0} \\ \mathit{0} & -\mathit{1} \end{pmatrix} \tag{200.3}$$

with $\boldsymbol{s}_n$ the three Pauli matrices and $\mathit{1}$ and $\mathit{0}$ standing, respectively, for the 2×2 matrices of unity and zero. Then,

$$\alpha_n = i\gamma_4 \gamma_n = \begin{pmatrix} \mathit{0} & \boldsymbol{s}_n \\ \boldsymbol{s}_n & \mathit{0} \end{pmatrix} \tag{200.4}$$

so that the hamiltonian (200.2) splits up in the form

$$H = \begin{pmatrix} V + mc^2; & -\hbar c i \sum_n \boldsymbol{s}_n D_n \\ -\hbar c i \sum_n \boldsymbol{s}_n D_n; & V - mc^2 \end{pmatrix}. \tag{200.5}$$

If here we introduce the two-component quantities ψ_a and ψ_b so that

$$\psi = \begin{pmatrix} \psi_a \\ \psi_b \end{pmatrix}$$

is the four-component Dirac spinor, the differential equation splits up into the wanted pair of two-component equations, viz.

$$\left.\begin{aligned} -\frac{\hbar}{i}\frac{\partial\psi_a}{\partial t} &= -i\hbar c(\vec{s}\cdot\vec{D})\psi_b+(V+mc^2)\psi_a, \\ -\frac{\hbar}{i}\frac{\partial\psi_b}{\partial t} &= -i\hbar c(\vec{s}\cdot\vec{D})\psi_a+(V-mc^2)\psi_b \end{aligned}\right\} \tag{200.6}$$

or, for a stationary state of positive energy, E,

$$\left.\begin{aligned} (\vec{s}\cdot\vec{D})\psi_b - i\frac{E-V-mc^2}{\hbar c}\psi_a &= 0; \\ (\vec{s}\cdot\vec{D})\psi_a - i\frac{E-V+mc^2}{\hbar c}\psi_b &= 0. \end{aligned}\right\} \tag{200.7}$$

Neutrino theory: If $m=0$, the two equations become identical so that

$$\psi_b = \lambda\psi_a \quad \text{with } \lambda = \pm 1 \tag{200.8}$$

are two possible solutions with ψ_a to be determined from

$$\left\{(\vec{s}\cdot\vec{D}) - \lambda i\frac{E-V}{\hbar c}\right\}\psi_a = 0. \tag{200.9}$$

Since the two systems decouple, there emerge two independent two-component theories of particles of rest-mass zero. It can easily be seen that, in the force-free case, the parameter λ becomes identical with the helicity quantum number. For this purpose we study a plane wave in z direction,

$$\psi_a = C\mathrm{e}^{ikz}$$

with C a constant two-component spinor. The first term of (200.9) then becomes

$$s_3\partial_3\psi_a = ik(s_3\psi_a)\mathrm{e}^{ikz}$$

and its second term

$$-\lambda i\frac{E}{\hbar c} = -ik\lambda$$

so that we have

$$s_3 C = \lambda C. \tag{200.10}$$

Therefore λ is the eigenvalue of the spin component (in units of $\hbar/2$) in the direction of propagation ("helicity"). With the Pauli matrix

$$s_3 = \begin{pmatrix} 1 & 0 \\ 0 & -1 \end{pmatrix}$$

Eq. (200.10) then is solved for $\lambda=+1$ by

$$C=\begin{pmatrix}1\\0\end{pmatrix} \quad \text{and} \quad \psi=\begin{pmatrix}C\\C\end{pmatrix}e^{ikz} \tag{200.11a}$$

whereas for the helicity $\lambda=-1$ it is solved by

$$C=\begin{pmatrix}0\\1\end{pmatrix} \quad \text{and} \quad \psi=\begin{pmatrix}C\\-C\end{pmatrix}e^{ikz}. \tag{200.11b}$$

NB. Experience shows that neutrinos always have helicity $h=-1$ so that only the second theory actually describes natural phenomena.

Now, the operator

$$\gamma_5=\gamma_1\gamma_2\gamma_3\gamma_4$$

in standard representation becomes

$$\gamma_5=\begin{pmatrix}0 & -1\\-1 & 0\end{pmatrix}$$

so that

$$1+\gamma_5=\begin{pmatrix}1 & -1\\-1 & 1\end{pmatrix}; \quad 1-\gamma_5=\begin{pmatrix}1 & 1\\1 & 1\end{pmatrix}.$$

If these operators act on any $\lambda=+1$ solution of (200.9), i.e. on

$$\psi_+=\begin{pmatrix}\psi_a\\\psi_a\end{pmatrix}$$

they lead to

$$(1+\gamma_5)\psi_+=0; \quad (1-\gamma_5)\psi_+=2\psi_+. \tag{200.12a}$$

Acting on any $\lambda=-1$ solution of (200.9), i.e. on

$$\psi_-=\begin{pmatrix}\psi_a\\-\psi_a\end{pmatrix}$$

they render

$$(1+\gamma_5)\psi_-=2\psi_-; \quad (1-\gamma_5)\psi_-=0. \tag{200.12b}$$

It cannot be decided whether, due to some unknown principle, only ψ_- is realized in nature, or whether the interaction operator producing neutrinos contains a factor $1+\gamma_5$, thus making creation of $\lambda=+1$ neutrinos impossible. It should be noted, however, that $1+\gamma_5$ is an operator without defined parity.

Problem 201. Central forces in Dirac theory

To use the splitting up of the Dirac equations in standard representation into a pair of two-component equations (Problem 200) in order to construct eigenspinors of a central potential field $V(r)$ which are simultaneously eigenspinors of the total angular momentum operators $\boldsymbol{J}^2$ and J_z. The calculations may be restricted to $m_j=+\frac{1}{2}$.

Solution. According to (200.7) we may start from the differential equations

$$\left.\begin{aligned}\sum_{n=1}^{3} s_n \partial_n \psi_b - i\,\frac{E-V(r)-mc^2}{\hbar c}\,\psi_a = 0,\\ \sum_{n=1}^{3} s_n \partial_n \psi_a - i\,\frac{E-V(r)+mc^2}{\hbar c}\,\psi_b = 0\end{aligned}\right\} \tag{201.1}$$

where s_n $(n=1,2,3)$ denote the three Pauli matrices, and

$$\psi = \begin{pmatrix}\psi_a\\ \psi_b\end{pmatrix} \tag{201.2}$$

is the Dirac spinor, composed of the two-component spinors ψ_a and ψ_b.

The angular momentum operators, deriving from $\boldsymbol{J} = \boldsymbol{L} + \frac{\hbar}{2}\boldsymbol{\sigma}$, have four-component standard representations which do not mix the first two components (ψ_a) with the other two (ψ_b) since the four-component extension of the spin matrices,

$$\sigma_n = \begin{pmatrix} s_n & 0\\ 0 & s_n\end{pmatrix},$$

is diagonal in the Pauli matrices. If therefore ψ_a and ψ_b are two-component eigenspinors of $\boldsymbol{J}^2$ and J_z, so will ψ, Eq. (201.2), be.

Now, in Problem 133, we have already constructed the two-component eigenspinors $u_{j,l}$ of $\boldsymbol{J}^2$ and J_z with quantum numbers $j = l \pm \frac{1}{2}$ and $m_j = +\frac{1}{2}$, viz.

$$u^{\mathrm{I}} = u_{j,j-\frac{1}{2}} = \frac{f_l(r)}{\sqrt{2l+1}}\begin{pmatrix}\sqrt{l+1}\,Y_{l,0}\\ -\sqrt{l}\,Y_{l,1}\end{pmatrix} = \frac{f_{j-\frac{1}{2}}(r)}{\sqrt{2j}}\begin{pmatrix}\sqrt{j+\frac{1}{2}}\,Y_{j-\frac{1}{2},0}\\ -\sqrt{j-\frac{1}{2}}\,Y_{j-\frac{1}{2},1}\end{pmatrix} \tag{201.3a}$$

and

$$u^{\mathrm{II}} = u_{j,j+\frac{1}{2}} = \frac{g_l(r)}{\sqrt{2l+1}}\begin{pmatrix}\sqrt{l}\,Y_{l,0}\\ \sqrt{l+1}\,Y_{l,1}\end{pmatrix} = \frac{g_{j+\frac{1}{2}}(r)}{\sqrt{2(j+1)}}\begin{pmatrix}\sqrt{j+\frac{1}{2}}\,Y_{j+\frac{1}{2},0}\\ \sqrt{j+\frac{3}{2}}\,Y_{j+\frac{1}{2},1}\end{pmatrix}. \tag{201.3b}$$

We shall try to solve our problem by combinations of these two spinors, taking either

$$\psi = \begin{pmatrix}u^{\mathrm{I}}\\ u^{\mathrm{II}}\end{pmatrix} \tag{201.4a}$$

or

$$\psi = \begin{pmatrix}u^{\mathrm{II}}\\ u^{\mathrm{I}}\end{pmatrix}, \tag{201.4b}$$

where the normalization of $f(r)$ and $g(r)$ is still left open and may be different in (201.4a) and (201.4b).

To follow this programme, according to (201.1), we need the expressions

$$S u^{\mathrm{I,II}} = \sum_{n=1}^{3} s_n \partial_n u^{\mathrm{I,II}} = \begin{pmatrix} \partial_z; & \partial_x - i\partial_y \\ \partial_x + i\partial_y; & -\partial_z \end{pmatrix} u^{\mathrm{I,II}}. \tag{201.5}$$

Using the well-known formulae

$$\begin{aligned} \pm(\partial_x \pm i\partial_y)(F(r)\,Y_{l,m}) = {} & \sqrt{\frac{(l\pm m+2)(l\pm m+1)}{(2l+3)(2l+1)}}\left(F' - \frac{l}{r}F\right) Y_{l+1,m\pm 1} \\ & - \sqrt{\frac{(l\mp m)(l\mp m-1)}{(2l+1)(2l-1)}}\left(F' + \frac{l+1}{r}F\right) Y_{l-1,m\pm 1} \end{aligned} \tag{201.6a}$$

and

$$\begin{aligned} \partial_z(F(r)\,Y_{l,m}) = {} & \sqrt{\frac{(l+m+1)(l-m+1)}{(2l+3)(2l+1)}}\left(F' - \frac{l}{r}F\right) Y_{l+1,m} \\ & + \sqrt{\frac{(l+m)(l-m)}{(2l+1)(2l-1)}}\left(F' + \frac{l+1}{r}F\right) Y_{l-1,m} \end{aligned} \tag{201.6b}$$

we arrive after some cumbersome but elementary reshaping of the expressions

$$S u^{\mathrm{I}} = \frac{1}{\sqrt{2l+1}} \begin{pmatrix} \sqrt{l+1}\,\partial_z(f_l Y_{l,0}) - \sqrt{l}\,(\partial_x - i\partial_y)(f_l Y_{l,1}) \\ \sqrt{l+1}\,(\partial_x + i\partial_y)(f_l Y_{l,0}) + \sqrt{l}\,\partial_z(f_l Y_{l,1}) \end{pmatrix}$$

and

$$S u^{\mathrm{II}} = \frac{1}{\sqrt{2l+1}} \begin{pmatrix} \sqrt{l}\,\partial_z(g_l Y_{l,0}) + \sqrt{l+1}\,(\partial_x - i\partial_y)(g_l Y_{l,1}) \\ \sqrt{l}\,(\partial_x + i\partial_y)(g_l Y_{l,0}) - \sqrt{l+1}\,\partial_z(g_l Y_{l,1}) \end{pmatrix}$$

at the results

$$\begin{aligned} S u^{\mathrm{I}} & = \frac{1}{\sqrt{2l+3}}\left(f_l' - \frac{l}{r} f_l\right) \begin{pmatrix} \sqrt{l+1}\,Y_{l+1,0} \\ \sqrt{l+2}\,Y_{l+1,1} \end{pmatrix} \\ & = \frac{1}{\sqrt{2(j+1)}}\left(f_{j-\frac{1}{2}}' - \frac{j-\frac{1}{2}}{r} f_{j-\frac{1}{2}}\right) \begin{pmatrix} \sqrt{j+\frac{1}{2}}\,Y_{j+\frac{1}{2},0} \\ \sqrt{j+\frac{3}{2}}\,Y_{j+\frac{1}{2},1} \end{pmatrix} \end{aligned} \tag{201.7a}$$

for $j = l + \frac{1}{2}$, and

$$\begin{aligned} S u^{\mathrm{II}} & = \frac{1}{\sqrt{2l-1}}\left(g_l' + \frac{l+1}{r} g_l\right) \begin{pmatrix} \sqrt{l}\,Y_{l-1,0} \\ -\sqrt{l-1}\,Y_{l-1,1} \end{pmatrix} \\ & = \frac{1}{\sqrt{2j}}\left(g_{j+\frac{1}{2}}' + \frac{j+\frac{3}{2}}{r} g_{j+\frac{1}{2}}\right) \begin{pmatrix} \sqrt{j+\frac{1}{2}}\,Y_{j-\frac{1}{2},0} \\ -\sqrt{j-\frac{1}{2}}\,Y_{j-\frac{1}{2},1} \end{pmatrix} \end{aligned} \tag{201.7b}$$

for $j = l - \frac{1}{2}$.

In order now to satisfy Eqs. (201.1), we first try the type of solution (201.4a):

$$S u^{\mathrm{II}} - i\frac{E-V-mc^2}{\hbar c} u^{\mathrm{I}} = 0; \qquad S u^{\mathrm{I}} - i\frac{E-V+mc^2}{\hbar c} u^{\mathrm{II}} = 0. \quad (201.8)$$

Putting u^{I} and u^{II} from (201.3a, b) and Su^{II}, Su^{I} from (201.7a, b) into (201.8), we find (omitting the subscripts of f and g)

$$S u^{\mathrm{II}} - i\frac{E-V-mc^2}{\hbar c} u^{\mathrm{I}}$$

$$= \frac{1}{\sqrt{2j}}\left(g' + \frac{j+\frac{3}{2}}{r} g - i\frac{E-V-mc^2}{\hbar c} f\right)\begin{pmatrix} \sqrt{j+\frac{1}{2}}\, Y_{j-\frac{1}{2},0} \\ -\sqrt{j-\frac{1}{2}}\, Y_{j-\frac{1}{2},1} \end{pmatrix}$$

and

$$S u^{\mathrm{I}} - i\frac{E-V+mc^2}{\hbar c} u^{\mathrm{II}}$$

$$= \frac{1}{\sqrt{2(j+1)}}\left(f' - \frac{j-\frac{1}{2}}{r} f - i\frac{E-V+mc^2}{\hbar c} g\right)\begin{pmatrix} \sqrt{j+\frac{1}{2}}\, Y_{j+\frac{1}{2},0} \\ \sqrt{j+\frac{3}{2}}\, Y_{j+\frac{1}{2},1} \end{pmatrix}.$$

These expressions vanish if the radial functions $f(r)$ and $g(r)$ satisfy the coupled differential equations

$$\left.\begin{aligned} g' + \frac{j+\frac{3}{2}}{r} g - i\frac{E-V(r)-mc^2}{\hbar c} f = 0, \\ f' - \frac{j-\frac{1}{2}}{r} f - i\frac{E-V(r)+mc^2}{\hbar c} g = 0. \end{aligned}\right\} \quad (201.9\,\mathrm{a})$$

The type of solution (201.4b), on the other hand, with u^{I} and u^{II} exchanged in (201.8) leads in the same way to formulae in which the role of the two equations (201.8) and thus the signs of the rest-mass terms are exchanged. Thus we arrive, not at (201.9a) but at the set of differential equations

$$\left.\begin{aligned} g' + \frac{j+\frac{3}{2}}{r} g - i\frac{E-V(r)+mc^2}{\hbar c} f = 0, \\ f' - \frac{j-\frac{1}{2}}{r} f - i\frac{E-V(r)-mc^2}{\hbar c} g = 0. \end{aligned}\right\} \quad (201.9\,\mathrm{b})$$

The differential equations (201.9a) and (201.9b) have still to be solved for any given potential $V(r)$, separately, thus providing the complete solution. By their coupling they determine the relative normalization of the two functions f and g.

It should be noted that, in contrast to the unrelativistic spin theory, contributions of different l values are mixed up in the four-component Dirac spinor so that l is no longer a good quantum number, but j and m_j of course still are.

Problem 202. Kepler problem in Dirac theory

To specialize the solutions of the central-force problem to the potential

$$V(r) = -\frac{Z e^2}{r} \tag{202.1}$$

and to determine the eigenvalues.

Solution. In the preceding problem we have found two sets of solutions for the general central-force problem, the differential equations for whose radial parts had been written up in Eqs. (201.9a) and (201.9b). Let us first discuss the system (201.9a). With the abbreviation

$$\beta = \frac{Z e^2}{\hbar c} = \frac{Z}{137}, \tag{202.2}$$

which, it should be borne in mind, generally is a very small number, and

$$\frac{\mu}{a} = \frac{mc^2 - E}{\hbar c}; \qquad \frac{1}{\mu a} = \frac{mc^2 + E}{\hbar c}$$

or

$$\mu = \sqrt{\frac{mc^2 - E}{mc^2 + E}}; \qquad a = \frac{\hbar c}{\sqrt{(mc^2 - E)(mc^2 + E)}} \tag{202.3}$$

the differential equations (201.9a) may be written for potential (202.1):

$$\left.\begin{aligned} g' + \frac{j+\frac{3}{2}}{r} g + i\left(\frac{\mu}{a} - \frac{\beta}{r}\right) f = 0; \\ f' - \frac{j-\frac{1}{2}}{r} f - i\left(\frac{1}{\mu a} + \frac{\beta}{r}\right) g = 0. \end{aligned}\right\} \tag{202.4}$$

These equations are to be solved.

We start by discussing their behaviour for very large and very small values of r. For $r \to \infty$ Eqs. (202.4) become

$$g' + \frac{i\mu}{a} f = 0; \qquad f' - \frac{i}{\mu a} g = 0$$

with normalizable solutions

$$g = C\,e^{-r/a}; \qquad f = -C\frac{i}{\mu}\,e^{-r/a}.$$

There is another set of type $e^{+r/a}$ which we need not discuss. For $r\to 0$, on the other hand, we expect regular solutions of the form

$$g = A\,r^{s-1}; \qquad f = B\,r^{s-1}.$$

Putting these in (202.4) we get

$$(s-1)A + (j+\tfrac{3}{2})A - i\beta B = 0,$$
$$(s-1)B - (j-\tfrac{1}{2})B - i\beta A = 0.$$

Vanishing of the determinant of this set of linear equations leads to

$$s = \sqrt{(j+\tfrac{1}{2})^2 - \beta^2}. \tag{202.5}$$

Combining these results, it seems reasonable to put

$$\left.\begin{aligned} g &= C\,r^{s-1}\,e^{-r/a}\,G(r);\\ f &= -\frac{i}{\mu}\,C\,r^{s-1}\,e^{-r/a}\,F(r); \end{aligned}\right\} \tag{202.6}$$

then, from (202.4), there follow the differential equations

$$\left.\begin{aligned} G' + \left(\frac{s+(j+\frac{1}{2})}{r} - \frac{1}{a}\right)G + \left(\frac{1}{a} - \frac{\beta}{\mu r}\right)F = 0,\\ F' + \left(\frac{s-(j+\frac{1}{2})}{r} - \frac{1}{a}\right)F + \left(\frac{1}{a} + \frac{\beta\mu}{r}\right)G = 0. \end{aligned}\right\} \tag{202.7}$$

Adding and substracting these equations, respectively, and putting

$$G + F = v(r); \qquad G - F = w(r) \tag{202.8}$$

we arrive at

$$\left.\begin{aligned} v' + \frac{s+p}{r}\,v &= -(\mathsf{k}+q)\frac{w}{r},\\ w' + \left(\frac{s-p}{r} - \frac{2}{a}\right)w &= -(\mathsf{k}-q)\frac{v}{r} \end{aligned}\right\} \tag{202.9}$$

with

$$p = \frac{\beta}{2}\left(\mu - \frac{1}{\mu}\right); \qquad q = \frac{\beta}{2}\left(\mu + \frac{1}{\mu}\right); \qquad \mathsf{k} = j + \frac{1}{2}. \tag{202.10}$$

From the first equation (202.9) we get

$$w = -\frac{1}{\mathsf{k}+q}\{r\,v' + (s+p)\,v\}; \qquad w' = -\frac{1}{\mathsf{k}+q}(r\,v'' + (s+p+1)\,v'\}. \tag{202.11}$$

This we put in the second equation (202.9) which thus becomes a second-order equation for only v, viz.

$$r v'' + \left[(2s+1) - \frac{2}{a} r\right] v' - \frac{2}{a}(s+p)v = 0. \tag{202.12}$$

This is a Kummer's equation; its solution, in arbitrary normalization and regular at the origin, is the confluent series

$$v = {}_1F_1\left(s+p, 2s+1; 2\frac{r}{a}\right). \tag{202.13}$$

From (202.11) we then derive $w(r)$ using the general formula

$$\left(z\frac{d}{dz} + a\right) {}_1F_1(a,c;z) = a\,{}_1F_1(a+1,c;z);$$

the result is

$$w = -\frac{s+p}{\mathrm{k}+q}\,{}_1F_1\left(s+p+1, 2s+1; 2\frac{r}{a}\right). \tag{202.14}$$

Putting the expressions (202.13) and (202.14) in (202.8) leads on to G and F and thence, using (202.6), we finally arrive at the radial functions

$$\left.\begin{aligned} g &= \frac{1}{2} C r^{s-1} \mathrm{e}^{-r/a} \left\{ {}_1F_1\left(s+p, 2s+1; 2\frac{r}{a}\right) \right. \\ &\quad \left. - \frac{s+p}{\mathrm{k}+q}\,{}_1F_1\left(s+p+1, 2s+1; 2\frac{r}{a}\right)\right\}; \\ f &= -\frac{i}{2\mu} C r^{s-1} \mathrm{e}^{-r/a} \left\{ {}_1F_1\left(s+p, 2s+1; 2\frac{r}{a}\right) \right. \\ &\quad \left. + \frac{s+p}{\mathrm{k}+q}\,{}_1F_1\left(s+p+1, 2s+1; 2\frac{r}{a}\right)\right\}. \end{aligned}\right\} \tag{202.15}$$

The two confluent series are asymptotically proportional to $\mathrm{e}^{+2r/a}$, thus destroying normalizability unless their first parameters are zero or negative integer:

$$s+p = -n_r; \quad n_r = 0,1,2,3,\ldots \tag{202.16}$$

If $s+p=0$, in the second confluent series we still have $s+p+1=+1$; however, a factor $s+p$ makes this part of Eqs. (202.15) vanish anyway in this particular case, so that (202.16) is the complete eigenvalue condition. Replacing in (202.16) p by (202.10) and taking μ from (202.3), we may introduce the energy and write

$$\frac{1}{2}\beta\left\{\sqrt{\frac{mc^2-E}{mc^2+E}}-\sqrt{\frac{mc^2+E}{mc^2-E}}\right\}=-(n_r+s).$$

This may be resolved for E:

$$E=\frac{mc^2}{\sqrt{1+\dfrac{\beta^2}{(n_r+s)^2}}}, \tag{202.17}$$

thus providing us with the energy level formula wanted.

We have so far not yet considered the second set of radial equations, (201.9b), which follows by replacing μ by $-1/\mu$ in (202.4) and consequently, q by $-q$ and p by $-p$ so that the eigenvalue condition (202.16) is changed into $s-p=-n_r$ which, however, alters nothing whatsoever in the energy level formula. Each energy term therefore is degenerate with two solutions to it.

If $\beta>1$ ($Z>137$) the exponent s, Eq. (202.5), will become imaginary for the ground state, so that the boundary condition at $r=0$ cannot be simply satisfied. For very large Z the potential hole may even become so deep that for the lowest bound state $E<-mc^2$. According to (202.3), a then becomes imaginary and the solutions g and f, Eq. (202.6), no longer decrease exponentially at large values of r. This is a consequence of the electron wave penetrating into the domain of negative energies (Klein's paradox), a phenomenon explained in some detail for a potential step in Problem 207 below (case c).

Problem 203. Hydrogen atom fine structure

For a hydrogen atom, the parameter β of the preceding problem becomes identical with Sommerfeld's fine-structure constant,

$$\alpha=\frac{e^2}{\hbar c}=\frac{1}{137}.$$

This parameter is small enough to justify power expansion of the results. This shall be performed, thus confirming the unrelativistic theory and adding its first relativistic correction.

Solution. The expansion of s, Eq. (202.5), leads to

$$s=\left(j+\frac{1}{2}\right)-\frac{\alpha^2}{2j+1}+O(\alpha^4). \tag{203.1}$$

If this is inserted into the energy formula (202.17), the principal quantum number,

$$n = n_r + j + \tfrac{1}{2}, \tag{203.2}$$

may be introduced with advantage. We then have

$$E = mc^2 \left\{1 + \left[\alpha^2 \Big/ \left(n - \frac{\alpha^2}{2j+1}\right)^2\right]\right\}^{-\frac{1}{2}}$$

or

$$E = mc^2 \left\{1 - \frac{\alpha^2}{2n^2}\left[1 + \frac{\alpha^2}{n^2}\left(\frac{n}{j+\frac{1}{2}} - \frac{3}{4}\right)\right] + O(\alpha^6)\right\}. \tag{203.3}$$

Since

$$mc^2\alpha^2 = \frac{me^4}{\hbar^2}$$

we arrive at the level formula for hydrogen, in first relativistic approximation:

$$E = mc^2 - \frac{me^4}{2\hbar^2 n^2}\left[1 + \frac{\alpha^2}{n^2}\left(\frac{n}{j+\frac{1}{2}} - \frac{3}{4}\right)\right], \tag{203.4}$$

where the first term is the rest energy, the second term, for $\alpha^2 = 0$, the unrelativistic Balmer term (cf. Problem 67), and the square bracket provides a first relativistic correction of the order of $\alpha^2 = 0{,}532 \times 10^{-4}$ or about $\frac{1}{200}$ percent of the binding energy. Since this correction depends upon j as well as upon n, each unrelativistic level will split up into several fine-structure components.

Next let us expand the length parameter a, Eq. (202.3). This leads to

$$a = \frac{\hbar^2}{me^2} n \left[1 - \frac{\alpha^2}{2n^2}\left(\frac{n}{j+\frac{1}{2}} - 1\right)\right] + O(\alpha^4). \tag{203.5}$$

Without the relativistic correction in the square bracket, this is the well-known Bohr radius of the n-th hydrogen orbital. Since the argument $2r/a$ occurs in the confluent series and $e^{-r/a}$ as a factor, in the radial wave functions (202.15), the atomic size is determined by a in a form quite similar to that in unrelativistic theory (cf. the following problem).

To perform the transition from the relativistic wave functions (202.15) to those of the unrelativistic Schrödinger theory we need the expansions of the parameters μ and q, Eqs. (202.3) and (202.10):

$$\mu = \frac{\alpha}{2n}\left[1 + \frac{\alpha^2}{n^2}\left(\frac{n}{2j+1} - \frac{1}{4}\right)\right]; \tag{203.6}$$

$$q = n\left[1 - \frac{\alpha^2}{n^2}\left(\frac{n}{2j+1} - \frac{1}{2}\right)\right]. \tag{203.7}$$

The factor in front of the second confluent series in (202.15), in unrelativistic approximation, is

$$\frac{s+p}{\mathsf{k}+q} = -\frac{n_r}{j+\frac{1}{2}+n} = -\frac{n_r}{n_r+(2j+1)}, \tag{203.8}$$

i. e. of the order of magnitude 1 (except in the case $n_r=0$ when it vanishes). The factor μ, according to (203.6) being of the order α, therefore makes f about two powers of 10 bigger than g. In unrelativistic approach and arbitrary normalization we entirely neglect g and conclude:

$$\left.\begin{aligned} g &= 0, \\ f &= r^{j-\frac{1}{2}} \mathrm{e}^{-\frac{r}{a}} \left\{ {}_1F_1\left(-n_r, 2j+2; 2\frac{r}{a}\right) \right. \\ &\quad \left. - \frac{n_r}{n_r+2j+1}\, {}_1F_1\left(-n_r+1, 2j+2; 2\frac{r}{a}\right) \right\}. \end{aligned}\right\} \tag{203.9}$$

If in the function f we put $j=l+\frac{1}{2}$, it is indeed transformed into the Schrödinger wave function [cf. (67.12)]

$$f_{\mathrm{Schr}} = r^l \mathrm{e}^{-\gamma r}\, {}_1F_1(l+1-n, 2l+2; 2\gamma r) \tag{203.9 S}$$

where $\gamma=1/a$. This is readily seen as follows. With $j=l+\frac{1}{2}$, Eq. (203.2) renders for the principal quantum number $n=n_r+l+1$ so that $n_r=n-l-1$ and

$$\begin{aligned} f = r^l \mathrm{e}^{-\gamma r} &\left\{ {}_1F_1(l+1-n, 2l+3; 2\gamma r) \right. \\ &\left. - \frac{n-l-1}{n+l-1}\, {}_1F_1(l+2-n, 2l+3; 2\gamma r) \right\}. \end{aligned}$$

Here we apply the general relation

$$a\, {}_1F_1(a+1, c+1; z) = (a-c)\, {}_1F_1(a, c+1; z) + c\, {}_1F_1(a, c; z)$$

which permits with $a=l+1-n$, $c=2l+2$, $z=2\gamma r$ transformation of the curly bracket

$${}_1F_1(a, c+1; z) + \frac{a}{c-a}\, {}_1F_1(a+1, c+1; z)$$

into

$$\frac{c}{c-a}\, {}_1F_1(a, c; z) = \frac{2l+2}{n+l+1}\, {}_1F_1(l+1-n, 2l+2; 2\gamma r)$$

as required in Eq. (203.9 S).

The second solution needs separate treatment. With μ replaced by $-1/\mu$, we now have f the small and g the big solution. We use, except for a normalization factor, the unrelativistic approach

$$\left.\begin{aligned} g &= r^{j-\frac{1}{2}} e^{-\gamma r} \{{}_1F_1(-n_r, 2j+2; 2\gamma r) - {}_1F_1(1-n_r, 2j+2; 2\gamma r)\}, \\ f &= 0 \end{aligned}\right\} \tag{203.10}$$

where we now have to put $j = l - \frac{1}{2}$:

$$g = r^{l-1} e^{-\gamma r} \{{}_1F_1(-n_r, 2l+1; 2\gamma r) - {}_1F_1(1-n_r, 2l+1; 2\gamma r)\}.$$

To show that this is again the function (203.9 S), we apply the general formulae

$$a\{{}_1F_1(a+1, c-1; z) - {}_1F_1(a, c-1; z)\} = z \frac{d}{dz} {}_1F_1(a, c-1; z)$$

and

$$(c-1) \frac{d}{dz} {}_1F_1(a, c-1; z) = a\, {}_1F_1(a+1, c; z)$$

which indeed perform the desired transformation into

$$g = r^{l} e^{-\gamma r} {}_1F_1(1-n_r, 2l+2; 2\gamma r)$$

except for a factor $2\gamma/(2l+1)$. Here, according to (203.2), $1-n_r = j + \frac{3}{2} - n$ which, with $j = l - \frac{1}{2}$, again becomes $l+1-n$ as in Eq. (203.9 S).

Hitherto we have treated l as a convenient parameter without referring to its physical significance. In order to check the latter, let us calculate the expectation values of the operator $\boldsymbol{L}^2$ for the two types of solution. Using just the simple relation

$$\boldsymbol{L}^2 Y_{l,m} = \hbar^2 l(l+1) Y_{l,m}$$

we get for both solutions the expectation value

$$\langle \boldsymbol{L}^2 \rangle = \hbar^2 \frac{\int_0^\infty dr\, r^2 \{(j-\frac{1}{2})(j+\frac{1}{2})|f|^2 + (j+\frac{1}{2})(j+\frac{3}{2})|g|^2\}}{\int_0^\infty dr\, r^2 \{|f|^2 + |g|^2\}}. \tag{203.11}$$

For the first solution we may neglect $|g|^2$, which is then of an order α^2 smaller than $|f|^2$; then it follows that

$$\langle \boldsymbol{L}^2 \rangle = (j-\tfrac{1}{2})(j+\tfrac{1}{2})\, \hbar^2 \tag{203.12a}$$

corresponding to $j - \frac{1}{2} = l$. For the second solution, inversely, $|f|^2$ may be neglected and we arrive at

$$\langle \boldsymbol{L}^2 \rangle = (j+\tfrac{1}{2})(j+\tfrac{3}{2})\, \hbar^2 \tag{203.12b}$$

leading to $j+\frac{1}{2}=l$. These are exactly the substitutions used above. In other words, in unrelativistic approximation, when f and g are no longer mixed in the same wave spinor, l again becomes a good quantum number.

Supplement. The eigenspinors are composed of

$$u^{\mathrm{I}} = \frac{1}{\sqrt{2j}} f(r) \begin{pmatrix} \sqrt{j+\frac{1}{2}}\, Y_{j-\frac{1}{2},0} \\ -\sqrt{j-\frac{1}{2}}\, Y_{j-\frac{1}{2},1} \end{pmatrix}$$

and

$$u^{\mathrm{II}} = \frac{1}{\sqrt{2(j+1)}} g(r) \begin{pmatrix} \sqrt{j+\frac{1}{2}}\, Y_{j+\frac{1}{2},0} \\ \sqrt{j+\frac{3}{2}}\, Y_{j+\frac{1}{2},1} \end{pmatrix}$$

in either of the two forms

$$\psi_a = \begin{pmatrix} u^{\mathrm{I}} \\ u^{\mathrm{II}} \end{pmatrix} \quad \text{or} \quad \psi_b = \begin{pmatrix} u^{\mathrm{II}} \\ u^{\mathrm{I}} \end{pmatrix}.$$

We have seen that for the first solution, ψ_a, $g \ll f$, for the other, ψ_b, $f \ll g$ so that they approximately simplify to

$$\psi_a = \begin{pmatrix} u^{\mathrm{I}} \\ 0 \end{pmatrix} \quad \text{and} \quad \psi_b = \begin{pmatrix} u^{\mathrm{II}} \\ 0 \end{pmatrix}.$$

Thus there remains the unrelativistic two-component spin theory in which now $j=l+\frac{1}{2}$ for ψ_a and $j=l-\frac{1}{2}$ for ψ_b without any mixing of different l-values for the same j.

We further know that the "big" function in ψ_a is

$$f = r^{j-\frac{1}{2}}\, e^{-r/n}\, {}_1F_1(j+\tfrac{1}{2}-n, 2j+1; 2\gamma r) \quad \text{with } j=l+\tfrac{1}{2}$$

and in ψ_b,

$$g = r^{j+\frac{1}{2}}\, e^{-r/n}\, {}_1F_1(j+\tfrac{3}{2}-n, 2j+3; 2\gamma r) \quad \text{with } j=l-\tfrac{1}{2}.$$

If, and only if, the first parameter of the confluent series is zero or a negative integer, there exists a normalizable solution. Thus we arrive at the lowest possible states given in the following table. Their radial wave functions, either f or g, are identical with those of the unrelativistic theory treated in problem 67; their two-component character agrees with the unrelativistic spin theory, cf. Problem 133.

j	Solution ψ_a	Solution ψ_b	Spectroscopic symbol for ψ_a	Spectroscopic symbol for ψ_b
$\frac{1}{2}$	$n \geqq 1$	$n \geqq 2$	$n\,S_{\frac{1}{2}}$	$n\,P_{\frac{1}{2}}$
$\frac{3}{2}$	$n \geqq 2$	$n \geqq 3$	$n\,P_{\frac{3}{2}}$	$n\,D_{\frac{3}{2}}$
$\frac{5}{2}$	$n \geqq 3$	$n \geqq 4$	$n\,D_{\frac{5}{2}}$	$n\,F_{\frac{5}{2}}$

The energy levels, according to (203.4), are, except for the residual energy,

$$E_{n,j} = -E_n^0 - \Delta E_{n,j}$$

with (in multiples of the atomic unit $me^4/\hbar^2 = 27.2$ eV)

$$E_n^0 = \frac{1}{2n^2}$$

the unrelativistic Balmer binding energy, and

$$\Delta E_{n,j} = \frac{\alpha^2}{2n^4}\left(\frac{n}{j+\frac{1}{2}} - \frac{3}{4}\right)$$

its relativistic correction. For $n=3$ the energy values and their corrections are tabulated. The table shows that the splitting is larger the lower the level. That is the reason why the red $H\alpha$ line ($n=3 \rightarrow n=2$) appears to be roughly a doublet of

$$(2.08-0.42)\times 10^{-6}\times 27.2\,\mathrm{eV}$$

or $0.365\,\mathrm{cm}^{-1}$ splitting.

n	E_n^0	$(2/\alpha^2)\Delta E_{n,j}$ for			$10^6\times\Delta E_{n,j}$ for		
		$j=\frac{1}{2}$	$j=\frac{3}{2}$	$j=\frac{5}{2}$	$j=\frac{1}{2}$	$j=\frac{3}{2}$	$j=\frac{5}{2}$
1	$\frac{1}{2}$	$\frac{1}{4}$	—	—	6.68	—	—
2	$\frac{1}{8}$	$\frac{5}{64}$	$\frac{1}{64}$	—	2.08	0.42	—
3	$\frac{1}{18}$	$\frac{1}{36}$	$\frac{1}{108}$	$\frac{1}{324}$	0.74	0.25	0.08

Problem 204. Radial Kepler solutions at positive kinetic energies

To determine the radial wave functions and their asymptotic behaviour if the electron in the Coulomb field has positive kinetic energy, $E-mc^2>0$, at infinite distances from the centre of attraction.

Solution. Since $mc^2-E<0$, we replace the relations (202.3) of Problem 202 by the definitions

$$\eta k = \frac{E-mc^2}{\hbar c}; \qquad \frac{k}{\eta} = \frac{E+mc^2}{\hbar c} \tag{204.1a}$$

where k is now the wave number at infinity. This is equivalent with

$$\eta = \sqrt{\frac{E-mc^2}{E+mc^2}}; \qquad k = \frac{\sqrt{(E-mc^2)(E+mc^2)}}{\hbar c}. \tag{204.1b}$$

The differential equations to be solved then run as follows,

$$\left.\begin{aligned} g' + \frac{j+\frac{3}{2}}{r} g + i\left(-\eta k - \frac{\beta}{r}\right) f = 0, \\ f' - \frac{j-\frac{1}{2}}{r} f - i\left(\frac{k}{\eta} + \frac{\beta}{r}\right) g = 0. \end{aligned}\right\} \tag{204.2}$$

We solve these equations in full analogy to Problem 202 by putting

$$g = \frac{1}{2}(w+v)\, r^{s-1}\, e^{ikr}; \qquad f = \frac{1}{2\eta}(w-v)\, r^{s-1}\, e^{ikr} \tag{204.3}$$

with

$$s=\sqrt{(j+\tfrac{1}{2})^2-\beta^2}\,. \tag{204.4}$$

After a straightforward calculation, using the abbreviations

$$P=\frac{\beta}{2}\left(\frac{1}{\eta}-\eta\right); \quad Q=\frac{\beta}{2}\left(\frac{1}{\eta}+\eta\right) \tag{204.5}$$

and the variable

$$z=-2ikr \tag{204.6}$$

we arrive at

$$z\frac{d^2w}{dz^2}+[(2s+1)-z]\frac{dw}{dz}-(s-iQ)\,w=0 \tag{204.7}$$

and

$$v=-\frac{1}{j+\frac{1}{2}+iP}\left[z\frac{dw}{dz}+(s-iQ)\,w\right]. \tag{204.8}$$

The solution of the differential equation (204.7), regular at the origin, in arbitrary normalization, is

$$w=C\,{}_1F_1(s-iQ,2s+1\,;z)\,. \tag{204.9}$$

From (204.8), using the formula

$$\left\{z\frac{d}{dz}+a\right\}{}_1F_1(a,c\,;z)=a\,{}_1F_1(a+1,c;z)\,,$$

we then find

$$v=-C\frac{s-iQ}{j+\frac{1}{2}+iP}\,{}_1F_1(1+s-iQ,2s+1;z)\,. \tag{204.10}$$

Eqs. (204.9) and (204.10) provide the complete solution of the radial problem in arbitrary normalization.

For positive real values of r, the variable z, Eq. (204.6), is negative imaginary. We therefore may apply without further precaution the asymptotic formula

$${}_1F_1(a,c;z)\to e^{-i\pi a}\frac{\Gamma(c)}{\Gamma(c-a)}z^{-a}+\frac{\Gamma(c)}{\Gamma(a)}e^z z^{a-c} \tag{204.11}$$

thus getting

$$w\to\frac{C\Gamma(2s+1)\,e^{-\frac{\pi Q}{2}}}{(2kr)^s}\left\{\frac{e^{-\frac{i\pi s}{2}+iQ\log 2kr}}{\Gamma(1+s+iQ)}+\frac{i\,e^{-\frac{i\pi s}{2}-iQ\log 2kr}}{\Gamma(s-iQ)}\cdot\frac{e^{-2ikr}}{2kr}\right\}$$

and

$$v \to -\frac{C\,\Gamma(2s+1)\,\mathrm{e}^{-\frac{\pi Q}{2}}}{(2kr)^s}\,\frac{s-iQ}{j+\frac{1}{2}+iP}\left\{\frac{\mathrm{e}^{-\frac{i\pi s}{2}+iQ\log 2kr}}{i\Gamma(s+iQ)}\cdot\frac{1}{2kr}\right.$$
$$\left.+\frac{\mathrm{e}^{\frac{i\pi s}{2}-iQ\log 2kr}}{\Gamma(1+s-iQ)}\,\mathrm{e}^{-2ikr}\right\}.$$

Here, the second term in the w bracket and the first term in the v bracket are of an order $1/kr$ smaller than the other terms, respectively, and may thus be neglected. Therefore, using (204.3), we arrive at the results

$$\begin{aligned} rg &\to C_1\,\mathrm{e}^{i(kr+Q\log 2kr)}+C_2\,\mathrm{e}^{-i(kr+Q\log 2kr)},\\ \eta r f &\to C_1\,\mathrm{e}^{i(kr+Q\log 2kr)}-C_2\,\mathrm{e}^{-i(kr+Q\log 2kr)} \end{aligned} \tag{204.12}$$

with the complex amplitude constants

$$\begin{aligned} C_1 &- \frac{C\Gamma(2s+1)\,\mathrm{e}^{-\frac{\pi Q}{2}}}{2(2k)^s}\cdot\frac{\mathrm{e}^{-\frac{i\pi s}{2}}}{\Gamma(1+s+iQ)};\\ C_2 &= -\frac{C\Gamma(2s+1)\,\mathrm{e}^{-\frac{\pi Q}{2}}}{2(2k)^s}\,\frac{s-iQ}{j+\frac{1}{2}+iP}\,\frac{\mathrm{e}^{\frac{i\pi s}{2}}}{\Gamma(1+s-iQ)}. \end{aligned} \tag{204.13}$$

The constants C_1 and C_2 can, of course, only differ by a phase factor since incoming and outgoing partial waves must have equal amplitudes. Indeed both the factors by which they differ from one another,

$$\frac{s-iQ}{j+\frac{1}{2}+iP}=\mathrm{e}^{2i\delta}\quad\text{and}\quad\frac{\Gamma(1+s+iQ)}{\Gamma(1+s-iQ)}=\mathrm{e}^{2i\zeta} \tag{204.14}$$

are phase factors. For the second this is obvious; the first one yields

$$\left|\frac{s-iQ}{j+\frac{1}{2}+iP}\right|^2=\frac{s^2+Q^2}{(j+\frac{1}{2})^2+P^2}=1$$

because of

$$Q^2-P^2=\beta^2\quad\text{and}\quad s^2=(j+\tfrac{1}{2})^2-\beta^2.$$

We therefore finally arrive at

$$g=|C_1|\,\mathrm{e}^{i\delta}\,2i\,\frac{\sin\xi}{r};\qquad f=|C_1|\,\mathrm{e}^{i\delta}\,2\,\frac{\cos\xi}{\eta r} \tag{204.15}$$

with

$$\xi=kr+Q\log 2kr-\frac{\pi s}{2}-\zeta-\delta. \tag{204.16}$$

NB. The result may be compared with the unrelativistic theory where $\eta \ll 1$ so that $f \gg g$. This gives

$$f \propto \frac{1}{r} \sin\left(kr + Q \log 2kr - \frac{\pi}{2}(s+1) - \zeta - \delta\right) \tag{204.17}$$

where, according to (204.5) and (204.1 b), Q may be expressed by E and further by the velocity v at infinity. From the well-known relation

$$E = mc^2 \left(1 - \frac{v^2}{c^2}\right)^{-\frac{1}{2}}$$

one easily deduces

$$Q = \frac{Ze^2}{\hbar v},$$

i.e. Q becomes identical with $\varkappa$, Problem 110. From (204.1 b) one further gathers that $\hbar k$ is the momentum of the electron, at infinity. The first two terms in the argument of (204.17) therefore agree with the classical expressions. The constant phase angles, in the same approximation, follow from

$$s \to j + \tfrac{1}{2} = l + 1 \quad \text{and} \quad P \to Q = \varkappa;$$

they become

$$\frac{\pi}{2}(s+1) + \zeta + \delta \to \frac{\pi}{2}(l+2) + \arg \Gamma(l+2+i\varkappa) - \arg(l+1+i\varkappa)$$

$$= \pi + \frac{\pi}{2} l + \arg \Gamma(l+1+i\varkappa),$$

and as the wave function is defined except for a sign (i.e., for a phase π) we arrive finally at

$$f \propto \frac{1}{r} \sin\left\{kr + \varkappa \log 2kr - \frac{\pi l}{2} - \arg \Gamma(l+1+i\varkappa)\right\}$$

in complete agreement with the results of Problem 111, but for an attractive Coulomb potential.

Problem 205. Angular momentum expansion of plane Dirac wave

The angular momentum eigenspinors of Problem 190 shall be used to construct the plane wave, of helicity $h = +1$, in z direction:

$$\psi = C \begin{pmatrix} 1 \\ 0 \\ \eta \\ 0 \end{pmatrix} e^{ikz} = C \begin{pmatrix} 1 \\ 0 \\ \eta \\ 0 \end{pmatrix} \sum_{l=0}^{\infty} \sqrt{4\pi(2l+1)}\, i^l\, Y_{l,0}\,. \tag{205.1}$$

Solution. We start by solving Eqs. (201.9a, b) in the force-free case. Using the abbreviation η, introduced for plane waves in Problem 190, we have with

$$\frac{E-mc^2}{\hbar c}=k\eta\,;\qquad \frac{E+mc^2}{\hbar c}=\frac{k}{\eta}, \tag{205.2}$$

instead of (201.9a)

$$g'+\frac{j+\frac{3}{2}}{r}g-ik\eta f=0\,; \tag{205.3a}$$

$$f'-\frac{j-\frac{1}{2}}{r}f-i\frac{k}{\eta}g=0\,. \tag{205.3b}$$

From (205.3b) we get

$$i\frac{k}{\eta}g=f'-\frac{j-\frac{1}{2}}{r}f \tag{205.4}$$

and by differentiation

$$i\frac{k}{\eta}g'=f''-\frac{j-\frac{1}{2}}{r}f'+\frac{j-\frac{1}{2}}{r^2}f\,.$$

Putting these expressions into (205.3a), a second-order equation for f results:

$$f''+\frac{2}{r}f'+\left[k^2-\frac{(j-\frac{1}{2})(j+\frac{1}{2})}{r^2}\right]f=0 \tag{205.5}$$

with the solution

$$f=\frac{1}{kr}j_{j-\frac{1}{2}}(kr) \tag{205.6}$$

regular at the origin, in arbitrary normalization. Putting then (205.6) into (205.4), we find

$$i\frac{k}{\eta}g=\frac{1}{kr}\left\{j'_{j-\frac{1}{2}}-\frac{j+\frac{1}{2}}{kr}j_{j-\frac{1}{2}}\right\}$$

with the prime denoting differentiation with respect to the argument kr. The general formula

$$j'_l(z)-\frac{l+1}{z}j_l(z)=-j_{l+1}(z)$$

then permits to write

$$g=\frac{i\eta}{kr}j_{j+\frac{1}{2}}(kr)\,. \tag{205.7}$$

Thus the normalization of g is fixed relative to that of f. Eqs. (205.6) and (205.7) allow us to write the full four-component spinor according to (201.4a) for given j and $m_j = +\frac{1}{2}$:

$$\psi_j^{\mathrm{I}} = \frac{1}{kr} \begin{pmatrix} \sqrt{\frac{j+\frac{1}{2}}{2j}}\, j_{j-\frac{1}{2}}(kr)\, Y_{j-\frac{1}{2},0} \\ -\sqrt{\frac{j-\frac{1}{2}}{2j}}\, j_{j-\frac{1}{2}}(kr)\, Y_{j-\frac{1}{2},1} \\ i\eta \sqrt{\frac{j+\frac{1}{2}}{2(j+1)}}\, j_{j+\frac{1}{2}}(kr)\, Y_{j+\frac{1}{2},0} \\ i\eta \sqrt{\frac{j+\frac{3}{2}}{2(j+1)}}\, j_{j+\frac{1}{2}}(kr)\, Y_{j+\frac{1}{2},1} \end{pmatrix} . \tag{205.8}$$

The second solution derived in Problem 201 is obtained by replacing η by $1/\eta$ in (205.3a, b) so that f is still given by (205.6), but in (205.7) the factor η has to be put in the denominator instead of in the numerator. From (201.4b) with an arbitrary normalization we get for the second solution,

$$\psi_j^{\mathrm{II}} = \frac{1}{kr} \begin{pmatrix} \frac{i}{\eta} \sqrt{\frac{j+\frac{1}{2}}{2(j+1)}}\, j_{j+\frac{1}{2}}(kr)\, Y_{j+\frac{1}{2},0} \\ \frac{i}{\eta} \sqrt{\frac{j+\frac{3}{2}}{2(j+1)}}\, j_{j+\frac{1}{2}}(kr)\, Y_{j+\frac{1}{2},1} \\ \sqrt{\frac{j+\frac{1}{2}}{2j}}\, j_{j-\frac{1}{2}}(kr)\, Y_{j-\frac{1}{2},0} \\ -\sqrt{\frac{j-\frac{1}{2}}{2j}}\, j_{j-\frac{1}{2}}(kr)\, Y_{j-\frac{1}{2},1} \end{pmatrix} . \tag{205.9}$$

In order to construct a solution like (205.1), we have to write

$$\psi = \sum_j (A_j \psi_j^{\mathrm{I}} + B_j \psi_j^{\mathrm{II}}) . \tag{205.10}$$

In this expansion it is, of course, possible to change the dummy j in a different way in different sums. We shall do this by again using the orbital momentum quantum number l so that in all sums over

$$j_{j-\frac{1}{2}}(kr)\, Y_{j-\frac{1}{2},m}$$

(with $m=0$ or 1) we shall put $j=l+\frac{1}{2}$, and in all sums over

$$j_{j+\frac{1}{2}}(kr)\,Y_{j+\frac{1}{2},m}$$

we shall put $j=l-\frac{1}{2}$. We then obtain from (205.8) and (205.9):

$$\psi=\sum_{l=0}^{\infty}\frac{j_l(kr)}{\sqrt{2l+1}\,kr}\begin{pmatrix}\left(A_{l+\frac{1}{2}}\sqrt{l+1}+B_{l-\frac{1}{2}}\frac{i}{\eta}\sqrt{l}\right)Y_{l,0}\\ \left(-A_{l+\frac{1}{2}}\sqrt{l}+B_{l-\frac{1}{2}}\frac{i}{\eta}\sqrt{l+1}\right)Y_{l,1}\\ (i\eta A_{l-\frac{1}{2}}\sqrt{l}+B_{l+\frac{1}{2}}\sqrt{l+1})\,Y_{l,0}\\ (i\eta A_{l-\frac{1}{2}}\sqrt{l+1}-B_{l+\frac{1}{2}}\sqrt{l})\,Y_{l,1}\end{pmatrix}. \tag{205.11}$$

Here it is correct to sum over all $l\geqq 0$. In the terms $j=l+\frac{1}{2}$, this follows directly from $j\geqq\frac{1}{2}$. On the other hand, for $j=l-\frac{1}{2}$ the sums would start with $l=1$. In the second and fourth line, however, the zeroth term vanishes because $Y_{0,1}=0$. There remain $j=l-\frac{1}{2}$ terms in the first and third line with $l=0$, but again these vanish due to the factor $\sqrt{l}$.

In order to make the sum (205.11) identical with (205.1) we need only put

$$\left.\begin{aligned}A_{l+\frac{1}{2}}&=\sqrt{4\pi}\,C\,i^l\sqrt{l+1},\\ B_{l+\frac{1}{2}}&=\eta\sqrt{4\pi}\,C\,i^l\sqrt{l+1}\end{aligned}\right\} \tag{205.12}$$

as can be directly checked.

Problem 206. Scattering by a central force potential

A plane Dirac wave of positive helicity is scattered by a central-force potential. To determine the asymptotic behaviour of the scattered wave if the phases are taken from the solutions of the radial wave equations.

Solution. In Problem 201 it has been shown that there are two sets of radial equations, viz.

$$\left.\begin{aligned}\textit{Set I}:\quad g_j'+\frac{j+\frac{3}{2}}{r}g_j-ik\eta f_j+iU(r)f_j&=0,\\ f_j'-\frac{j-\frac{1}{2}}{r}f_j-i\frac{k}{\eta}g_j+iU(r)g_j&=0\end{aligned}\right\} \tag{206.1a}$$

with $U(r)=V(r)/\hbar c$. It can easily be seen that, for $U(r)$ decreasing more steeply than $1/r$, the asymptotic solution may be written

$$g_j(r)\to\frac{i\eta}{r}\sin\sigma_j\,;\qquad f_j(r)\to\frac{1}{r}\cos\sigma_j \tag{206.2a}$$

with

$$\sigma_j=kr-(j+\tfrac{1}{2})\frac{\pi}{2}+\alpha_j\,. \tag{206.3a}$$

The two functions are mutually shifted by a phase angle $\pi/2$ and their respective amplitudes, given in arbitrary normalization, are coupled in such a way that, as the unrelativistic limit ($\eta\to 0$) is approached, f_j becomes the big and g_j the small wave function. If f_j is chosen to be real, g_j is purely imaginary. The phase α_j is determined by integrating the set (206.1a) with the boundary conditions $g_j(0)=0$, $f_j(0)=0$. In the classical approach, $j=l+\frac{1}{2}$ so that we get

$$f_j(r)\to\frac{1}{r}\sin\left(kr-l\frac{\pi}{2}+\alpha_{l+\frac{1}{2}}\right);\qquad g_j(r)\to 0$$

at large r.

$$\textit{Set II}:\quad \left.\begin{aligned} &g_j'+\frac{j+\frac{3}{2}}{r}g_j-i\frac{k}{\eta}f_j+iU(r)f_j=0,\\ &f_j'-\frac{j-\frac{1}{2}}{r}f_j-ik\eta g_j+iU(r)g_j=0\end{aligned}\right\} \tag{206.1b}$$

with the asymptotic solution

$$g_j\to\frac{i}{\eta r}\sin\tau_j\,;\qquad f_j\to\frac{1}{r}\cos\tau_j \tag{206.2b}$$

and

$$\tau_j=kr-(j+\tfrac{1}{2})\frac{\pi}{2}+\beta_j \tag{206.3b}$$

with other phase constants β_j than in case I. Since (206.1b) differs from (206.1a) by η being replaced by $1/\eta$, in the classical limit there will g_j become the big, and f_j the small wave function. This leads to the identification $j=l-\frac{1}{2}$ in this approach:

$$g_j(r)\to\frac{1}{r}\sin\left(kr-l\frac{\pi}{2}+\beta_{l-\frac{1}{2}}\right);\qquad f_j(r)\to 0\,.$$

We have further seen in Problem 201 that there exist two solutions ψ_j^{I} and ψ_j^{II} for each value of j belonging to spin in positive z direction, constructed from these two sets. They behave asymptotically as

$$\psi_j^{\mathrm{I}} \to \frac{1}{kr}\begin{pmatrix} \sqrt{\frac{j+\frac{1}{2}}{2j}}\cos\sigma_j\, Y_{j-\frac{1}{2},0} \\ -\sqrt{\frac{j-\frac{1}{2}}{2j}}\cos\sigma_j\, Y_{j-\frac{1}{2},1} \\ i\eta\sqrt{\frac{j+\frac{1}{2}}{2(j+1)}}\sin\sigma_j\, Y_{j+\frac{1}{2},0} \\ i\eta\sqrt{\frac{j+\frac{3}{2}}{2(j+1)}}\sin\sigma_j\, Y_{j+\frac{1}{2},1} \end{pmatrix};$$

$$\psi_j^{\mathrm{II}} \to \frac{1}{kr}\begin{pmatrix} \frac{i}{\eta}\sqrt{\frac{j+\frac{1}{2}}{2(j+1)}}\sin\tau_j\, Y_{j+\frac{1}{2},0} \\ \frac{i}{\eta}\sqrt{\frac{j+\frac{3}{2}}{2(j+1)}}\sin\tau_j\, Y_{j+\frac{1}{2},1} \\ \sqrt{\frac{j+\frac{1}{2}}{2j}}\cos\tau_j\, Y_{j-\frac{1}{2},0} \\ -\sqrt{\frac{j-\frac{1}{2}}{2j}}\cos\tau_j\, Y_{j-\frac{1}{2},1} \end{pmatrix}. \tag{206.4}$$

The most general solution is

$$\psi = \sum_j (A_j \psi_j^{\mathrm{I}} + B_j \psi_j^{\mathrm{II}}) \tag{206.5}$$

which, by changing the summation subscript j into $l \pm \frac{1}{2}$ everywhere so that spherical harmonics of the order l emerge in the sum, may be written

$$\psi \to \frac{1}{kr}\sum_{l=0}^{\infty}\frac{1}{\sqrt{2l+1}}$$

$$\begin{pmatrix} \left[\sqrt{l+1}\,A_{l+\frac{1}{2}}\cos\sigma_{l+\frac{1}{2}} + \frac{i}{\eta}\sqrt{l}\,B_{l-\frac{1}{2}}\sin\tau_{l-\frac{1}{2}}\right] Y_{l,0} \\ \left[-\sqrt{l}\,A_{l+\frac{1}{2}}\cos\sigma_{l+\frac{1}{2}} + \frac{i}{\eta}\sqrt{l+1}\,B_{l-\frac{1}{2}}\sin\tau_{l-\frac{1}{2}}\right] Y_{l,1} \\ \left[i\eta\sqrt{l}\,A_{l-\frac{1}{2}}\sin\sigma_{l-\frac{1}{2}} + \sqrt{l+1}\,B_{l+\frac{1}{2}}\cos\tau_{l+\frac{1}{2}}\right] Y_{l,0} \\ \left[i\eta\sqrt{l+1}\,A_{l-\frac{1}{2}}\sin\sigma_{l-\frac{1}{2}} - \sqrt{l}\,B_{l+\frac{1}{2}}\cos\tau_{l+\frac{1}{2}}\right] Y_{l,1} \end{pmatrix}. \tag{206.6}$$

This has the same structure as the plane wave (205.11, 12) with which it becomes identical if $\alpha_j=0$ and $\beta_j=0$ for all j. Let us write the plane wave

$$\psi^0=\sum_j\left(A_j^0\,\psi_j^{0\,\mathrm{I}}+B_j^0\,\psi_j^{0\,\mathrm{II}}\right) \tag{206.7}$$

with

$$\left.\begin{aligned} &A^0_{l+\frac12}=\sqrt{4\pi}\,C\,i^l\sqrt{l+1}\,; \qquad B^0_{l+\frac12}=\eta\,A^0_{l+\frac12}\,;\\ &\sigma^0_{l-\frac12}=\tau^0_{l-\frac12}=k\,r-l\,\frac{\pi}{2}\,. \end{aligned}\right\} \tag{206.8}$$

The boundary conditions of a scattering problem, for $r\to\infty$, are then satisfied if, and only if, the difference

$$\psi_s=\psi-\psi^0 \tag{206.9}$$

contains no incoming spherical wave parts of the form $\mathrm{e}^{-ikr}/k\,r$ but only outgoing waves so that it may be identified with the scattered wave. This, according to (206.6), leads to four coefficient relations, viz.

$$\begin{aligned}
&\sqrt{l+1}\,A_{l+\frac12}\,\mathrm{e}^{i\left(\frac{\pi}{2}-\alpha_{l+\frac12}\right)}+\frac{1}{\eta}\sqrt{l}\,B_{l-\frac12}\,\mathrm{e}^{-i\beta_{l-\frac12}}=\sqrt{l+1}\,A^0_{l+\frac12}\,\mathrm{e}^{i\frac{\pi}{2}}+\frac{1}{\eta}\sqrt{l}\,B^0_{l-\frac12}\,,\\
&-\sqrt{l}\,A_{l+\frac12}\,\mathrm{e}^{i\left(\frac{\pi}{2}-\alpha_{l+\frac12}\right)}+\frac{1}{\eta}\sqrt{l+1}\,B_{l-\frac12}\,\mathrm{e}^{-i\beta_{l-\frac12}}=-\sqrt{l}\,A^0_{l+\frac12}\,\mathrm{e}^{i\frac{\pi}{2}}+\frac{1}{\eta}\sqrt{l+1}\,B^0_{l-\frac12}\,,\\
&-\eta\sqrt{l}\,A_{l-\frac12}\,\mathrm{e}^{-i\alpha_{l-\frac12}}+\sqrt{l+1}\,B_{l+\frac12}\,\mathrm{e}^{i\left(\frac{\pi}{2}-\beta_{l+\frac12}\right)}=-\eta\sqrt{l}\,A^0_{l-\frac12}+\sqrt{l+1}\,B^0_{l+\frac12}\,\mathrm{e}^{i\frac{\pi}{2}}\,,\\
&-\eta\sqrt{l+1}\,A_{l-\frac12}\,\mathrm{e}^{-i\alpha_{l-\frac12}}-\sqrt{l}\,B_{l+\frac12}\,\mathrm{e}^{i\left(\frac{\pi}{2}-\beta_{l+\frac12}\right)}=-\eta\sqrt{l+1}\,A^0_{l-\frac12}-\sqrt{l}\,B^0_{l+\frac12}\,\mathrm{e}^{i\frac{\pi}{2}}\,.
\end{aligned} \tag{206.10}$$

These equations are satisfied if, and only if,

$$A_j=A_j^0\,\mathrm{e}^{i\alpha_j}\,; \qquad B_j=B_j^0\,\mathrm{e}^{i\beta_j}\,. \tag{206.11}$$

The scattered wave then can be shown by straightforward calculation to behave asymptotically as

$$\psi_s\to\sqrt{4\pi}\,C\,\frac{\mathrm{e}^{ikr}}{2ikr}\sum_{l=0}^{\infty}\frac{1}{\sqrt{2l+1}}\begin{pmatrix}[(l+1)(\mathrm{e}^{2i\alpha_{l+\frac12}}-1)+l(\mathrm{e}^{2i\beta_{l-\frac12}}-1)]\;Y_{l,0}\\ \sqrt{l(l+1)}\,(\mathrm{e}^{2i\beta_{l-\frac12}}-\mathrm{e}^{2i\alpha_{l+\frac12}})\;Y_{l,1}\\ \eta\,[l(\mathrm{e}^{2i\alpha_{l-\frac12}}-1)+(l+1)(\mathrm{e}^{2i\beta_{l+\frac12}}-1)]\;Y_{l,0}\\ \eta\sqrt{l(l+1)}\,(\mathrm{e}^{2i\alpha_{l-\frac12}}-\mathrm{e}^{2i\beta_{l+\frac12}})\;Y_{l,1}\end{pmatrix}. \tag{206.12}$$

Problem 207. Continuous potential step

A plane wave of positive helicity $h=+1$ is falling perpendicularly from negative z upon a potential step described by

$$V(z)=\tfrac{1}{2}V_0\left(1+\tanh\frac{z}{l}\right), \tag{207.1}$$

i.e. with the potential increasing from $V=0$ at $z=-\infty$ to $V=+V_0$ at $z=+\infty$, within a layer around $z=0$ of a thickness of the order l. The coefficient of transmission of the step shall be investigated for the cases of different step heights, viz.

$$\left.\begin{array}{ll}\text{Case a:} & V_0<E-mc^2,\\ \text{Case b:} & E-mc^2<V_0<E+mc^2,\\ \text{Case c:} & E+mc^2<V_0.\end{array}\right\} \tag{207.2}$$

The three cases are sketched in Fig. 72.

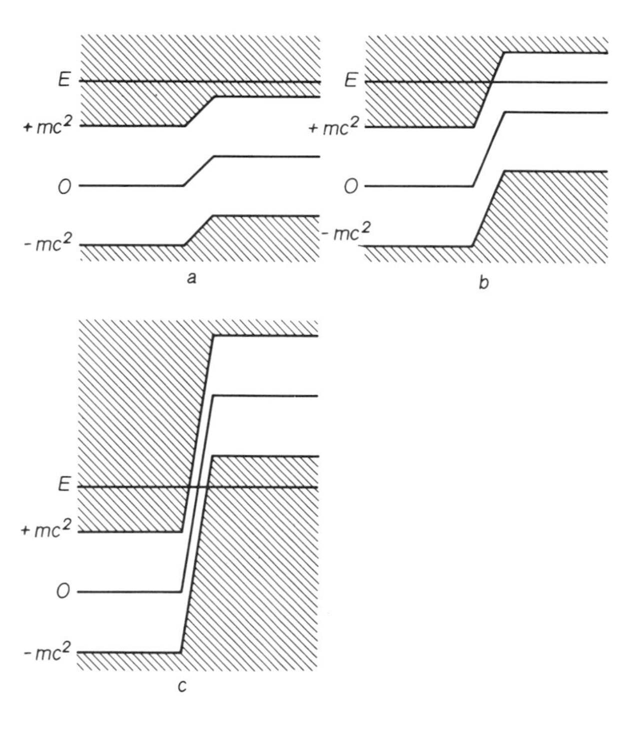

Fig. 72a—c. Three potential steps of different heights. The energy domains in which the particle can move are indicated by hatching

Solution. This is a special case of Problem 197. For helicity $h=+1$, the component functions u_2 and u_4 vanish so that we are left with a set of only two differential equations,

$$\left.\begin{aligned} -i\frac{du_3}{dz}+(\varkappa+Q)\,u_1=0,\\ i\frac{du_1}{dz}+(\varkappa-Q)\,u_3=0 \end{aligned}\right\} \tag{207.3}$$

with

$$Q=\frac{V(z)-E}{\hbar c}; \qquad \varkappa=\frac{mc^2}{\hbar c}. \tag{207.4}$$

Let us introduce, instead of u_1 and u_3, their symmetric and antisymmetric combinations,

$$\varphi_s=u_1+u_3; \qquad \varphi_a=u_1-u_3, \tag{207.5}$$

then, instead of (207.3), we have the equations

$$-\varkappa\varphi_s=i\varphi_a'+Q\varphi_a; \qquad \varkappa\varphi_a=i\varphi_s'-Q\varphi_s \tag{207.6}$$

which permit a simpler elimination of one of the two functions than do the original equations (207.3). We find

$$\varphi_s''+[Q^2-\varkappa^2+iQ']\,\varphi_s=0. \tag{207.7}$$

We have to solve this differential equation with fitting boundary conditions and afterwards to derive φ_a from the second equation (207.6) in order to solve the problem completely.

The differential equation (207.7) can be written with rational coefficient functions if we use the variable

$$x=(1+e^{2z/l})^{-1} \tag{207.8}$$

instead of z with

$$\frac{du}{dz}=-\frac{2}{l}x(1-x)\frac{du}{dx}; \qquad V(z)=V_0(1-x). \tag{207.9}$$

Further, we shall introduce an energy unit $2\hbar c/l$ and use the dimensionless abbreviations

$$\varepsilon=\frac{El}{2\hbar c}; \qquad \varepsilon_0=\frac{lmc^2}{2\hbar c}; \qquad v_0=\frac{lV_0}{2\hbar c}. \tag{207.10}$$

Eq. (207.7) then is transformed into

$$x(1-x)\frac{d}{dx}\left[x(1-x)\frac{d\varphi_s}{dx}\right]+\{[v_0(1-x)-\varepsilon]^2-\varepsilon_0^2+iv_0x(1-x)\}\,\varphi_s=0. \tag{207.11}$$

If we put

$$\varphi_s = x^{\nu}(1-x)^{\mu} f(x) \tag{207.12}$$

with

$$\nu^2 = \varepsilon_0^2 - (\varepsilon - v_0)^2\,; \qquad \mu^2 = \varepsilon_0^2 - \varepsilon^2 \tag{207.13}$$

this reduces to the hypergeometrical equation

$$x(1-x) f'' + [(2\nu+1) - (2\nu+2\mu+2)x] f' \\ - (\mu+\nu-i v_0)(\mu+\nu-i v_0+1) f = 0 \tag{207.14}$$

of which we need, as we shall immediately see, only the solution regular at $x=0$,

$$f(x) = {}_2F_1(\mu+\nu-i v_0, \mu+\nu+i v_0+1, 2\nu+1; x)\,. \tag{207.15}$$

In order to see this, we next study the *boundary conditions*. According to (207.8) we have $x=1$ for $z=-\infty$, and $x=0$ for $z=+\infty$. Further, according to (207.10) and (207.13), we have

$$\mu = -i\frac{l}{2}k\,; \qquad k^2 = \frac{E^2-(mc^2)^2}{(\hbar c)^2} = \left(\frac{p}{\hbar}\right)^2; \tag{207.16}$$

hence μ is always an imaginary parameter proportional to the particle momentum $p=\hbar k$ for $V=0$, i.e. on the far left. The hypergeometric series (207.15), in the neighbourhood of $x=1$, may be transformed according to the rule

$${}_2F_1(a,b,c;x) = \frac{\Gamma(c)\Gamma(c-a-b)}{\Gamma(c-a)\Gamma(c-b)}\,{}_2F_1(a,b,a+b-c+1;1-x) \\ +(1-x)^{c-a-b}\frac{\Gamma(c)\Gamma(a+b-c)}{\Gamma(a)\Gamma(b)}\,{}_2F_1(c-a,c-b,c-a-b+1;1-x)$$

which, with (207.8) and (207.15), leads for $z\to-\infty$, $x\simeq 1$ to

$$\varphi_s \to (1-x)^{\mu}\left\{\frac{\Gamma(2\nu+1)\Gamma(-2\mu)}{\Gamma(\nu-\mu+i v_0+1)\Gamma(\nu-\mu-i v_0)} \right. \\ \left. + (1-x)^{-2\mu}\frac{\Gamma(2\nu+1)\Gamma(2\mu)}{\Gamma(\nu+\mu-i v_0)\Gamma(\nu+\mu+i v_0+1)}\right\}$$

or, with

$$1-x = \frac{e^{2z/l}}{1+e^{2z/l}} \to e^{2z/l}$$

and (207.16), to

$$\varphi_s \to A\,e^{ikz} + B\,e^{-ikz}; \tag{207.17}$$

$$A = \frac{\Gamma(2\nu+1)\,\Gamma(2\mu)}{\Gamma(\nu+\mu-i v_0)\,\Gamma(\nu+\mu+i v_0+1)}; \quad B = \frac{\Gamma(2\nu+1)\,\Gamma(-2\mu)}{\Gamma(\nu-\mu-i v_0)\,\Gamma(\nu-\mu+i v_0+1)} \tag{207.18}$$

where B differs from A just by the sign of μ. The function φ_s therefore is, at large negative z, composed of an incoming wave with amplitude A and a reflected wave with amplitude B, conforming to the physical problem. Hence the special solution (207.15) satisfies the boundary condition at large negative values of z. The same composition of two waves holds for φ_a if we apply the second equation (207.6) to the asymptotic function (207.17),

$$\varphi_a \to A\left(\frac{E}{mc^2} - \frac{k}{\varkappa}\right) e^{ikz} + B\left(\frac{E}{mc^2} + \frac{k}{\varkappa}\right) e^{-ikz}. \tag{207.19}$$

The electrical current density (cf. Problem 198, Eq. (198.13))

$$j_z = ec(u_1^* u_3 + u_3^* u_1) = \tfrac{1}{2} ec(|\varphi_s|^2 - |\varphi_a|^2) \tag{207.20}$$

is then, except for interference terms,

$$j_z = j_{\text{in}} - j_{\text{refl}} \tag{207.21}$$

with the incident current

$$j_{\text{in}} = \tfrac{1}{2} ec|A|^2 \left\{1 - \left(\frac{E}{mc^2} - \frac{k}{\varkappa}\right)^2\right\} = ec|A|^2 \frac{pc(E-pc)}{(mc^2)^2} \tag{207.22}$$

and the reflected current

$$j_{\text{refl}} = \tfrac{1}{2} ec|B|^2 \left\{1 - \left(\frac{E}{mc^2} + \frac{k}{\varkappa}\right)^2\right\} = ec|B|^2 \frac{pc(E+pc)}{(mc^2)^2} \tag{207.23}$$

where the energy formula

$$E = \sqrt{(pc)^2 + (mc^2)^2}$$

is to be remembered.

Let us now pass on to a discussion of the behaviour of the wave function on the right-hand side of the potential step, near $x=0$ or for $z \to +\infty$. It then follows from (207.12) and (207.15) directly that

$$\varphi_s \to x^\nu = e^{-\frac{2}{l}\nu z} \tag{207.24}$$

where, according to (207.10) and (207.13)

$$\nu = -i\frac{l}{2}k'; \quad k'^2 = \frac{(E-V_0)^2 - (mc^2)^2}{(\hbar c)^2}. \tag{207.25}$$

Here we have to distinguish three cases according to Eq. (207.2). If $E-V_0>mc^2$ or $V_0-E>mc^2$ (cases a and c), $k'^2>0$ and k' is real. If, on the other hand, $|E-V_0|<mc^2$ (case b), ν will be real and k' imaginary. In the last case, with $\nu>0$, Eq. (207.24) describes total reflection of the incident wave so that the *coefficient of reflection*,

$$R=j_{\mathrm{refl}}/j_{\mathrm{in}}=\left|\frac{B}{A}\right|^2\frac{E+pc}{E-pc}, \tag{207.26}$$

must become $=1$.

This can easily be derived from A and B, Eq. (207.18), by using the identity $\Gamma(z+1)=z\Gamma(z)$,

$$\frac{B}{A}=\frac{\nu+\mu+i v_0}{\nu-\mu+i v_0}\cdot\frac{\Gamma(\nu+\mu-i v_0)\Gamma(\nu+\mu+i v_0)}{\Gamma(\nu-\mu+i v_0)\Gamma(\nu-\mu-i v_0)}\cdot\frac{\Gamma(-2\mu)}{\Gamma(2\mu)}. \tag{207.27}$$

Since

$$\mu=-i\sigma$$

is always imaginary, the last of the three fractions in (207.27) never contributes to the absolute square $|B/A|^2$. If ν is real (case b), the second factor also is the ratio of two conjugate numbers not contributing. There then remains

$$\left|\frac{B}{A}\right|^2=\frac{\nu^2+(v_0-\sigma)^2}{\nu^2+(v_0+\sigma)^2}=\frac{2v_0(\varepsilon-\sigma)}{2v_0(\varepsilon+\sigma)}=\frac{E-pc}{E+pc},$$

so that (207.26) indeed yields $R=1$.

In cases a and c, on the other hand, ν is imaginary so that a running wave exists on the far right,

$$\varphi_s=e^{ik'z}$$

and according to (207.6),

$$\varphi_a=\left(-\frac{k'}{\varkappa}+\frac{E-V_0}{mc^2}\right)e^{ik'z}=\frac{E'-p'c}{mc^2}e^{ik'z}$$

where $p'=\hbar k'$ is the particle momentum for $z\to+\infty$, and $E'=E-V_0$. The electrical current transmitted is then, according to (207.20),

$$j_{\mathrm{trans}}=\tfrac{1}{2}ec\left\{1-\left(\frac{E'-p'c}{mc^2}\right)^2\right\}=ec\frac{p'c(E'-p'c)}{(mc^2)^2}. \tag{207.28}$$

This yields, with (207.22), a *coefficient of transmission*,

$$T=j_{\mathrm{trans}}/j_{\mathrm{in}}=\frac{1}{|A|^2}\cdot\frac{p'c(E'-p'c)}{pc(E-pc)}. \tag{207.29}$$

In now computing $|A|^2$ from Eq. (207.18) we use, besides $\Gamma(z+1)=z\,\Gamma(z)$, the general formula

$$|\Gamma(\pm i y)|^2=\frac{\pi}{y\sinh\pi y}\ (y\ \text{real});$$

we then have, with

$$\mu=-i\sigma,\qquad \nu=-i\sigma',$$

$$1/|A|^2=\frac{\sigma}{\sigma'}\cdot\frac{\sigma+\sigma'-v_0}{\sigma+\sigma'+v_0}\cdot\frac{\sinh 2\pi\sigma\,\sinh 2\pi\sigma'}{\sinh\pi(\sigma+\sigma'+v_0)\,\sinh\pi(\sigma+\sigma'-v_0)}. \tag{207.30}$$

Combining (207.30) with (207.29), there occurs a factor

$$\frac{p'c(E'-p'c)}{pc(E-pc)}\cdot\frac{\sigma}{\sigma'}\cdot\frac{\sigma+\sigma'-v_0}{\sigma+\sigma'+v_0}=\frac{(E'-p'c)(pc+p'c-V_0)}{(E-pc)(pc+p'c+V_0)}$$

which can easily be shown to be $=1$: if we replace V_0 by $E-E'$, the expression may be written

$$\frac{(E'-p'c)(E'+p'c)-(E'-p'c)(E-pc)}{(E-pc)(E+pc)-(E-pc)(E'-p'c)}=1$$

since $E'^2-(p'c)^2=(mc^2)^2=E^2-(pc)^2$ in the first terms of numerator and denominator. We therefore finally obtain the coefficient of transmission,

$$T=\frac{\sinh 2\pi\sigma\,\sinh 2\pi\sigma'}{\sinh\pi(\sigma+\sigma'+v_0)\,\sinh\pi(\sigma+\sigma'-v_0)}. \tag{207.31}$$

The denominator can be suitably reshaped so that the characteristic quantity v_0, proportional to the product of breadth and height of the step and independent of the particle energy, is isolated from the momentum quantities σ and σ':

$$T=\frac{\sinh 2\pi\sigma\,\sinh 2\pi\sigma'}{\sinh^2\pi(\sigma+\sigma')\cosh^2\pi v_0-\cosh^2\pi(\sigma+\sigma')\sinh^2\pi v_0}. \tag{207.32}$$

In case a of Eq. (207.2), we still have $\sigma+\sigma'-v_0>0$ or $V_0<(p+p')c$. This is the normal case which also occurs in unrelativistic theory. In case c, on the other hand, we find $\sigma+\sigma'-v_0<0$ and therefore $T<0$. The wave then penetrates, for large positive z, into the domain of negative energies (cf. Fig. 72) where negative electrical current accompanies positive momentum. If $\pi v_0\gg 1$ or

$$V_0\gg\frac{2\pi\hbar c}{l},$$

the expression (207.32) becomes

$$T=-4\sinh\frac{\pi l p}{\hbar}\sinh\frac{\pi l p'}{\hbar}\,e^{-\frac{\pi l V_0}{\hbar c}},$$

i.e. the penetrability of the potential step from positive to negative energies rapidly becomes very small with increasing "step size" $V_0 l$. Since, in case c, $V_0 > mc^2$, the exponential in T contributes a factor smaller than

$$e^{-\frac{\pi l m c}{\hbar}} = e^{-\pi l/\lambda}$$

with $\lambda = \hbar/mc$ the Compton wavelength.

Literature. Klein, O.: Z. f. Physik **53**, 157 (1929); Sauter, F.: Z. f. Physik **69**, 742; **73**, 547 (1931).

Problem 208. Plane wave at a potential jump

A plane Dirac wave of arbitrary polarization falls obliquely on a potential jump smaller than its kinetic energy. The laws of reflection and refraction shall be derived, and the state of polarization of the transmitted wave calculated.

Solution. Let the incident wave ψ be described by a wave vector $\boldsymbol{k}$ in the direction defined by polar angles ϑ, φ, the reflected wave ψ' by $\boldsymbol{k}'$ in the direction ϑ', φ', and the transmitted wave ψ'' by $\boldsymbol{k}''$ in the direction ϑ'', φ''. Let further $z=0$ be the refracting surface and ψ and ψ' be defined for $z<0$, ψ'' for $z>0$ (see Fig. 73).

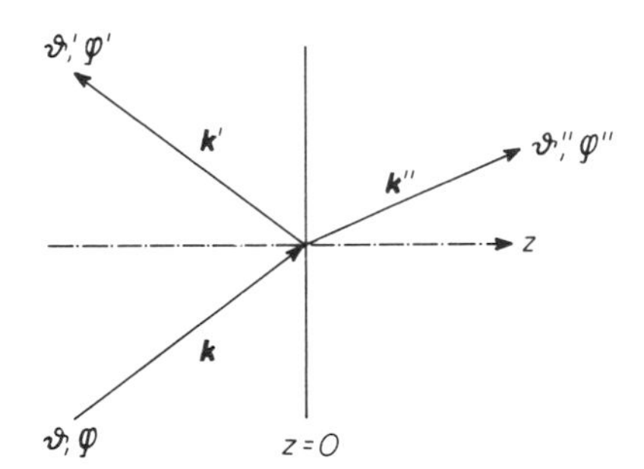

Fig. 73. The three wave vectors at a potential jump

On the surface $z=0$ there must hold for all values of x and y the relation

$$\psi + \psi' = \psi'' \tag{208.1}$$

according to which the three $\boldsymbol{k}$ vectors have the same x and y components:

$$k \sin\vartheta \cos\varphi = k \sin\vartheta' \cos\varphi' = k'' \sin\vartheta'' \cos\varphi'' ;$$

$$k \sin\vartheta \sin\varphi = k \sin\vartheta' \sin\varphi' = k'' \sin\vartheta'' \sin\varphi'' .$$

These relations are satisfied if

$$\varphi=\varphi'=\varphi'' ; \tag{208.2}$$

$$\vartheta'=\pi-\vartheta ; \tag{208.3}$$

$$k \sin\vartheta=k'' \sin\vartheta'' . \tag{208.4}$$

Eq. (208.2) shows that the three $\boldsymbol{k}$ vectors are lying in the same meridional plane, which we may choose to be the x, z plane, so that the y components of the three vectors vanish along with $\varphi=0$, $\varphi'=0$, $\varphi''=0$. Eq. (208.3) is the law of reflection, and Eq. (208.4) the law of refraction, the index of refraction being $n=k''/k$. These two laws are very much the same as those holding for unrelativistic Schrödinger waves (Problem 45).

Additional information arises for the *polarization*. Using $\varphi=0$ and $\vartheta'=\pi-\vartheta$, the three wave functions in standard representation run as follows:

$$\psi=\frac{e^{i\boldsymbol{k}\boldsymbol{r}}}{\sqrt{V(1+\eta^2)}}\begin{pmatrix} A\cos\frac{\vartheta}{2}+B\sin\frac{\vartheta}{2} \\ A\sin\frac{\vartheta}{2}-B\cos\frac{\vartheta}{2} \\ \eta(A\cos\frac{\vartheta}{2}-B\sin\frac{\vartheta}{2}) \\ \eta(A\sin\frac{\vartheta}{2}+B\cos\frac{\vartheta}{2}) \end{pmatrix};$$

$$\psi'=\frac{e^{i\boldsymbol{k}'\boldsymbol{r}}}{\sqrt{V(1+\eta^2)}}\begin{pmatrix} C\sin\frac{\vartheta}{2}+D\cos\frac{\vartheta}{2} \\ C\cos\frac{\vartheta}{2}-D\sin\frac{\vartheta}{2} \\ \eta(C\sin\frac{\vartheta}{2}-D\cos\frac{\vartheta}{2}) \\ \eta(C\cos\frac{\vartheta}{2}+D\sin\frac{\vartheta}{2}) \end{pmatrix};$$

$$\psi''=\frac{e^{i\boldsymbol{k}''\boldsymbol{r}}}{\sqrt{V(1+\eta''^2)}}\begin{pmatrix} E\cos\frac{\vartheta''}{2}+F\sin\frac{\vartheta''}{2} \\ E\sin\frac{\vartheta''}{2}-F\cos\frac{\vartheta''}{2} \\ \eta''(E\cos\frac{\vartheta''}{2}-F\sin\frac{\vartheta''}{2}) \\ \eta''(E\sin\frac{\vartheta''}{2}+F\cos\frac{\vartheta''}{2}) \end{pmatrix}. \tag{208.5}$$

Here, the first parts of the spinors, proportional to the constants A, C, and E, respectively, belong to waves with helicity $+1$, the parts with B, D, and F to helicity -1.

The boundary condition (208.1) applied to the amplitudes then leads to the following four-component relations:

$$\begin{aligned} A+pB+pC+D&=r(E+qF), \\ pA-B+C-pD&=r(qE-F), \\ A-pB+pC-D&=r\lambda(E-qF), \\ pA+B+C+pD&=r\lambda(qE+F) \end{aligned} \tag{208.6}$$

with the abbreviations

$$p=\tan\frac{\vartheta}{2};\quad q=\tan\frac{\vartheta''}{2};\quad r=\sqrt{\frac{1+\eta^2}{1+\eta''^2}\;\frac{\cos\frac{\vartheta''}{2}}{\cos\frac{\vartheta}{2}}};\quad \lambda=\frac{\eta''}{\eta}. \tag{208.7}$$

By combination we find from Eqs. (208.6)

$$\begin{aligned} A+pC&=\tfrac{1}{2}r[(1+\lambda)E+(1-\lambda)qF],\\ pA+C&=\tfrac{1}{2}r[(1+\lambda)qE-(1-\lambda)F] \end{aligned} \tag{208.8a}$$

and

$$\begin{aligned} B+pD&=\tfrac{1}{2}r[-(1-\lambda)qE+(1+\lambda)F],\\ pB+D&=\tfrac{1}{2}r[(1-\lambda)E+(1+\lambda)qF]. \end{aligned} \tag{208.8b}$$

From the first pair of equations we then eliminate C and from the other pair D so that there remain relations connecting the amplitude coefficients E and F of the transmitted wave with A and B of the incident wave, viz.

$$\begin{aligned} (1-p^2)A&=\tfrac{1}{2}r[(1+\lambda)(1-pq)E+(1-\lambda)(p+q)F];\\ (1-p^2)B&=\tfrac{1}{2}r[-(1-\lambda)(p+q)E+(1+\lambda)(1-pq)F]. \end{aligned} \tag{208.9}$$

The expectation value of the helicity (or, briefly, the polarization) of the incident wave is defined by

$$h=\frac{A^2-B^2}{A^2+B^2}=\frac{1-(B/A)^2}{1+(B/A)^2} \tag{208.10}$$

and that of the transmitted wave by

$$h''=\frac{E^2-F^2}{E^2+F^2}=\frac{1-(F/E)^2}{1+(F/E)^2}. \tag{208.11}$$

From (208.9) we have

$$B/A=\frac{-(1-\lambda)(p+q)+(1+\lambda)(1-pq)(F/E)}{(1+\lambda)(1-pq)+(1-\lambda)(p+q)(F/E)}. \tag{208.12}$$

The expression can be much simplified by using the abbreviation

$$u=\frac{1-\lambda}{1+\lambda}\cdot\frac{p+q}{1-pq}=\frac{\eta-\eta''}{\eta+\eta''}\tan\frac{\vartheta+\vartheta''}{2} \tag{208.13}$$

which allows us to write

$$B/A=\frac{(F/E)-u}{1+u(F/E)}. \tag{208.14}$$

If, into this latter expression, h instead of B/A and h'' instead of F/E are introduced according to (208.10) and (208.11), a simple calculation finally leads to

$$h'' = \frac{1-u^2}{1+u^2}\,h - \frac{2u}{1+u^2}\sqrt{1-h^2}\,. \qquad (208.15)$$

If the incident wave is completely polarized in either of the two directions ($h = \pm 1$) we find

$$h'' = \pm\frac{1-u^2}{1+u^2},$$

i.e. a partial depolarization will occur at the potential jump which is, however, of second order only in the parameter u. An entirely unpolarized beam ($h=0$), on the other hand, leads to

$$h'' = -\frac{2u}{1+u^2}$$

the potential jump giving rise to at least partial polarization. The latter effect is of first order in u. Since u in reasonable arrangements generally turns out to be rather small (about 0.1), the partial polarization of an unpolarized beam will be of greater interest than the partial depolarization of a beam of well-defined helicity. It should further be remarked that u vanishes for perpendicular incidence, so that grazing incidence will favour the effect.

Problem 209. Reflected intensity at a potential jump

To calculate the reflection coefficient of the plane wave of mixed helicity investigated in the preceding problem, and to derive the law of continuity of the electrical current at the surface.

Solution. The electrical current density

$$j_l = iec\,\psi^\dagger\gamma_4\gamma_l\psi = ec\,\psi^\dagger\alpha_l\psi$$

can be expressed by the spinor components ψ_μ in standard representation as follows:

$$\begin{aligned} j_x &= ec(\psi_1^*\psi_4+\psi_2^*\psi_3+\psi_3^*\psi_2+\psi_4^*\psi_1),\\ j_y &= eci(-\psi_1^*\psi_4+\psi_2^*\psi_3-\psi_3^*\psi_2+\psi_4^*\psi_1),\\ j_z &= ec(\psi_1^*\psi_3-\psi_2^*\psi_4+\psi_3^*\psi_1-\psi_4^*\psi_2). \end{aligned}$$

Since the exponential factors of ψ_μ^* and ψ_ν in a plane wave cancel out, there only remain the spinor amplitudes which, using the Dirac wave functions of Eq. (208.5), are all real according to our choice of $\varphi=0$. Indeed, this leads to $j_y=0$ above. The two other components become for the incident wave

$$j_x=\frac{2ec}{V(1+\eta^2)}\left\{\left(A\cos\frac{\vartheta}{2}+B\sin\frac{\vartheta}{2}\right)\eta\left(A\sin\frac{\vartheta}{2}+B\cos\frac{\vartheta}{2}\right)\right.$$
$$\left.+\left(A\sin\frac{\vartheta}{2}-B\cos\frac{\vartheta}{2}\right)\eta\left(A\cos\frac{\vartheta}{2}-B\sin\frac{\vartheta}{2}\right)\right\};$$

$$j_z=\frac{2ec}{V(1+\eta^2)}\left\{\left(A\cos\frac{\vartheta}{2}+B\sin\frac{\vartheta}{2}\right)\eta\left(A\cos\frac{\vartheta}{2}-B\sin\frac{\vartheta}{2}\right)\right.$$
$$\left.-\left(A\sin\frac{\vartheta}{2}-B\cos\frac{\vartheta}{2}\right)\eta\left(A\sin\frac{\vartheta}{2}+B\cos\frac{\vartheta}{2}\right)\right\}$$

which may be simplified to

$$j_x=\frac{2ec\eta}{V(1+\eta^2)}(A^2+B^2)\sin\vartheta\,; \quad j_z=\frac{2ec\eta}{V(1+\eta^2)}(A^2+B^2)\cos\vartheta\,. \tag{209.1}$$

By an analogous procedure we arrive at

$$j'_x=\frac{2ec\eta}{V(1+\eta^2)}(C^2+D^2)\sin\vartheta\,; \; j'_z=-\frac{2ec\eta}{V(1+\eta^2)}(C^2+D^2)\cos\vartheta \tag{209.2}$$

for the reflected[2], and at

$$j''_x=\frac{2ec\eta''}{V(1+\eta''^2)}(E^2+F^2)\sin\vartheta''\,; \quad j''_z=\frac{2ec\eta''}{V(1+\eta''^2)}(E^2+F^2)\cos\vartheta'' \tag{209.3}$$

for the transmitted wave.

In order to calculate the three currents we first express the amplitude constants C, D, E, F by those, A and B, of the incident wave by solving the linear system (208.6) or (208.8a, b). The result of this rather cumbersome but elementary computation is

$$C=\frac{-\alpha A+\beta B}{\Delta}; \quad D=-\frac{\beta A+\alpha B}{\Delta}; \tag{209.4}$$

$$E=\frac{\rho A+\sigma B}{2r\lambda}; \quad F=\frac{-\sigma A+\rho B}{2r\lambda} \tag{209.5}$$

---

[2] Interference terms of incident and reflected waves may be omitted for the present purpose since $\int d^3x\, e^{i(\boldsymbol{k}-\boldsymbol{k}')\boldsymbol{r}}=0$. When dealing only with local dependence of densities, these interferences are of interest, cf. Problem 23.

with the abbreviations

$$\begin{aligned}
\alpha &= 2(\lambda^2+1)p(1+q^2)-4\lambda q(1+p^2);\\
\beta &= (\lambda^2-1)(1-p^2)(1+q^2);\\
\Delta &= (\lambda+1)^2(1-pq)^2+(\lambda-1)^2(p+q)^2;\\
\rho\Delta &= (\lambda+1)(\Delta-\alpha p)-(\lambda-1)\beta=4\lambda(\lambda+1)(1-pq)(1-p^2);\\
\sigma\Delta &= (\lambda-1)(p\Delta-\alpha)+(\lambda+1)\beta p=4\lambda(\lambda-1)(p+q)(1-p^2).
\end{aligned} \tag{209.6}$$

From Eqs. (209.4) and (209.5) it further follows that

$$\begin{aligned}
\Delta^2\cdot(C^2+D^2)&=(\alpha^2+\beta^2)(A^2+B^2);\\
4r^2\lambda^2(E^2+F^2)&=(\rho^2+\sigma^2)(A^2+B^2).
\end{aligned} \tag{209.7}$$

If the expressions (209.6) are put into (209.7) we find by elementary though lengthy calculation,

$$\Delta\cdot(\rho^2+\sigma^2)=16\lambda^2(1-p^2)^2 \tag{209.8}$$

and

$$\Delta^2-(\alpha^2+\beta^2)=\Delta\cdot 4\lambda(1-p^2)(1-q^2). \tag{209.9}$$

It is now easy to express the reflected and transmitted currents by the incident one. We find for the z components perpendicular to the potential jump surface,

$$j_z' = -\frac{C^2+D^2}{A^2+B^2}j_z = -\frac{\alpha^2+\beta^2}{\Delta^2}j_z \tag{209.10}$$

and

$$\begin{aligned}
j_z'' &= \frac{\eta''}{1+\eta''^2}\cdot\frac{1+\eta^2}{\eta}\cdot\frac{\cos\vartheta''}{\cos\vartheta}\cdot\frac{E^2+F^2}{A^2+B^2}j_z\\
&= \frac{\eta''}{\eta}\cdot\frac{1+\eta^2}{1+\eta''^2}\cdot\frac{\cos\vartheta''}{\cos\vartheta}\cdot\frac{\rho^2+\sigma^2}{4r^2\lambda^2}j_z.
\end{aligned}$$

In the last formula we introduce the definitions (208.7) which render

$$\frac{\eta''}{\eta}\cdot\frac{1+\eta^2}{1+\eta''^2}\cdot\frac{\cos\vartheta''}{\cos\vartheta}\cdot\frac{1}{r^2}=\lambda\frac{1-q^2}{1-p^2};$$

hence

$$j_z''=\frac{1-q^2}{1-p^2}\cdot\frac{\rho^2+\sigma^2}{4\lambda}j_z. \tag{209.11}$$

Eqs. (209.10) and (209.11) when combined with (209.8) and (209.9) then yield

$$j'_z = -\left\{1 - \frac{4\lambda(1-p^2)(1-q^2)}{\Delta}\right\} j_z ; \qquad j''_z = \frac{4\lambda(1-p^2)(1-q^2)}{\Delta} j_z \tag{209.12}$$

with Δ defined in (209.6). From (209.12) there follows immediately the equation of continuity,

$$j_z + j'_z = j''_z . \tag{209.13}$$

The quantity

$$R = 1 - \frac{4\lambda(1-p^2)(1-q^2)}{\Delta} \tag{209.14}$$

is the coefficient of reflection. At normal incidence ($p=0$, $q=0$) it simply becomes

$$R = \left(\frac{\lambda-1}{\lambda+1}\right)^2 . \tag{209.15}$$

VII. Radiation Theory

Problem 210. Quantization of Schrödinger field

The quantization of a force-free Schrödinger wave field into particles obeying either Bose or Fermi statistics shall be discussed using suitable expressions for energy, momentum and electric charge of the field.

Solution. Let us start by treating the force-free Schrödinger equation as a classical wave field, ψ being simply a scalar function of space coordinates and time. Then, in the usual normalization, we have the following integral expressions (cf. Problems 3 and 5): the total field energy is

$$W = -\frac{\hbar^2}{2m}\int d^3x\, \psi^* \nabla^2 \psi\,, \tag{210.1}$$

the total momentum of the field is

$$\boldsymbol{P} = \frac{\hbar}{i}\int d^3x\, \psi^* \nabla \psi\,, \tag{210.2}$$

and the total electric charge of the field is

$$Q = e\int d^3x\, \psi^* \psi\,. \tag{210.3}$$

Here m and e are phenomenological constants not yet explained as particle properties since, so far, the field does not yet consist of particles.

We further know that a Schrödinger field must satisfy two conjugate wave equations,

$$-\frac{\hbar}{i}\frac{\partial \psi}{\partial t} = -\frac{\hbar^2}{2m}\nabla^2\psi\,; \qquad \frac{\hbar}{i}\frac{\partial \psi^*}{\partial t} = -\frac{\hbar^2}{2m}\nabla^2\psi^*\,. \tag{210.4}$$

They may be solved by plane waves which we normalize within an arbitrary periodicity cube of volume $\mathscr{V}$; the complete solution then may be written in the form

$$\psi(\boldsymbol{r},t) = \mathscr{V}^{-\frac{1}{2}} \sum_{\boldsymbol{k}} c_{\boldsymbol{k}}\, e^{i(\boldsymbol{k}\boldsymbol{r}-\omega t)} \tag{210.5}$$

with the law of dispersion following from (210.4),

$$\hbar\omega = \frac{\hbar^2 k^2}{2m}. \tag{210.6}$$

If we put (210.5) in the expressions (210.1) to (210.3) and use the orthonormality relation of the plane waves,

$$\frac{1}{\mathscr{V}} \int_{\mathscr{V}} d^3x \, e^{i(\boldsymbol{k}-\boldsymbol{k}')\boldsymbol{r}} = \delta_{\boldsymbol{k},\boldsymbol{k}'}$$

we get

$$W = \frac{\hbar^2}{2m} \sum_{\boldsymbol{k}} k^2 c_{\boldsymbol{k}}^* c_{\boldsymbol{k}}; \qquad \boldsymbol{P} = \sum_{\boldsymbol{k}} \hbar \boldsymbol{k} c_{\boldsymbol{k}}^* c_{\boldsymbol{k}}; \qquad Q = e \sum_{\boldsymbol{k}} c_{\boldsymbol{k}}^* c_{\boldsymbol{k}}. \tag{210.7}$$

This classical theory now we shall quantize. The wave function ψ shall be replaced by an operator ψ which operates on Hilbert vectors χ of particle numbers. The same then holds for the coefficients $c_{\boldsymbol{k}}$ of the Fourier series (210.5). They have to be replaced by such operators $c_{\boldsymbol{k}}$ and their hermitian conjugates $c_{\boldsymbol{k}}^\dagger$ as to make the eigenvalues of $c_{\boldsymbol{k}}^\dagger c_{\boldsymbol{k}}$ integers, viz.

$$\begin{aligned} &\text{either } N_{\boldsymbol{k}} = 0, 1, 2, 3, \ldots \text{ in the Bose case} \\ &\text{or} \quad N_{\boldsymbol{k}} = 0, 1 \text{ only} \quad \text{in the Fermi case}. \end{aligned} \tag{210.8}$$

If this is done, the three expressions (210.7), too, become operators with the eigenvalues

$$W = \sum_{\boldsymbol{k}} N_{\boldsymbol{k}} E_k; \qquad E_k = \frac{\hbar^2 k^2}{2m}; \tag{210.9a}$$

$$\boldsymbol{P} = \sum_{\boldsymbol{k}} N_{\boldsymbol{k}} \boldsymbol{p}_k; \qquad \boldsymbol{p}_k = \hbar \boldsymbol{k}; \tag{210.9b}$$

$$Q = e \sum_{\boldsymbol{k}} N_{\boldsymbol{k}}, \tag{210.9c}$$

describing a system of particles without interactions of which $N_{\boldsymbol{k}}$ are in the state $\boldsymbol{k}$ and have each the energy E_k, the momentum $\boldsymbol{p}_{\boldsymbol{k}}$ and the charge e.

Quantization leading to the required eigenvalues (210.8) is performed if the coefficients satisfy the following commutation rules[1,2]:

$$\begin{aligned} &[c_{\boldsymbol{k}}; c_{\boldsymbol{k}'}^\dagger]_- \equiv c_{\boldsymbol{k}} c_{\boldsymbol{k}'}^\dagger - c_{\boldsymbol{k}'}^\dagger c_{\boldsymbol{k}} = \delta_{\boldsymbol{k}\boldsymbol{k}'} \text{ in the Bose case} \\ \text{or} \quad &[c_{\boldsymbol{k}}; c_{\boldsymbol{k}'}^\dagger]_+ \equiv c_{\boldsymbol{k}} c_{\boldsymbol{k}'}^\dagger + c_{\boldsymbol{k}'}^\dagger c_{\boldsymbol{k}} = \delta_{\boldsymbol{k}\boldsymbol{k}'} \text{ in the Fermi case}. \end{aligned} \tag{210.10}$$

---

[1] The commutator notation $[a,b] = ab - ba$ used in this chapter differs by a factor $i/\hbar$ from the one used in chapter I.

[2] For the Bose case, it has been shown in Problem 31 that the eigenvalues (210.8) are the consequence of the commutation rule (210.10). The same method may be applied in the Fermi case.

From these relations the commutator of the wave functions can easily be constructed:

$$[\psi(\boldsymbol{r},t);\psi^{\dagger}(\boldsymbol{r}',t)]_{\pm}=\frac{1}{\mathcal{V}}\sum_{\boldsymbol{k}}\sum_{\boldsymbol{k}'}[c_{\boldsymbol{k}};c_{\boldsymbol{k}'}^{\dagger}]_{\pm}\,\mathrm{e}^{i(\boldsymbol{k}\boldsymbol{r}-\boldsymbol{k}'\boldsymbol{r}')-i(\omega-\omega')t}$$

$$=\frac{1}{\mathcal{V}}\sum_{\boldsymbol{k}}\mathrm{e}^{i\boldsymbol{k}(\boldsymbol{r}-\boldsymbol{r}')}=\delta(\boldsymbol{r}-\boldsymbol{r}')\,. \tag{210.11}$$

This commutation relation apparently holds for both commutator signs.

Problem 211. Scattering in Born approximation

The quantized Schrödinger theory shall be applied to the elastic scattering of a particle in a central force potential, $V(r)$.

Solution. Let the force-free quantized field of the preceding problem be disturbed by the potential $V(r)$; then the hamiltonian W of the field has to be supplemented by the perturbation cncrgy,

$$W'=\int d^3x\,\psi^{\dagger}\,V\,\psi\,. \tag{211.1}$$

If we stick to the first approximation, we may put the plane-wave decomposition (210.5) of the preceding problem into W' for ψ and $\psi^{\dagger}$, thus getting

$$W'=\frac{1}{\mathcal{V}}\sum_{\boldsymbol{k}}\sum_{\boldsymbol{k}'}c_{\boldsymbol{k}'}^{\dagger}c_{\boldsymbol{k}}\int d^3x\,V(r)\,\mathrm{e}^{i(\boldsymbol{k}-\boldsymbol{k}')\boldsymbol{r}}\,\mathrm{e}^{i(\omega'-\omega)t}\,. \tag{211.2}$$

The integrals in (211.2) are well known from the Born approximation (cf. Problem 105); with the abbreviation

$$\boldsymbol{k}-\boldsymbol{k}'=\boldsymbol{K} \tag{211.3}$$

we again arrive at the expression (note that its dimension is erg·cm^3)

$$\langle\boldsymbol{k}'|\,V\,|\boldsymbol{k}\rangle=\int d^3x\,V(r)\,\mathrm{e}^{i\boldsymbol{K}\boldsymbol{r}}=4\pi\int_0^{\infty}dr\,r^2\,V(r)\frac{\sin Kr}{Kr} \tag{211.4}$$

and may briefly write

$$W'=\frac{1}{\mathcal{V}}\sum_{\boldsymbol{k}}\sum_{\boldsymbol{k}'}c_{\boldsymbol{k}'}^{\dagger}c_{\boldsymbol{k}}\langle\boldsymbol{k}'|\,V\,|\boldsymbol{k}\rangle\,\mathrm{e}^{i(\omega'-\omega)t}\,. \tag{211.5}$$

We now turn to describing the scattering process. If in the initial state there is just one particle in the state $\boldsymbol{k}_0$ and all other one-particle states are empty, so that the initial Hilbert vector is

$$\chi_i = |0 \ldots 1_{\boldsymbol{k}_0} \ldots 0_{\boldsymbol{k}_f} \ldots\rangle\,, \tag{211.6a}$$

and if the final state is defined by there being just one particle in the state $\boldsymbol{k}_f$,

$$\chi_f = |0 \ldots 0_{\boldsymbol{k}_0} \ldots 1_{\boldsymbol{k}_f} \ldots\rangle\,, \tag{211.6b}$$

then we need the matrix element

$$\langle \chi_f | \, W' \, | \chi_i \rangle \tag{211.7}$$

to determine the transition probability between these two states.

Using the general rules

$$c_{\boldsymbol{k}} |0_{\boldsymbol{k}}\rangle = 0\,; \quad c_{\boldsymbol{k}} |1_{\boldsymbol{k}}\rangle = |0_{\boldsymbol{k}}\rangle$$

we find

$$c_{\boldsymbol{k}} \chi_i = \delta_{\boldsymbol{k}\boldsymbol{k}_0} |0_{\boldsymbol{k}_0}\rangle\,,$$

i.e. all terms of the sum over $\boldsymbol{k}$ in Eq. (211.5) vanish when W' is applied to χ_i except the term $\boldsymbol{k} = \boldsymbol{k}_0$ and here the operator $c_{\boldsymbol{k}_0}$ exactly annihilates the initial particle, thus generating the state vector of vacuum, $|0\rangle$. If now $c_{\boldsymbol{k}'}$ is applied, all sum terms will contribute, viz.

$$W' |1_{\boldsymbol{k}_0}\rangle = \frac{1}{\mathscr{V}} \sum_{\boldsymbol{k}'} \langle \boldsymbol{k}' | \, V \, | \boldsymbol{k}_0 \rangle \, \mathrm{e}^{i(\omega' - \omega_0)t} |1_{\boldsymbol{k}'}\rangle\,.$$

The matrix element (211.7) then again leaves only one term of this sum, in consequence of the orthogonality,

$$\langle \chi_f | 1_{\boldsymbol{k}'} \rangle = \langle 1_{\boldsymbol{k}_f} | \, 1_{\boldsymbol{k}'} \rangle = \delta_{\boldsymbol{k}_f \boldsymbol{k}'}\,,$$

therefore

$$\langle \chi_f | \, W' \, | \chi_i \rangle = \frac{1}{\mathscr{V}} \langle \boldsymbol{k}_f | \, V \, | \boldsymbol{k}_0 \rangle \, \mathrm{e}^{i(\omega' - \omega_0)t}\,. \tag{211.8}$$

From the matrix element we now finally proceed to the differential cross section using the Golden Rule,

$$d\sigma = \frac{2\pi}{\hbar} \rho_f \, |\langle \chi_f | \, W' \, | \chi_i \rangle|^2 \cdot \frac{\mathscr{V}}{v} \tag{211.9}$$

with

$$\rho_f = \frac{\mathscr{V} p^2 \, dp \, d\Omega}{(2\pi\hbar)^3 \, dE} = \frac{m^2 v \mathscr{V}}{8\pi^3 \hbar^3} d\Omega \tag{211.10}$$

where all the quantities refer to the final state. Putting (211.10) in (211.9) and using the matrix element formula (211.8), the (large but otherwise arbitrary) periodicity volume $\mathscr{V}$ cancels out and we arrive at the result,

$$\frac{d\sigma}{d\Omega}=\left(\frac{m}{2\pi\hbar^2}\right)^2 |\langle \boldsymbol{k}_f|V|\boldsymbol{k}_0\rangle|^2 . \tag{211.11}$$

With the expression (211.4), this completely agrees with Born's first approximation formula, cf. Problems 105 and 184.

Problem 212. Quantization of classical radiation field

A classical Maxwell field in vacuo, underlying the condition of periodicity within a cube $\mathscr{V}=L^3$, shall be quantized into photons, using the classical expressions of field energy and momentum.

Solution. The classical radiation field is described by a vector potential $\boldsymbol{A}$ satisfying the differential equations

$$\Box^2 \boldsymbol{A}=0 \quad \text{and} \quad \operatorname{div}\boldsymbol{A}=0 \tag{212.1}$$

in the usual gauge leading to transverse waves, The physical meaning of these equations can either be explained by coupling $\boldsymbol{A}$ with the field strengths,

$$\mathscr{E}=-\frac{1}{c}\dot{\boldsymbol{A}}; \quad \mathscr{H}=\operatorname{rot}\boldsymbol{A}, \tag{212.2}$$

or by the mechanical expressions of total field energy,

$$W=\frac{1}{8\pi}\int d^3x(\mathscr{E}^2+\mathscr{H}^2)=\frac{1}{8\pi}\int d^3x\left\{\frac{1}{c^2}\dot{\boldsymbol{A}}^2+(\operatorname{rot}\boldsymbol{A})^2\right\}, \tag{212.3}$$

and of total field momentum,

$$\boldsymbol{P}=\frac{1}{4\pi c}\int d^3x(\mathscr{E}\times\mathscr{H})=-\frac{1}{4\pi c^2}\int d^3x(\dot{\boldsymbol{A}}\times\operatorname{rot}\boldsymbol{A}). \tag{212.4}$$

The differential equations (212.1) are solved by plane waves in the usual standard form,

$$\boldsymbol{A}=\sqrt{\frac{4\pi c^2}{\mathscr{V}}}\sum_{\boldsymbol{k}\lambda}\boldsymbol{u}_{\boldsymbol{k}\lambda}(q_{\boldsymbol{k}\lambda}\,e^{i(\boldsymbol{kr}-\omega t)}+q^*_{\boldsymbol{k}\lambda}\,e^{-i(\boldsymbol{kr}-\omega t)}) \tag{212.5}$$

with $\lambda=1,2$ denoting the two states of transverse polarization and $\boldsymbol{u}_{\boldsymbol{k}\lambda}$ a unit vector. There hold three orthogonality relations,

$$(\boldsymbol{u}_{\boldsymbol{k}1}\cdot\boldsymbol{k})=0; \quad (\boldsymbol{u}_{\boldsymbol{k}2}\cdot\boldsymbol{k})=0; \quad (\boldsymbol{u}_{\boldsymbol{k}1}\cdot\boldsymbol{u}_{\boldsymbol{k}2})=0. \tag{212.6}$$

The normalization factor in front of the summation sign in (212.5) is arbitrary and merely conventional. The $\boldsymbol{k}$ vectors are restricted by the periodicity within the cube $\mathscr{V}=L^3$ to

$$\boldsymbol{k}=\frac{2\pi}{L}\boldsymbol{n} \tag{212.7}$$

with integer components $n_i=0, \pm 1, \pm 2, \ldots$ The frequency ω is connected with k through the law of dispersion,

$$\omega=kc\,. \tag{212.8}$$

Since each term in the sum (212.5) consists of two complex conjugate members, the vector potential $\boldsymbol{A}$ is a real function of $\boldsymbol{r}$ and t, as necessary in the classical Maxwell theory.

Putting (212.5) in the energy integral (212.3), we get

$$\begin{aligned} W=\frac{c^2}{2\mathscr{V}}\sum_{\boldsymbol{k}\lambda}\sum_{\boldsymbol{k}'\lambda'}\int d^3x\left\{-\frac{\omega\omega'}{c}\boldsymbol{u}_{\boldsymbol{k}'\lambda'}\boldsymbol{u}_{\boldsymbol{k}\lambda}-(\boldsymbol{k}'\times\boldsymbol{u}_{\boldsymbol{k}'\lambda'})(\boldsymbol{k}\times\boldsymbol{u}_{\boldsymbol{k}\lambda})\right\}\\ \times\left\{q_{\boldsymbol{k}'\lambda'}\,\mathrm{e}^{i(\boldsymbol{k}'\boldsymbol{r}-\omega' t)}-q^*_{\boldsymbol{k}'\lambda'}\,\mathrm{e}^{-i(\boldsymbol{k}'\boldsymbol{r}-\omega' t)}\right\}\\ \times\left\{q_{\boldsymbol{k}\lambda}\,\mathrm{e}^{i(\boldsymbol{k}\boldsymbol{r}-\omega t)}-q^*_{\boldsymbol{k}\lambda}\,\mathrm{e}^{-i(\boldsymbol{k}\boldsymbol{r}-\omega t)}\right\}. \end{aligned}$$

Multiplication of the two last brackets and integration leaves either $\boldsymbol{k}'=-\boldsymbol{k}$ in the products of type qq and q^*q^*, or $\boldsymbol{k}'=\boldsymbol{k}$ for the types qq^* and q^*q. The first bracket then becomes, when use is yet made of (212.6),

$$-\frac{\omega^2}{c}-(\boldsymbol{k}'\cdot\boldsymbol{k})=\begin{cases}0 & \text{if } \boldsymbol{k}'=-\boldsymbol{k},\\ -2\dfrac{\omega^2}{c^2}\delta_{\lambda\lambda'} & \text{if } \boldsymbol{k}'=+\boldsymbol{k}.\end{cases}$$

Hence, there remains

$$W=\sum_{\boldsymbol{k}\lambda}\omega^2(q_{\boldsymbol{k}\lambda}q^*_{\boldsymbol{k}\lambda}+q^*_{\boldsymbol{k}\lambda}q_{\boldsymbol{k}\lambda})\,. \tag{212.9}$$

For the momentum, Eq. (212.4), a similar calculation leads to

$$\boldsymbol{P}=\sum_{\boldsymbol{k}\lambda}\omega\,\boldsymbol{k}(q_{\boldsymbol{k}\lambda}q^*_{\boldsymbol{k}\lambda}+q^*_{\boldsymbol{k}\lambda}q_{\boldsymbol{k}\lambda})\,. \tag{212.10}$$

Now we are ready to proceed to quantizing the classical radiation field by replacing the amplitudes $q_{\boldsymbol{k}\lambda}$ and $q^*_{\boldsymbol{k}\lambda}$ by operators and their hermitian conjugates. Written in the form

$$q_{\boldsymbol{k}\lambda}=C_k b_{\boldsymbol{k}\lambda} \quad\text{and}\quad q^\dagger_{\boldsymbol{k}\lambda}=C_k b^\dagger_{\boldsymbol{k}\lambda} \tag{212.11}$$

with a real normalization factor C_k, the expressions (212.9) and (212.10) turn into

$$W = \sum_{k\lambda} \omega^2 C_k^2 (b_{k\lambda} b^\dagger_{k\lambda} + b^\dagger_{k\lambda} b_{k\lambda});$$

and (212.12)

$$\boldsymbol{P} = \sum_{k\lambda} \omega \boldsymbol{k} C_k^2 (b_{k\lambda} b^\dagger_{k\lambda} + b^\dagger_{k\lambda} b_{k\lambda}).$$

With the operators $b_{k\lambda}$ and $b^\dagger_{k\lambda}$ chosen to satisfy the commutation rules

$$b_{k\lambda} b^\dagger_{k'\lambda'} - b^\dagger_{k'\lambda'} b_{k\lambda} = \delta_{kk'} \delta_{\lambda\lambda'}, \tag{212.13}$$

and all other combinations commuting, the eigenvalues of the operators $b^\dagger_{k\lambda} b_{k\lambda}$ become integers, $N_{k\lambda}$, including zero,

$$N_{k\lambda} = 0, 1, 2, 3, \ldots \tag{212.14}$$

and those of $b_{k\lambda} b^\dagger_{k\lambda}$ become $N_{k\lambda}+1$. (Cf. Problem 31). If we further put

$$C_k = \sqrt{\frac{\hbar}{2\omega}}, \tag{212.15}$$

the operators (212.12) become

$$W = \sum_{k\lambda} \frac{\hbar\omega}{2} (b_{k\lambda} b^\dagger_{k\lambda} + b^\dagger_{k\lambda} b_{k\lambda})$$

and (212.16)

$$\boldsymbol{P} = \sum_{k\lambda} \hbar \boldsymbol{k} (b_{k\lambda} b^\dagger_{k\lambda} + b^\dagger_{k\lambda} b_{k\lambda})$$

with eigenvalues

$$W = \sum_{k\lambda} \hbar\omega (N_{k\lambda} + \tfrac{1}{2}); \qquad \boldsymbol{P} = \sum_{k\lambda} \hbar \boldsymbol{k} (N_{k\lambda} + \tfrac{1}{2}). \tag{212.17}$$

This permits interpretation of $N_{k\lambda}$ as the number of photons in a state defined by $\boldsymbol{k}$ and λ, each of the photons having the energy $\hbar\omega$ and the momentum in the wave propagation direction $\hbar k = \hbar\omega/c$.

There occurs a zero-point energy, i.e. an energy of the vacuum,

$$W_0 = \sum_{k\lambda} \frac{\hbar\omega}{2}. \tag{212.18}$$

In spite of its being infinitely large, it has no serious physical consequences and may be normalized away by using

$$W - W_0 = \sum_{k\lambda} \hbar\omega N_{k\lambda}, \tag{212.19}$$

i.e. the difference of the actual state from vacuum. In the momentum the zero-point contribution vanishes anyway, because in the sum pairs of terms with opposite $\boldsymbol{k}$ vectors cancel each other out.

The vector potential $\boldsymbol{A}$ then becomes an operator, creating or annihilating photons. It may be gathered from Eqs. (212.5), (212.11) and (212.15) that we shall write in quantum theory:

$$\boldsymbol{A} = \sum_{\boldsymbol{k}\lambda} \sqrt{\frac{2\pi c\hbar}{\mathscr{V} k}} \{b_{\boldsymbol{k}\lambda} \mathrm{e}^{i(\boldsymbol{kr}-\omega t)} + b^{\dagger}_{\boldsymbol{k}\lambda} \mathrm{e}^{-i(\boldsymbol{kr}-\omega t)}\} \boldsymbol{u}_{\boldsymbol{k}\lambda}. \tag{212.20}$$

Problem 213. Emission probability of a photon

The probability that an electron from an excited state in a central potential field $V(r)$ will jump to a lower level and emit a photon shall be determined. Retardation effects may be neglected.

Solution. The interaction of matter (the electron) with radiation is given by the classical Maxwell expression,

$$H' = \frac{1}{c}\int d^3x (\boldsymbol{A}\cdot\boldsymbol{j}) \tag{213.1}$$

with $\boldsymbol{A}$ the vector potential and $\boldsymbol{j}$ the electrical current density of matter. The first has been translated into the theory of quantized fields in Problem 212[3],

$$\boldsymbol{A} = \sum_{\boldsymbol{k}\lambda} \sqrt{\frac{2\pi\hbar c}{k\mathscr{V}}}\, \boldsymbol{u}^{(\lambda)}_{\boldsymbol{k}} (b_{\boldsymbol{k}\lambda} \mathrm{e}^{i\boldsymbol{kr}} + b^{\dagger}_{\boldsymbol{k}\lambda} \mathrm{e}^{-i\boldsymbol{kr}}). \tag{213.2}$$

The current can be taken from the quantized Schrödinger field,

$$\psi = \sum_n c_n u_n(\boldsymbol{r}); \qquad -\frac{\hbar^2}{2m}\nabla^2 u_n + V u_n = E_n u_n \tag{213.3}$$

using the formula

$$\boldsymbol{j} = -\frac{e\hbar}{2mi}(\psi^{\dagger}\nabla\psi - \nabla\psi^{\dagger}\cdot\psi)$$

(with the electron charge $-e$), hence

$$\boldsymbol{j} = -\frac{e\hbar}{2mi}\sum_n\sum_{n'} (u^*_{n'}\nabla u_n - u_n\nabla u^*_{n'})\, c^{\dagger}_{n'} c_n. \tag{213.4}$$

[3] Omission of the time factors of (212.20) and (210.5) in (213.2) and (213.3) corresponds to the transition from the Schrödinger to the Heisenberg picture.

The functions $u_n(\boldsymbol{r})$ and $u^*_{n'}(\boldsymbol{r})$ are one-particle wave functions according to (213.3), the subscript n (or n') being a comprehensive notation for a set of three quantum numbers. The c_n and $c^\dagger_{n'}$ are the operators used in Problem 210 and may satisfy either of the commutation rules

$$\begin{aligned} c_n c^\dagger_{n'} \pm c^\dagger_{n'} c_n &= \delta_{nn'}, \\ c_n c_{n'} \pm c_{n'} c_n &= 0. \end{aligned} \tag{213.5}$$

Spontaneous emission of a photon is a process in which the electron jumps from an initial state n_i to a final state n_f or, in the wording of quantized wave theory, an electron in the initial state n_i is annihilated and an electron in the final state n_f created instead. At the same time a photon in a state $(\boldsymbol{k}, \lambda)$ is created. Such a process will be originated by a term with the operator product

$$b^\dagger_{\boldsymbol{k}\lambda} c^\dagger_{n_f} c_{n_i}$$

in the hamiltonian. If we put (213.2) and (213.4) in (213.1) such a term will indeed occur; we write it in the form

$$\langle f|H'|i\rangle\, b^\dagger_{\boldsymbol{k}\lambda} c^\dagger_{n_f} c_{n_i} \tag{213.6a}$$

with

$$\langle f|H'|i\rangle = \frac{1}{c}\int d^3x \sqrt{\frac{2\pi\hbar c}{k\mathscr{V}}}\,\frac{e\hbar}{2m}\, i\,\mathrm{e}^{-i\boldsymbol{k}\boldsymbol{r}}\, \boldsymbol{u}^{(\lambda)}_{\boldsymbol{k}}\cdot(u^*_{n_f}\nabla u_{n_i} - u_{n_i}\nabla u^*_{n_f}) \tag{213.6b}$$

the matrix element of H' between initial and final state. The transition probability then follows from the Golden Rule (cf. Problem 183),

$$P = \frac{2\pi}{\hbar}\rho_f |\langle f|H'|i\rangle|^2. \tag{213.7}$$

The density ρ_f of final states in the energy scale is completely determined by the photon:

$$\rho_f = \frac{k^2\, dk\, d\Omega_{\boldsymbol{k}}\, \mathscr{V}}{(2\pi)^3 \hbar c\, dk} = \frac{\mathscr{V}}{8\pi^3\hbar c} k^2\, d\Omega_{\boldsymbol{k}} \tag{213.8}$$

with $d\Omega_{\boldsymbol{k}}$ the solid angle element into which the photon is emitted.

There remains the evaluation of the integral in (213.6b),

$$I = \int d^3x\, \mathrm{e}^{-i\boldsymbol{k}\boldsymbol{r}} (u^*_{n_f}\nabla u_{n_i} - u_{n_i}\nabla u^*_{n_f}).$$

Retardation may be neglected if the wavelength of the emitted light is large compared with the atomic integration domain, i.e., if in the integrand

$e^{-i\boldsymbol{k}\boldsymbol{r}}=1$ is a good approximation. Of the two terms of the integral, the second then may be partially integrated, thus yielding

$$I=2\int d^3x\, u_{n_f}^* \nabla u_{n_i}\,.$$

From the Schrödinger equations for u_{n_i} and $u_{n_f}^*$ an identity may be derived (cf. Problem 187),

$$I=\frac{2m}{\hbar^2}(E_i-E_f)\int d^3x\, u_{n_f}^* \boldsymbol{r}\, u_{n_i}$$

and from the conservation of energy there follows

$$E_i-E_f=\hbar\omega$$

so that the matrix element (213.6b) becomes

$$\langle f|H'|i\rangle=\frac{e}{c}\,i\sqrt{\frac{2\pi\hbar c}{k\mathscr{V}}}\,\omega\langle f|\boldsymbol{u}_{\boldsymbol{k}}^{(\lambda)}\cdot\boldsymbol{r}|i\rangle \tag{213.9}$$

with

$$\langle f|\boldsymbol{u}_{\boldsymbol{k}}^{(\lambda)}\cdot\boldsymbol{r}|i\rangle=\int d^3x\, u_{n_f}^*(\boldsymbol{u}_{\boldsymbol{k}}^{(\lambda)}\cdot\boldsymbol{r})\,u_{n_i}\,. \tag{213.10}$$

By gathering the expressions (213.8) and (213.9) into (213.7), we then finally obtain

$$P_{\boldsymbol{k},\lambda}=\frac{d\Omega_{\boldsymbol{k}}}{2\pi}\,\frac{e^2}{\hbar c}\,\frac{\omega^3}{c^2}\,|\langle f|\boldsymbol{u}_{\boldsymbol{k}}^{(\lambda)}\cdot\boldsymbol{r}|i\rangle|^2 \tag{213.11a}$$

or, if we introduce $\nu=\omega/2\pi$ instead of ω, in the usual notation,

$$P_{\boldsymbol{k},\lambda}=\frac{e^2}{\hbar c}\,\frac{4\pi^2\nu^3}{c^2}\,d\Omega_{\boldsymbol{k}}\,|\langle f|\boldsymbol{u}_{\boldsymbol{k}}^{(\lambda)}\cdot\boldsymbol{r}|i\rangle|^2\,. \tag{213.11b}$$

The remaining matrix element in this formula may still be decomposed into two factors according to

$$\langle f|\boldsymbol{u}_{\boldsymbol{k}}^{(\lambda)}\cdot\boldsymbol{r}|i\rangle=\boldsymbol{u}_{\boldsymbol{k}}^{(\lambda)}\cdot\boldsymbol{r}_{if} \tag{213.12}$$

with the first factor depending only upon the direction and polarization of the photon emitted, and the second only upon atomic parameters.

Problem 214. Angular distribution of radiation

The final formulae of the preceding problem shall be used to investigate the angular distribution of photons emitted in a $P\rightarrow S$ transition of one electron.

Solution. Let Θ and Φ be polar angles defining the direction of $\boldsymbol{k}$. We then may define two polarization states, one with $\boldsymbol{u}_{\boldsymbol{k}}^{(1)}$ in the meridional plane, the other with $\boldsymbol{u}_{\boldsymbol{k}}^{(2)}$ perpendicular to it. The two unit vectors then have the components

$$u_x^{(1)}=\cos\Theta\cos\Phi\,; \quad u_y^{(1)}=\cos\Theta\sin\Phi\,; \quad u_z^{(1)}=-\sin\Theta \tag{214.1a}$$

and

$$u_x^{(2)}=-\sin\Phi\,; \quad u_y^{(2)}=\cos\Phi\,; \quad u_z^{(2)}=0\,. \tag{214.1b}$$

To construct the components of $\boldsymbol{r}_{if}$ we first write the three components x, y, z of $\boldsymbol{r}$ in terms of spherical harmonics:

$$\left.\begin{aligned} x&=r\sqrt{\frac{2\pi}{3}}\,(Y_{1,1}+Y_{1,-1})\,;\\ y&=-ir\sqrt{\frac{2\pi}{3}}\,(Y_{1,1}-Y_{1,-1})\,;\\ z&=r\sqrt{\frac{4\pi}{3}}\,Y_{1,0}\,. \end{aligned}\right\} \tag{214.2}$$

Of these three expressions we then form matrix elements between the two electrons states,

$$|i\rangle=v(r)\,Y_{1,m}(\vartheta,\varphi)\,; \qquad |f\rangle=u(r)\,Y_{0,0}(\vartheta,\varphi)\,. \tag{214.3}$$

It should be noted that $Y_{0,0}=(4\pi)^{-\frac{1}{2}}$ is a constant. Thence we get,

$$\left.\begin{aligned} x_{if}&=\frac{1}{\sqrt{6}}\,R(\delta_{m,1}+\delta_{m,-1}),\\ y_{if}&=-\frac{i}{\sqrt{6}}\,R(\delta_{m,1}-\delta_{m,-1}),\\ z_{if}&=\frac{1}{\sqrt{3}}\,R\,\delta_{m,0} \end{aligned}\right\} \tag{214.4}$$

with the abbreviation

$$R=\int_0^\infty dr\,r^3\,u^*(r)\,v(r)=\sqrt{3}\,|\boldsymbol{r}_{if}| \tag{214.5}$$

for the radial integral.

The probability of emission of a photon of polarization λ in the solid angle element $d\Omega_{\boldsymbol{k}}$ is, according to (213.11a) of the preceding problem,

$$P_{\boldsymbol{k},\lambda} = \frac{e^2 \omega^3}{\hbar c^3} \frac{d\Omega_{\boldsymbol{k}}}{2\pi} |\boldsymbol{u}_{\boldsymbol{k}}^{(\lambda)} \cdot \boldsymbol{r}_{if}|^2. \tag{214.6}$$

With (214.1a, b) for $\boldsymbol{u}_{\boldsymbol{k}}^{(\lambda)}$ and (214.4) for $\boldsymbol{r}_{if}$ we find for the scalar product $\boldsymbol{u}_{\boldsymbol{k}}^{(\lambda)} \cdot \boldsymbol{r}_{if}$ in the case $\lambda = 1$

$$\left.\begin{array}{ll} \text{for } m = +1: & \dfrac{R}{\sqrt{6}} \cos\Theta \, \mathrm{e}^{-i\Phi}, \\ \text{for } m = 0: & -\dfrac{R}{\sqrt{3}} \sin\Theta, \\ \text{for } m = -1: & \dfrac{R}{\sqrt{6}} \cos\Theta \, \mathrm{e}^{i\Phi} \end{array}\right\} \tag{214.7}$$

and in the case $\lambda = 2$

$$\left.\begin{array}{ll} \text{for } m = +1: & -i \dfrac{R}{\sqrt{6}} \mathrm{e}^{-i\Phi}, \\ \text{for } m = 0: & 0, \\ \text{for } m = -1: & +i \dfrac{R}{\sqrt{6}} \mathrm{e}^{i\Phi}. \end{array}\right\} \tag{214.8}$$

Thus, if we write

$$P_{k,\lambda} = \frac{e^2 \omega^3}{\hbar c^3} \frac{R^2}{6\pi} d\Omega_{\boldsymbol{k}} \cdot D_{\boldsymbol{k}\lambda}(\Theta, \Phi), \tag{214.9}$$

the directional factor $D_{\boldsymbol{k}\lambda}$ will be given by the expressions assembled in the accompanying table. In the case $\lambda = 2$, all directions of photon

Directional factors $D_{\boldsymbol{k}\lambda}$ for

m	$\lambda = 1$	$\lambda = 2$
$+1$	$\frac{1}{2}\cos^2\Theta$	$\frac{1}{2}$
0	$\sin^2\Theta$	0
-1	$\frac{1}{2}\cos^2\Theta$	$\frac{1}{2}$

emission have equal probabilities; only from the initial state $m = 0$ are no photons of this polarization emitted at all. The P state with $m = 0$, only decaying therefore under emission of a photon in polarization state $\lambda = 1$, has an angular distribution as $\sin^2\Theta$, the main direction of emission for $m=0$ thus falling in the equatorial plane $\Theta=90°$. For $\lambda=1$ and $m=\pm 1$, however, the distribution becomes proportional to $\cos^2\Theta$ and the emission occurs preferably in the directions $\Theta=0°$ and $\Theta=180°$.

Problem 215. Transition probability

What is the transition probability of an electron from a higher P to a lower S state by emission of a photon of any direction or polarization whatsoever? As an example, the mean lifetime of the $2P$ state in a hydrogen atom shall be computed.

Solution. In the preceding problem, the differential emission probability for the photon going into the solid angle element $d\Omega_{\mathbf{k}}$ in the direction Θ, Φ was computed for $m=+1, 0, -1$ and both states of polarization. Gathering these formulae and performing summation over both polarization states, we obtain

$$\sum_\lambda P_{\mathbf{k}\lambda} = \frac{e^2\omega^3}{\hbar c^3}\cdot\frac{R^2}{6\pi}d\Omega_{\mathbf{k}}\cdot\sin^2\Theta \qquad \text{for } m=0; \tag{215.1a}$$

$$\sum_\lambda P_{\mathbf{k}\lambda} = \frac{e^2\omega^3}{\hbar c^3}\cdot\frac{R^2}{6\pi}d\Omega_{\mathbf{k}}\cdot\frac{1}{2}(1+\cos^2\Theta) \quad \text{for } m=\pm 1. \tag{215.1b}$$

Integration of these expressions over all directions of the photon then yields the same transition probability from P to S state for all three initial values of m, viz.

$$P = \frac{e^2\omega^3}{\hbar c^3}\cdot\frac{R^2}{6\pi}\cdot\frac{8\pi}{3} = \frac{4}{9}\,\frac{e^2\omega^3}{\hbar c^3}R^2. \tag{215.2}$$

Since R has been defined in some detail in the preceding problem, we may now immediately turn to the example. From the $2P$ state of a hydrogen atom transitions are possible only to the $1S$ ground state. The eigenfunctions of the two states are

$$|i\rangle = \tfrac{1}{12}\sqrt{6}\,r\,e^{-\frac{r}{2}}\,Y_{1,m} \tag{215.3a}$$

and

$$|f\rangle = 2e^{-r}\,Y_{0,0} \tag{215.3b}$$

in atomic units (unit length: $\hbar^2/me^2$). It then follows from (214.5) that

$$R = \int_0^\infty dr\,r^3\cdot\frac{\sqrt{6}}{12}\,r\,e^{-\frac{r}{2}}\cdot 2e^{-r} = \frac{1}{\sqrt{6}}\left(\frac{2}{3}\right)^5\cdot 24; \qquad R^2 = \frac{2^{15}}{3^9}$$

or, if we make the notation independent of units again,

$$R^2 = \frac{2^{15}}{3^9} \frac{\hbar^4}{m^2 e^4}. \tag{215.4}$$

The light frequency ω, in our example, follows from the hydrogen level formula

$$\omega = \frac{3}{8} \frac{m e^4}{\hbar^3}. \tag{215.5}$$

Putting R^2 from (215.4) and ω from (215.5) in (215.2) and suitably arranging the factors, we arrive at the result

$$P = \left(\frac{2}{3}\right)^8 \left(\frac{e^2}{\hbar c}\right)^3 \frac{m e^4}{\hbar^2}. \tag{215.6}$$

The reciprocal value of this transition probability is the mean lifetime τ of the $2P$ state, since there are no other ways of decaying in this special case. Hence,

$$\tau = \left(\frac{3}{2}\right)^8 \left(\frac{\hbar c}{e^2}\right)^3 \frac{\hbar^3}{m e^4}. \tag{215.7}$$

The quantity

$$\frac{\hbar^3}{m e^4} = 2.4187 \times 10^{-17}\,\text{sec} \tag{215.8}$$

is a suitable time unit for lifetimes of excited electron states. The factor

$$\frac{\hbar c}{e^2} = 137.0373$$

is the reciprocal fine-structure constant. The numerical value of the mean lifetime of the hydrogen $2P$ state then becomes

$$\tau = 1.5953 \times 10^{-9}\,\text{sec}.$$

Problem 216. Selection rules for dipole radiation

It has been shown in Problem 213 that the transition probabilities between one-electron states in an atom depend upon the matrix elements of the electrical dipole moment, if the photon wavelength is large compared to atomic dimensions. Selection rules shall be derived from this fact. What can be concluded for the normal Zeeman effect?

Solution. The probability that a photon is emitted into a solid angle element $d\Omega_{\boldsymbol{k}}$ in the direction $\boldsymbol{k}$ (polar angles Θ, Φ) in a state λ of polarization is, according to (213.11b),

$$P_{\boldsymbol{k},\lambda} = \frac{e^2}{\hbar c} \cdot \frac{4\pi^2 \nu^3}{c^2} d\Omega_{\boldsymbol{k}} |\langle f | \boldsymbol{u}_{\boldsymbol{k}}^{(\lambda)} \cdot \boldsymbol{r} | i \rangle|^2 \tag{216.1}$$

in dipole approximation. Here $\boldsymbol{u}_{\boldsymbol{k}}^{(\lambda)}$ is a unit vector in the direction of polarization, viz. according to (214.1a, b), either with the components

$$u_x^{(1)} = \cos\Theta \cos\Phi; \quad u_y^{(1)} = \cos\Theta \sin\Phi; \quad u_z^{(1)} = -\sin\Theta \tag{216.2a}$$

or

$$u_x^{(2)} = -\sin\Phi; \quad u_y^{(2)} = \cos\Phi; \quad u_z^{(2)} = 0. \tag{216.2b}$$

The two atomic states have wave functions

$$|i\rangle = \varphi_i(r)\, Y_{l,m}(\vartheta,\varphi); \qquad \langle f| = \varphi_f(r)\, Y^*_{l',m'}(\vartheta,\varphi)$$

so that the matrix element of $\boldsymbol{r}$ has the components

$$\left.\begin{aligned} \langle f | x \pm i y | i \rangle &= \int_0^\infty dr\, r^3 \varphi_i \varphi_f \oint d\Omega\, Y^*_{l',m'} \sin\vartheta\, e^{\pm i\varphi} Y_{l,m}, \\ \langle f | z | i \rangle &= \int_0^\infty dr\, r^3 \varphi_i \varphi_f \oint d\Omega\, Y^*_{l',m'} \cos\vartheta\, Y_{l,m}. \end{aligned}\right\} \tag{216.3}$$

We may now use the relations

$$\left.\begin{aligned} \sin\vartheta\, e^{\pm i\varphi} Y_{l,m} &= \pm A_{l+1,\pm m+1} Y_{l+1,m\pm 1} \mp A_{l,\mp m} Y_{l-1,m\pm 1}, \\ \cos\vartheta\, Y_{l,m} &= B_{l+1,m} Y_{l+1,m} + B_{l,m} Y_{l-1,m} \end{aligned}\right\} \tag{216.4}$$

with coefficients

$$A_{l,m} = \sqrt{\frac{(l+m)(l+m-1)}{(2l+1)(2l-1)}}; \qquad B_{l,m} = \sqrt{\frac{(l+m)(l-m)}{(2l+1)(2l-1)}} \tag{216.5}$$

which permit angular integrations in (216.3),

$$\begin{aligned} \langle f | x \pm i y | i \rangle = R_{if} \{ &\pm A_{l+1,\pm m+1} \delta_{l',l+1} \delta_{m',m\pm 1} \\ &\mp A_{l,\mp m} \delta_{l',l-1} \delta_{m',m\pm 1} \}, \end{aligned} \tag{216.6a}$$

$$\langle f | z | i \rangle = R_{if} \{ B_{l+1,m} \delta_{l',l+1} + B_{l,m} \delta_{l',l-1} \} \delta_{m',m} \tag{216.6b}$$

where the abbreviation R_{if} stands for the radial part,

$$R_{if} = \int_0^\infty dr\, r^3 \varphi_i(r) \varphi_f(r). \tag{216.7}$$

These matrix elements vanish if not $l' = l \pm 1$, thus providing us with the first basic selection rule for dipole transitions. Further, it is seen that for $m'=m\pm 1$ only the matrix element of $x\pm iy$, and for $m'=m$ of z, do not vanish. Any other changes of the quantum numbers l and m cannot occur in dipole transitions.

Combining these selection rules with the polarization vectors (216.2a, b), we may compute the matrix elements

$$\langle f|\boldsymbol{u}_{\boldsymbol{k}}^{(\lambda)}\cdot\boldsymbol{r}|i\rangle$$

occurring in (216.1), for the two polarization states $\lambda=1$ and $\lambda=2$. The results are summarized in the accompanying table.

m'	λ	$l'=l+1$	$l'=l-1$
$m+1$	1	$\frac{1}{2}\cos\Theta\, e^{-i\Phi} R_{if} A_{l+1,m+1}$	$-\frac{1}{2}\cos\Theta\, e^{-i\Phi} R_{if} A_{l,-m}$
	2	$-\frac{i}{2}e^{-i\Phi} R_{if} A_{l+1,m+1}$	$\frac{i}{2}e^{-i\Phi} R_{if} A_{l,-m}$
m	1	$-\sin\Theta\, R_{if} B_{l+1,m}$	$-\sin\Theta\, R_{if} B_{l,m}$
	2	0	0
$m-1$	1	$-\frac{1}{2}\cos\Theta\, e^{i\Phi} R_{if} A_{l+1,-m+1}$	$\frac{1}{2}\cos\Theta\, e^{i\Phi} R_{if} A_{l,m}$
	2	$-\frac{i}{2}e^{i\Phi} R_{if} A_{l+1,-m+1}$	$\frac{i}{2}e^{i\Phi} R_{if} A_{l,m}$

If the radiating atoms are not oriented in space, observation gives only intensity averages over all directions. The orientation, though, which is performed under Zeeman effect conditions gives more infor-

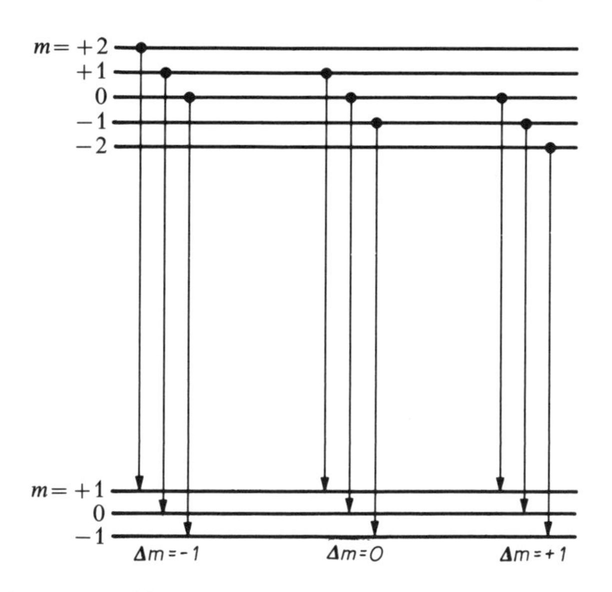

Fig. 74. Zeeman transitions $D\to P$. There are three lines of different polarization according to $\Delta m=+1, 0, -1$

mation. The polar axis $\Theta=0$ then coincides with the direction of the magnetic field. If we observe light emission in field direction, therefore, the matrix elements for $m'=m$ will vanish, so that only the lines $m'=m+1$ and $m'=m-1$ are observed. In any direction perpendicular to the field, on the other hand, we have $\cos\Theta=0$ so that $m'=m\pm1$ lines appear in polarization state 2, but $m'=m$ in state 1, only.

As an example let us take the Zeeman transition from a D state (5 components) to a P state (3 components). If the frequency of the line emitted in absence of the magnetic field is ω_0, we then find three possible frequencies in the magnetic field (Fig. 74), viz.

$$\left.\begin{array}{ll}\omega_1=\omega_0+\omega_L & \text{for } m'=m-1,\\ \omega_0 & \text{for } m'=m,\\ \omega_{-1}=\omega_0-\omega_L & \text{for } m'=m+1\end{array}\right\}\text{ with } \omega_L=\frac{e\mathscr{H}}{2mc}.$$

Observing in field direction ($\Theta=0$), we find the middle line (ω_0) does not occur and there is a doublet of frequencies $\omega_0+\omega_L$ and $\omega_0-\omega_L$. In perpendicular observation all three lines occur, forming the full Zeeman triplet, though in different states of polarization.

Problem 217. Intensities of Lyman lines

To compare the intensities of emission of the two first Lyman lines of atomic hydrogen, Lyα and Lyβ.

Solution. We are concerned with the two transitions

$$\text{Ly}\,\alpha\text{: } 2p\to1s \quad\text{and}\quad \text{Ly}\,\beta\text{: } 3p\to1s.$$

The emission probability, integrated over all directions and summed up over both polarizations, is

$$P=\frac{4}{3}\,\frac{e^2\omega^3}{\hbar c^3}\,|\langle f|\boldsymbol{r}|i\rangle|^2; \tag{217.1}$$

the intensity of a spectral line (energy per second) is proportional to $\omega\cdot P$ so that for the two lines under consideration we have the intensity ratio

$$\frac{I_\alpha}{I_\beta}=\left(\frac{E_\alpha}{E_\beta}\right)^4\left|\frac{\langle 1s|\boldsymbol{r}|2p\rangle}{\langle 1s|\boldsymbol{r}|3p\rangle}\right|^2. \tag{217.2}$$

Here the energy differences in atomic units are

$$E_\alpha=\tfrac{1}{2}-\tfrac{1}{8}=\tfrac{3}{8} \quad\text{and}\quad E_\beta=\tfrac{1}{2}-\tfrac{1}{18}=\tfrac{4}{9}. \tag{217.3}$$

We still have to calculate the two matrix elements.

From Problem 67 we know the wave function of the final state to be

$$|1s\rangle = \frac{1}{\sqrt{\pi}} e^{-r}, \tag{217.4a}$$

whereas for the initial state we have either in the Lyα case

$$|2p\rangle = \frac{1}{4\sqrt{2\pi}} r e^{-\frac{1}{2}r} \cos\vartheta \tag{217.4b}$$

or in the Lyβ case

$$|3p\rangle = \frac{4}{27\sqrt{2\pi}} \left(r - \frac{1}{6} r^2\right) e^{-\frac{1}{3}r} \cos\vartheta. \tag{217.4c}$$

Here we have arbitrarily used the p states with $m=0$, a choice which does not restrict generality as long as no directional effects are discussed. Since the vector $\boldsymbol{r}$ has components

$$x \pm i y = r \sin\vartheta \, e^{\pm i\varphi} \quad \text{and} \quad z = r\cos\vartheta,$$

it is immediately seen that the matrix elements of $x \pm iy$ vanish in consequence of the integration over φ. Therefore only $\langle f|z|i\rangle$ remains to be computed. We obtain,

$$\langle 1s|z|2p\rangle = \frac{1}{4\pi\sqrt{2}} \oint d\Omega \cos^2\vartheta \int_0^\infty dr\, r^4 e^{-\frac{3}{2}r}$$

and

$$\langle 1s|z|3p\rangle = \frac{4}{27\pi\sqrt{2}} \oint d\Omega \cos^2\vartheta \int_0^\infty dr\, r^3 \left(r - \frac{1}{6} r^2\right) e^{-\frac{4}{3}r}.$$

Elementary evaluation leads to

$$\langle 1s|z|2p\rangle = \frac{1}{\sqrt{2}} \cdot \frac{256}{243}; \qquad \langle 1s|z|3p\rangle = \frac{1}{\sqrt{2}} \cdot \frac{27}{64}. \tag{217.5}$$

Thus, Eq. (217.2) with (217.3) and (217.5) yields the final result

$$\frac{I_\alpha}{I_\beta} = \left(\frac{27}{32}\right)^4 \cdot \left(\frac{256}{243} \cdot \frac{64}{27}\right)^2 = 0.510 \times 6.23 \tag{217.6}$$

or

$$I_\alpha / I_\beta = 3.18.$$

Literature. Radial matrix elements for other pairs of hydrogen states are given by Bethe, H. A., Salpeter, E. E., in: *Encyclopedia of Physics*, vol. **35** (1956), cf. their section 63, and especially table 13.

Problem 218. Compton effect

The scattering of a photon by a free electron at rest shall be investigated in the unrelativistic frame.

Solution. In the presence of a radiation field, the electric current density of an electron field is described by[4]

$$\boldsymbol{j} = -\frac{e\hbar}{2mi}(\psi^\dagger \nabla\psi - \nabla\psi^\dagger \cdot \psi) - \frac{e^2}{mc}\boldsymbol{A}\,\psi^\dagger\psi = \boldsymbol{j}' + \boldsymbol{j}''. \tag{218.1}$$

The interaction energy of the two fields ψ and $\boldsymbol{A}$ is

$$W = \frac{1}{c}\int d^3x\, \boldsymbol{j}\cdot\boldsymbol{A} = W' + W''. \tag{218.2}$$

If for ψ the quantized Schrödinger field,

$$\psi = \frac{1}{\sqrt{\mathscr{V}}}\sum_{\boldsymbol{q}} c_{\boldsymbol{q}}\,\mathrm{e}^{i\boldsymbol{q}\boldsymbol{r}} \tag{218.3}$$

and for $\boldsymbol{A}$ the quantized radiation field

$$\boldsymbol{A} = \sum_{\boldsymbol{k}\lambda}\sqrt{\frac{2\pi\hbar c}{\mathscr{V}k}}\,\boldsymbol{u}_{\boldsymbol{k}}^{(\lambda)}(b_{\boldsymbol{k}\lambda}\mathrm{e}^{i\boldsymbol{k}\boldsymbol{r}} + b_{\boldsymbol{k}\lambda}^\dagger \mathrm{e}^{-i\boldsymbol{k}\boldsymbol{r}}) \tag{218.4}$$

are put in (218.2), it can easily be seen that W'' (arising from $\boldsymbol{j}''$) will contribute to scattering in first-order, whereas W' (arising from $\boldsymbol{j}'$) does so only in second-order approximation. We shall therefore in what follows confine our attention to the term

$$W'' = -\frac{e^2}{mc^2}\int d^3x\, \boldsymbol{A}^2\,\psi^\dagger\psi. \tag{218.5}$$

This interaction term can easily be understood from the viewpoint of the classical picture in which the electric field strength $\mathscr{E} = -\frac{1}{c}\dot{\boldsymbol{A}}$ of the incident light wave plucks at the electron according to its equation of motion,

$$m\ddot{\boldsymbol{r}} = -e\mathscr{E} = \frac{e}{c}\dot{\boldsymbol{A}}; \quad \text{hence } \dot{\boldsymbol{r}} = \frac{e}{mc}\boldsymbol{A}.$$

This originates an induced current density $\boldsymbol{j}'' = \rho\dot{\boldsymbol{r}} = \frac{e}{mc}\rho\,\boldsymbol{A}$, if ρ denotes the

[4] In the radiation problems previously dealt with, the last supplement term in (218.1) would not have contributed in first order.

charge density, and the interaction between the current and the radiation field is, according to Maxwell's theory,

$$W'' = \frac{1}{c} \int d^3x \, \boldsymbol{j}'' \cdot \boldsymbol{A} = \frac{e}{mc^2} \int d^3x \, \rho \, \boldsymbol{A}^2 .$$

If here we set $\rho = -e\psi^\dagger \psi$, Eq. (218.5) results.

The Compton scattering process is, in first order, described by a hamiltonian term with the operator combination

$$c^\dagger_{\boldsymbol{q}'} c_{\boldsymbol{q}} b^\dagger_{\boldsymbol{k}'\lambda'} b_{\boldsymbol{k}\lambda} \tag{218.6}$$

if a photon in the state $(\boldsymbol{k}, \lambda)$ and an electron of momentum $\hbar \boldsymbol{q}$ of the initial state are both annihilated and replaced by the newly created photon in the state $(\boldsymbol{k}', \lambda')$ and electron of momentum $\hbar \boldsymbol{q}'$.

If we pick out of (218.5) the factor of the operator product (218.6), we have the matrix element,

$$\langle f | W'' | i \rangle = -\frac{e^2}{mc^2} \int d^3x \frac{2\pi\hbar c}{\mathscr{V}^2 \sqrt{kk'}} (\boldsymbol{u}_{\boldsymbol{k}}^{(\lambda)} \cdot \boldsymbol{u}_{\boldsymbol{k}'}^{(\lambda')}) \, e^{i(\boldsymbol{k}+\boldsymbol{q}-\boldsymbol{k}'-\boldsymbol{q}')\boldsymbol{r}} . \tag{218.7}$$

The integral in (218.7) vanishes unless

$$\boldsymbol{k} + \boldsymbol{q} = \boldsymbol{k}' + \boldsymbol{q}', \tag{218.8}$$

i. e. unless the law of momentum conservation holds for the process. If it holds, (218.7) becomes

$$\langle f | W'' | i \rangle = -\frac{2\pi e^2 \hbar}{mc\mathscr{V}} \frac{(\boldsymbol{u}_{\boldsymbol{k}}^{(\lambda)} \cdot \boldsymbol{u}_{\boldsymbol{k}'}^{(\lambda')})}{\sqrt{kk'}} . \tag{218.9}$$

For determining the cross section we apply the Golden Rule,

$$d\sigma(\boldsymbol{k}', \lambda') = \frac{2\pi}{\hbar} \frac{\mathscr{V}}{c} \rho_f \, |\langle f | W'' | i \rangle|^2 \tag{218.10}$$

where the final density of states follows from

$$\rho_f = \frac{k'^2 \, dk' \, d\Omega' \, \mathscr{V}}{8\pi^3 \, dE_f} \tag{218.11}$$

and the final energy is

$$E_f = \hbar c k' + \frac{\hbar^2}{2m} \boldsymbol{q}'^2 = \hbar c \left\{ k' + \frac{1}{2\varkappa} (\boldsymbol{k}' - \boldsymbol{k} - \boldsymbol{q})^2 \right\} \tag{218.12}$$

with $\varkappa = mc/\hbar$.

Let us now deal with polarization. In Fig. 75 the directions $\boldsymbol{k}$ and $\boldsymbol{k}'$ of the photon before and after the collision are lying in the plane of drawing. The vectors $\boldsymbol{u}_{\boldsymbol{k}}^{(1)}$ and $\boldsymbol{u}_{\boldsymbol{k}'}^{(1)}$ both lie in this plane, $\boldsymbol{u}_{\boldsymbol{k}}^{(2)}$ and $\boldsymbol{u}_{\boldsymbol{k}'}^{(2)}$ (not drawn) are perpendicular to it. The scalar products in (218.9) can then be read directly from the figure:

$$(\boldsymbol{u}_{\boldsymbol{k}}^{(1)} \cdot \boldsymbol{u}_{\boldsymbol{k}'}^{(1)}) = \cos\vartheta; \qquad (\boldsymbol{u}_{\boldsymbol{k}}^{(1)} \cdot \boldsymbol{u}_{\boldsymbol{k}'}^{(2)}) = 0; \tag{218.13a}$$

$$(\boldsymbol{u}_{\boldsymbol{k}}^{(2)} \cdot \boldsymbol{u}_{\boldsymbol{k}'}^{(1)}) = 0; \qquad (\boldsymbol{u}_{\boldsymbol{k}}^{(2)} \cdot \boldsymbol{u}_{\boldsymbol{k}'}^{(2)}) = 1. \tag{218.13b}$$

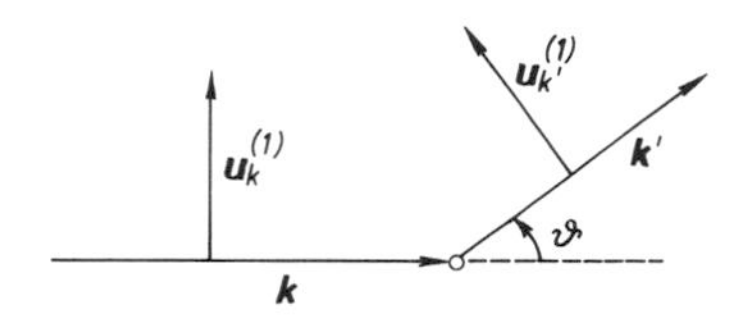

Fig. 75. Compton effect. Definition of polarization vectors in the initial state ($\boldsymbol{k}$) and final state ($\boldsymbol{k}'$) of the photon. Vectors $\boldsymbol{u}_{\boldsymbol{k}}^{(1)}$ and $\boldsymbol{u}_{\boldsymbol{k}'}^{(1)}$ in the $\boldsymbol{k}$, $\boldsymbol{k}'$ plane, vectors $\boldsymbol{u}_{\boldsymbol{k}}^{(2)}$ and $\boldsymbol{u}_{\boldsymbol{k}'}^{(2)}$ (not drawn) perpendicular to it

These relations show that there can only be transitions with both polarization vectors, before and after the collision, lying in the $\boldsymbol{k}, \boldsymbol{k}'$ plane, or with both perpendicular to it. In the first case the transition probability becomes proportional to $\cos^2\vartheta$, in the second case it is independent of the scattering angle. If the incident light is unpolarized we have to average over λ and to sum up over the final states λ' thus getting

$$\sum_{\lambda'} \overline{(\boldsymbol{u}_{\boldsymbol{k}}^{(\lambda)} \cdot \boldsymbol{u}_{\boldsymbol{k}'}^{(\lambda')})^2} = \tfrac{1}{2}(1+\cos^2\vartheta). \tag{218.14}$$

For the further discussion, let us suppose the electron to be at rest in the initial state,

$$\boldsymbol{q} = 0. \tag{218.15}$$

Then the law of energy conservation reads, according to (218.12),

$$\hbar c \left\{ k' + \frac{1}{2\varkappa} (\boldsymbol{k}' - \boldsymbol{k})^2 \right\} = \hbar c k. \tag{218.16}$$

This is a quadratic equation for the determination of k' because

$$(\boldsymbol{k}' - \boldsymbol{k})^2 = k'^2 + k^2 - 2kk'\cos\vartheta.$$

Its solution is

$$k' = k\cos\vartheta - \varkappa + \sqrt{\varkappa^2 + 2\varkappa k(1-\cos\vartheta) - k^2\sin^2\vartheta}. \tag{218.17}$$

To evaluate ρ_f, Eq. (218.11), we derive from (218.16)

$$\frac{dE_f}{dk'} = \hbar c \left\{1 + \frac{1}{\varkappa}(k' - k\cos\vartheta)\right\};$$

therefore,

$$\rho_f = \frac{\mathscr{V}}{8\pi^3 \hbar c} \cdot \frac{k'^2}{1 + \frac{1}{\varkappa}(k' - k\cos\vartheta)} d\Omega', \tag{218.18}$$

and the differential scattering cross section, according to (218.10), (218.9), (218.14) and (218.18) becomes

$$d\sigma = \left(\frac{e^2}{mc^2}\right)^2 \frac{k'^2}{1 + \frac{1}{\varkappa}(k' - k\cos\vartheta)} \frac{\frac{1}{2}(1 + \cos^2\vartheta)}{k k'} d\Omega', \tag{218.19}$$

where k' still may be replaced by the full expression (218.17).

So far, these are all rigorous unrelativistic formulae. They can, of course, be used only as long as the electron does not receive a kinetic energy comparable with mc^2:

$$E_{\mathrm{kin}} = \hbar c(k - k') \ll mc^2 \quad \text{or} \quad k - k' \ll \varkappa.$$

It is therefore reasonable to expand (218.17) and (218.19) in powers of $k/\varkappa$:

$$\frac{k'}{\varkappa} = \frac{k}{\varkappa} - \frac{k^2}{\varkappa^2}(1 - \cos\vartheta) + \cdots$$

and

$$d\sigma = \frac{1}{2}\left(\frac{e^2}{mc^2}\right)^2 \left\{1 - 2\frac{k}{\varkappa}(1 - \cos\vartheta)\right\}(1 + \cos^2\vartheta)\, d\Omega'. \tag{218.20}$$

The second-order contribution from W', Eq. (218.2) vanishes for $q=0$. In the relativistic treatment, the Dirac expression (199.1) for the current density is generally used which only in second order can originate Compton transitions. The expression, however, may be split up according to Problem 199, so that the relativistic treatment may be performed in complete analogy to the unrelativistic method of the present Problem.

The total cross section then follows by elementary integration of (218.20) over all directions and may be written

$$\sigma = \frac{8\pi}{3}\left(\frac{e^2}{mc^2}\right)^2 \frac{1}{1 + 2\frac{k}{\varkappa}}. \tag{218.21}$$

It is well known from classical electrodynamics that the so-called Thomson cross section,

$$\sigma_0 = \frac{8\pi}{3}\left(\frac{e^2}{mc^2}\right)^2 = 6.652 \times 10^{-25}\,\mathrm{cm}^2, \tag{218.22}$$

represents the long-wavelength limit to our problem. The additional factor in (218.21), lowering the cross section with increasing photon energy ($k/\varkappa = \hbar\omega/mc^2$), is a first quantum theoretical correction, sufficient as long as $k/\varkappa \ll 1$, or as long as the wavelength is still large compared with the Compton wavelength $1/\varkappa = \hbar/mc$. (For $\hbar\omega = mc^2 = 0.51$ MeV or $k = \varkappa$, the wavelength is $\lambda = 2\pi\hbar/mc$.)

NB 1. The second-order contribution from W', Eq. (218.2) vanishes for $q=0$. In the relativistic treatment, usually the current density expression (199.1) is used, which can only in second order originate Compton transitions. If this expression is split up according to Problem 199, we may formulate the relativistic treatment in complete analogy to the unrelativistic one of the present problem.

NB 2. At higher photon energies the electron field has to be treated by the Dirac theory. The result then is the Klein-Nishina formula instead of (218.21). That our approximation can be used in rather a wide domain of energies may be seen from the following figures. For $k/\varkappa = 0.2$, Eq. (218.21) gives $\sigma/\sigma_0 = 0.714$ whereas the rigorous Klein-Nishina formula leads to 0.737. At $k/\varkappa = 1$ the two values are $\sigma/\sigma_0 = 0.333$ from (218.21) and 0.431 (Kl.-N.). The real values of cross sections decrease much more slowly with increasing energy than those of our approximation, e.g. at $k/\varkappa = 1000$ we find $\sigma/\sigma_0 = 0.0050$ instead of the exact value 0.0215.

Problem 219. Bremsstrahlung

In an unrelativistic treatment, the production of an x ray photon by an electron passing a heavy nucleus may be dealt with as a second-order process in which the nucleus is simply described by its electrostatic field and its mass is supposed to be infinitely large. The bremsstrahlung spectrum shall be calculated in this approximation.

Solution. Reproduced in Fig. 76 are the two simplest possible graphs of this process. In the initial state there is the nucleus at rest and an electron of momentum $\hbar\boldsymbol{q}$, in the final state the infinitely heavy nucleus is still at rest, the electron has a smaller momentum $\hbar\boldsymbol{q}'$ and a photon $(\boldsymbol{k},\lambda)$ has been created. This creation process, together with simple Rutherford scattering of the electron at the nucleus, gives rise to the two vertices in one or the other order of succession. In consequence of the infinite mass of the nucleus, not energy but momentum may be transferred to it ($M=\infty$, p finite, $p^2/2M=0$, $v=0$) so that between the initial and final states of the other particles energy conservation will still hold, but conservation of momentum will not.

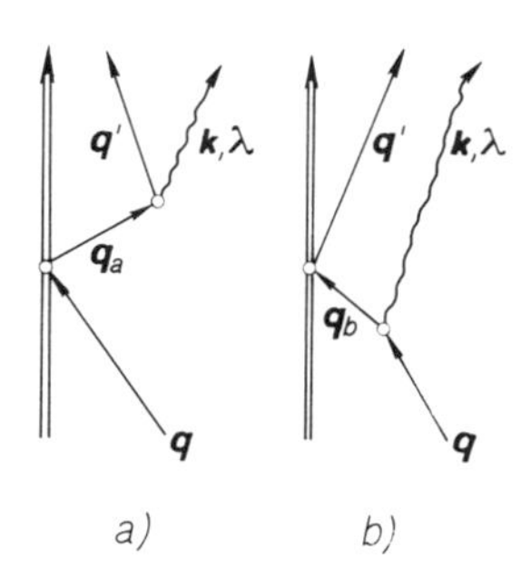

Fig. 76a and b. Lowest Feynman graphs for bremsstrahlung in unrelativistic approach. Double line nucleus, single line electron, wavy line photon

The perturbation energy consists of two terms,

$$H'=H_1+H_2 \tag{219.1}$$

viz.

$$H_1=-Ze^2\int d^3x\,\frac{1}{r}\,\psi^\dagger\psi, \tag{219.2}$$

the Coulomb interaction of the nucleus (charge Ze) and electron (charge density $\rho=-e\psi^\dagger\psi$), and

$$H_2=\frac{1}{c}\int d^3x\,(\boldsymbol{A}\cdot\boldsymbol{j});\quad \boldsymbol{j}=-\frac{e\hbar}{2mi}(\psi^\dagger\nabla\psi-\nabla\psi^\dagger\cdot\psi), \tag{219.3}$$

the radiation interaction. The field operators are

$$\boldsymbol{A}=\sum_{\boldsymbol{k},\lambda}\sqrt{\frac{2\pi\hbar c}{k\mathscr{V}}}\,\boldsymbol{u}_{\boldsymbol{k}}^{(\lambda)}(b_{\boldsymbol{k}\lambda}e^{i\boldsymbol{k}\boldsymbol{r}}+b_{\boldsymbol{k}\lambda}^\dagger e^{-i\boldsymbol{k}\boldsymbol{r}}) \tag{219.4}$$

and

$$\psi=\frac{1}{\sqrt{\mathscr{V}}}\sum_\nu c_\nu e^{i\boldsymbol{q}_\nu\boldsymbol{r}};\quad \nabla\psi=\frac{i}{\sqrt{\mathscr{V}}}\sum_\nu \boldsymbol{q}_\nu c_\nu e^{i\boldsymbol{q}_\nu\boldsymbol{r}}. \tag{219.5}$$

Putting (219.4) and (219.5) in (219.2) and (219.3) and performing the space integrations, we arrive at

$$H_1=-\frac{Ze^2}{\mathscr{V}}\sum_{\mu\nu}c_\mu^\dagger c_\nu\frac{4\pi}{|\boldsymbol{q}_\mu-\boldsymbol{q}_\nu|^2} \tag{219.6}$$

and

$$H_2 = -\frac{e\hbar}{2mc}\sum_{\boldsymbol{k},\lambda}\sum_{\mu,\nu}\sqrt{\frac{2\pi\hbar c}{k\mathscr{V}}}\,\boldsymbol{u}_{\boldsymbol{k}\lambda}\cdot(\boldsymbol{q}_\mu+\boldsymbol{q}_\nu)c_\mu^\dagger c_\nu\{b_{\boldsymbol{k}\lambda}\delta_{\boldsymbol{k},\boldsymbol{q}_\mu-\boldsymbol{q}_\nu}+b_{\boldsymbol{k}\lambda}^\dagger\delta_{\boldsymbol{k},\boldsymbol{q}_\nu-\boldsymbol{q}_\mu}\}. \tag{219.7}$$

For graph a in Fig. 76, we have to look up the factors of $c_{\boldsymbol{q}}c_{\boldsymbol{q}_a}^\dagger$ in H_1 and of $c_{\boldsymbol{q}_a}c_{\boldsymbol{q}'}^\dagger b_{\boldsymbol{k}\lambda}^\dagger$ in H_2 to obtain the matrix elements

$$\langle\boldsymbol{q}_a|H_1|\boldsymbol{q}\rangle = -\frac{4\pi Ze^2}{\mathscr{V}}\cdot\frac{1}{|\boldsymbol{q}-\boldsymbol{q}_a|^2} \tag{219.8a}$$

and

$$\langle\boldsymbol{q}',\boldsymbol{k}\lambda|H_2|\boldsymbol{q}_a\rangle = -\frac{e\hbar}{2mc}\sqrt{\frac{2\pi\hbar c}{k\mathscr{V}}}\,\boldsymbol{u}_{\boldsymbol{k}\lambda}\cdot(\boldsymbol{q}_a+\boldsymbol{q}'). \tag{219.8b}$$

No law of conservation holds at the first vertex but, at the second, conservation of momentum yields

$$\boldsymbol{k}+\boldsymbol{q}'=\boldsymbol{q}_a \tag{219.9a}$$

so that, with $\boldsymbol{u}_{\boldsymbol{k}\lambda}$ perpendicular to $\boldsymbol{k}$,

$$\boldsymbol{u}_{\boldsymbol{k}\lambda}\cdot(\boldsymbol{q}_a+\boldsymbol{q}') = 2\boldsymbol{u}_{\boldsymbol{k}\lambda}\cdot\boldsymbol{q}'. \tag{219.10a}$$

For the graph b in Fig. 76 we want the factors of $c_{\boldsymbol{q}}c_{\boldsymbol{q}_b}^\dagger b_{\boldsymbol{k}\lambda}^\dagger$ in H_2 and of $c_{\boldsymbol{q}_b}c_{\boldsymbol{q}'}^\dagger$ in H_1, viz.

$$\langle\boldsymbol{q}_b,\boldsymbol{k}\lambda|H_2|\boldsymbol{q}\rangle = -\frac{e\hbar}{2mc}\sqrt{\frac{2\pi\hbar c}{k\mathscr{V}}}\,\boldsymbol{u}_{\boldsymbol{k}\lambda}\cdot(\boldsymbol{q}+\boldsymbol{q}_b) \tag{219.11a}$$

and

$$\langle\boldsymbol{q}'|H_1|\boldsymbol{q}_b\rangle = -\frac{4\pi Ze^2}{\mathscr{V}}\cdot\frac{1}{|\boldsymbol{q}'-\boldsymbol{q}_b|^2}. \tag{219.11b}$$

In this case, momentum conservation holds at the first vertex,

$$\boldsymbol{k}-\boldsymbol{q}+\boldsymbol{q}_b=0 \tag{219.9b}$$

and therefore

$$\boldsymbol{u}_{\boldsymbol{k}\lambda}\cdot(\boldsymbol{q}+\boldsymbol{q}_b) = 2\boldsymbol{u}_{\boldsymbol{k}\lambda}\cdot\boldsymbol{q}. \tag{219.10b}$$

The energy of the initial state,

$$E_i = \frac{\hbar^2 q^2}{2m} \tag{219.12}$$

must be equal to the energy of the final state,

$$E_f = \frac{\hbar^2 q'^2}{2m} + \hbar ck, \tag{219.13}$$

hence

$$q^2 = q'^2 + 2\varkappa k; \qquad \varkappa = \frac{mc}{\hbar}. \tag{219.14}$$

Using (219.9a, b) we find for the two intermediate states

$$E_a = \frac{\hbar^2 q_a^2}{2m} = \frac{\hbar^2}{2m}(\boldsymbol{q}' + \boldsymbol{k})^2; \tag{219.15}$$

$$E_b = \frac{\hbar^2 q_b^2}{2m} + \hbar c k = \frac{\hbar^2}{2m}\{(\boldsymbol{q} - \boldsymbol{k})^2 + 2\varkappa k\}. \tag{219.16}$$

With these abbreviations we now may write the second-order matrix element in the form

$$\langle f|H'|i\rangle = \frac{\langle f|H_2|a\rangle\langle a|H_1|i\rangle}{E_i - E_a} + \frac{\langle f|H_1|b\rangle\langle b|H_2|i\rangle}{E_i - E_b},$$

which, collecting the matrix elements from (219.8a, b) and (219.11a, b) and the expressions for $\boldsymbol{q}_a$ and $\boldsymbol{q}_b$, may be written in more detail

$$\langle f|H'|i\rangle = \frac{4\pi Z e^2}{\mathscr{V}} \cdot \frac{e\hbar}{mc} \sqrt{\frac{2\pi\hbar c}{\mathscr{V} k}} \frac{1}{|\boldsymbol{q} - \boldsymbol{q}' - \boldsymbol{k}|^2} \left\{ \frac{(\boldsymbol{u}_{k\lambda} \cdot \boldsymbol{q}')}{E_i - E_a} + \frac{(\boldsymbol{u}_{k\lambda} \cdot \boldsymbol{q})}{E_i - E_b} \right\}. \tag{219.17}$$

To find the cross section, we must apply the Golden Rule and therefore need the final state density ρ_f. This is a little difficult to determine because, in consequence of non-conservation of momentum, the two final particles are emitted in independent directions. For one particle (1) we know that

$$\rho_1 = \frac{d^3 p_1 \mathscr{V}}{h^3 dE_1} = \frac{p_1^2}{v_1} \frac{\mathscr{V}}{h^3} d\Omega_1.$$

The other particle (2), for which ρ_2 must be a similar expression, is bound to lie within an interval dE_2 whose width and position are already determined by energy conservation when the interval for the first particle is given. Hence,

$$\rho_f = \rho_1 \rho_2 \, dE_2$$

or, in our special case, if 1 is the unrelativistic electron ($\boldsymbol{q}'$) and 2 the photon ($\boldsymbol{k}$), so that

$$\frac{p_1^2}{v_1} = m\hbar q'; \qquad \frac{p_2^2}{v_2} = \frac{1}{c}\hbar^2 k^2; \qquad dE_2 = \hbar c\, dk,$$

we have

$$\rho_f = \frac{m\hbar q'}{8\pi^3\hbar^3}\mathscr{V}\,d\Omega' \cdot \frac{\hbar^2 k^2}{8\pi^3\hbar^3 c}\mathscr{V}\,d\Omega_k \cdot \hbar c\,dk. \tag{219.18}$$

From the general formula

$$d\sigma = \frac{\mathscr{V}}{v_1}\cdot\frac{2\pi}{\hbar}\rho_f |\langle f|H|i\rangle|^2,$$

giving the differential cross section of the electron being scattered into $d\Omega'$ and the photon of polarization λ falling in the interval dk and within the solid-angle element $d\Omega_k$, we then find by inserting (219.17) and (219.18):

$$d\sigma = \frac{e^2}{\hbar c}\cdot\frac{Z^2 e^4}{\pi^2}\cdot\frac{q' k}{q}\cdot\frac{1}{|\boldsymbol{q}-\boldsymbol{q}'-\boldsymbol{k}|^4}\left\{\frac{\boldsymbol{u}_{k\lambda}\cdot\boldsymbol{q}'}{E_i-E_a}+\frac{\boldsymbol{u}_{k\lambda}\cdot\boldsymbol{q}}{E_i-E_b}\right\}^2 d\Omega_k\,d\Omega'\,dk. \tag{219.19}$$

There remains the problem of determining the photon energy spectrum, whatever the directions of both particles or the polarization of the photon emitted. This means integration over the directions and summation over λ. The integration procedures in such problems are often rather laborious. In the present case, however, they become very simple, as shall now be shown.

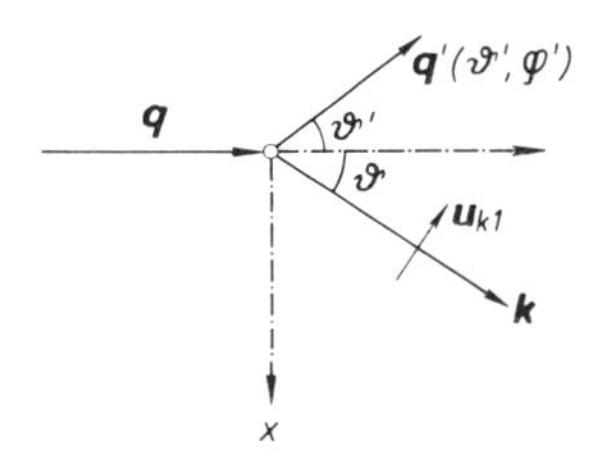

Fig. 77. Notations for bremsstrahlung

In Fig. 77 the three momenta have been drawn in a coordinate system. They are not coplanar, i.e. if the xz plane is chosen so that $\boldsymbol{q}$ and $\boldsymbol{k}$ fall in this plane, $\boldsymbol{q}'$ will have a y component. The components of the three momenta thus become

$$\boldsymbol{q} = q(0,0,1);$$
$$\boldsymbol{k} = k(\sin\vartheta, 0, \cos\vartheta);$$
$$\boldsymbol{q}' = q'(\sin\vartheta'\cos\varphi', \sin\vartheta'\sin\varphi', \cos\vartheta')$$

and of the two polarization vectors

$$\boldsymbol{u}_{k1}=(-\cos\vartheta, 0, \sin\vartheta); \qquad \boldsymbol{u}_{k2}=(0,1,0).$$

For further evaluation of the cross section, we now make the tyipically unrelativistic approximation to neglect the photon momentum compared to that of the electron, since from (219.14) it may be concluded that $k \ll q$ and $k \ll q'$ because $\varkappa$ is large. This allows us to simplify the energy denominators in (219.19). Making use of Eqs. (219.12), (219.15) and (219.16), we get

$$E_i - E_a = \frac{\hbar^2}{2m}(q^2 - q'^2 - 2\boldsymbol{q}'\cdot\boldsymbol{k} - k^2);$$

$$E_i - E_b = \frac{\hbar^2}{2m}(-2\varkappa k + 2\boldsymbol{q}\cdot\boldsymbol{k} - k^2).$$

In both expressions the two last terms then may be neglected, and $2\varkappa k$ replaced by $q^2 - q'^2$ according to (219.14) so that

$$E_i - E_a \simeq \frac{\hbar^2}{2m}(q^2 - q'^2) \simeq -(E_i - E_b). \qquad (219.20)$$

The two energy denominators in (219.19) thus becoming opposite and equal, we may simply subtract the two numerators for either $\lambda=1$ or $\lambda=2$:

$$\begin{aligned} \boldsymbol{u}_{k1}\cdot\boldsymbol{q}' - \boldsymbol{u}_{k1}\cdot\boldsymbol{q} &= q'(-\cos\vartheta\sin\vartheta'\cos\varphi' + \sin\vartheta\cos\vartheta') - q\sin\vartheta; \\ \boldsymbol{u}_{k2}\cdot\boldsymbol{q}' - \boldsymbol{u}_{k2}\cdot\boldsymbol{q} &= q'\sin\vartheta'\sin\varphi'. \end{aligned} \qquad (219.21)$$

To perform summation over λ we then square and add these two expressions.

Finally, the Rutherford denominator in (219.19) may be written in the same approximation

$$(\boldsymbol{q} - \boldsymbol{q}' - \boldsymbol{k})^4 \simeq (\boldsymbol{q} - \boldsymbol{q}')^4 = (q^2 + q'^2 - 2qq'\cos\vartheta')^2. \qquad (219.22)$$

Assembling all these factors, it is seen at once that the angles ϑ and φ' occur in the sum of the squares of (219.21) only, so that integration over these angles can be performed separately in these terms by a straightforward elementary calculation:

$$\int_0^{2\pi} d\varphi' \oint d\Omega_k \sum_\lambda (\boldsymbol{u}_{k\lambda}\cdot\boldsymbol{q}' - \boldsymbol{u}_{k\lambda}\cdot\boldsymbol{q})^2 = \frac{16\pi^2}{3}(q^2 + q'^2 - 2qq'\cos\vartheta'). \qquad (219.23)$$

Fortunately we find here the same bracket in the numerator of the cross section that according to (219.22) occurs twice in its denominator. Gathering up all these factors, we then arrive at

$$d\sigma(k) = Z^2 \left(\frac{e^2}{\hbar c}\right)^3 \cdot \frac{16}{3} \frac{q'}{q} \frac{dk}{k} \int\limits_{-1}^{+1} \frac{d\cos\vartheta'}{q^2 + q'^2 - 2qq'\cos\vartheta'}$$

where $d\sigma(k)$ stands for the partial cross section of a photon created in the energy interval dk, whatever else may have happened. The integral permits elementary evaluation,

$$\int\limits_{-1}^{+1} \frac{d\cos\vartheta'}{q^2 + q'^2 - 2qq'\cos\vartheta'} = \frac{1}{qq'} \log \frac{q+q'}{q-q'},$$

leaving us finally with the result

$$d\sigma(k) = \frac{16}{3} Z^2 \left(\frac{e^2}{\hbar c}\right)^3 \left(\frac{1}{q^2} \log \frac{q+q'}{q-q'}\right) \frac{dk}{k}. \tag{219.24}$$

Using (219.14), the momenta may be eliminated from this formula and q' be expressed by the energies E of the incident electron, and $E_k = \hbar c k$ of the outgoing photon:

$$\log \frac{q+q'}{q-q'} = \log \frac{(q+q')^2}{q^2 - q'^2} = \log \left\{ \frac{(\sqrt{E} + \sqrt{E - E_k})^2}{E_k} \right\}.$$

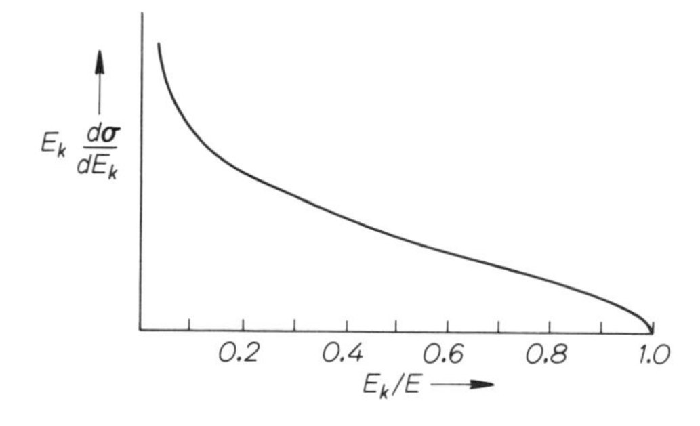

Fig. 78. Bremsstrahlung intensity distribution in unrelativistic theory. The logarithmic divergence at $E_k = 0$ does not occur if screening of the Coulomb field is taken into account

This bremsstrahlung spectrum has been shown in Fig. 78. It shows a remarkable singularity for the production of photons of very small energy, the so-called infrared divergence.

Literature. For relativistic treatment and for screening problems cf. Heitler, W.: Quantum Theory of Radiation, 3rd ed., Oxford 1954, pp. 242—256.

Mathematical Appendix

Coordinate systems

We start with rectangular coordinates x, y, z and list the transformation formulae for some frequently occurring curvilinear coordinate systems. We add the formulae of transformation for the distance from the coordinate centre,

$$r = \sqrt{x^2 + y^2 + z^2}$$

and of the Laplacian,

$$\nabla^2 = \frac{\partial^2}{\partial x^2} + \frac{\partial^2}{\partial y^2} + \frac{\partial^2}{\partial z^2}.$$

a) Spherical polar coordinates. Let the z axis be the polar axis and denote by ϑ the angle between the vector $\boldsymbol{r}$ and this axis, and by φ the azimuth angle about this axis (cf. Fig. 33 on p. 145, vol. I).

$$x = r \sin\vartheta \cos\varphi; \qquad y = r \sin\vartheta \sin\varphi; \qquad z = r\cos\vartheta;$$

$$\nabla^2 u = \frac{\partial^2 u}{\partial r^2} + \frac{2}{r}\frac{\partial u}{\partial r} + \frac{1}{r^2}\left\{\frac{1}{\sin\vartheta}\frac{\partial}{\partial\vartheta}\left(\sin\vartheta\frac{\partial u}{\partial\vartheta}\right) + \frac{1}{\sin^2\vartheta}\frac{\partial^2 u}{\partial\varphi^2}\right\}.$$

b) Circular cylindrical coordinates. Let the z axis be the common axis of circular cylinders of radii $\rho = \text{constans}$. The angle φ again shall be the azimuth angle about the z axis, and ρ, φ, z be chosen as coordinates.

$$x = \rho\cos\varphi; \qquad y = \rho\sin\varphi; \qquad z = z; \qquad r = \sqrt{\rho^2 + z^2};$$

$$\nabla^2 u = \frac{\partial^2 u}{\partial\rho^2} + \frac{1}{\rho}\frac{\partial u}{\partial\rho} + \frac{1}{\rho^2}\frac{\partial^2 u}{\partial\varphi^2} + \frac{\partial^2 u}{\partial z^2}.$$

c) Parabolic coordinates. Let the z axis be the common axis of rotation for two sets of paraboloids $\xi = \text{constans}$ and $\eta = \text{constans}$, all having their foci in the coordinate centre ($z = 0$) and opened the one towards the positive, the other towards the negative z axis. Again φ

shall be the azimuth angle about the z axis. There are two customary choices then of the coordinates ξ, η, φ.

First system. $x=\sqrt{\xi\eta}\cos\varphi$; $y=\sqrt{\xi\eta}\sin\varphi$; $z=\frac{1}{2}(\xi-\eta)$;
$r=\frac{1}{2}(\xi+\eta)$; $\xi=r+z$; $\eta=r-z$; $\rho=\sqrt{\xi\eta}$;

$$\nabla^2 u = \frac{4}{\xi+\eta}\left\{\frac{\partial}{\partial\xi}\left(\xi\frac{\partial u}{\partial\xi}\right)+\frac{\partial}{\partial\eta}\left(\eta\frac{\partial u}{\partial\eta}\right)+\frac{\xi+\eta}{4\xi\eta}\frac{\partial^2 u}{\partial\varphi^2}\right\}.$$

Second system. $x=\xi\eta\cos\varphi$; $y=\xi\eta\sin\varphi$; $z=\frac{1}{2}(\xi^2-\eta^2)$;
$r=\frac{1}{2}(\xi^2+\eta^2)$; $\xi^2=r+z$; $\eta^2=r-z$; $\rho=\xi\eta$;

$$\nabla^2 u = \frac{1}{\xi^2+\eta^2}\left\{\frac{1}{\xi}\frac{\partial}{\partial\xi}\left(\xi\frac{\partial u}{\partial\xi}\right)+\frac{1}{\eta}\frac{\partial}{\partial\eta}\left(\eta\frac{\partial u}{\partial\eta}\right)+\left(\frac{1}{\xi^2}+\frac{1}{\eta^2}\right)\frac{\partial^2 u}{\partial\varphi^2}\right\}.$$

d) Ellipsoidal coordinates. Two points lying on the z axis at $z=\pm c$ shall be chosen as common foci to a set of prolate ellipsoids of rotation described by $\xi=$constans. There exists a set of two-sheet hyperboloids of rotation orthogonally intersecting them and having the same foci, $\eta=$constans. Again φ shall be the azimuth angle about the z axis. Let r_1 and r_2 be the distances from the two foci $z=-c$ and $z=+c$, respectively.

$x=c\sqrt{(\xi^2-1)(1-\eta^2)}\cos\varphi$; $y=c\sqrt{(\xi^2-1)(1-\eta^2)}\sin\varphi$; $z=c\xi\eta$;

$r_1=c(\xi+\eta)$; $r_2=c(\xi-\eta)$; $\xi=\frac{1}{2c}(r_1+r_2)$; $\eta=\frac{1}{2c}(r_1-r_2)$;

$r=c\sqrt{\xi^2+\eta^2-1}$; $\rho=c\sqrt{(\xi^2-1)(1-\eta^2)}$.

Domains of values: $1\leq\xi<\infty$; $-1\leq\eta\leq+1$; $0\leq\varphi<2\pi$.

$$\nabla^2 u = \frac{1}{c^2(\xi^2-\eta^2)}\left\{\frac{\partial}{\partial\xi}(\xi^2-1)\frac{\partial u}{\partial\xi}+\frac{\partial}{\partial\eta}(1-\eta^2)\frac{\partial u}{\partial\eta}+\frac{\xi^2-\eta^2}{(1-\eta^2)(\xi^2-1)}\frac{\partial^2 u}{\partial\varphi^2}\right\}.$$

Γ function

The Γ function is a generalization of the *factorial*, the latter being defined only for positive integers by

$$n!=1\cdot 2\cdot 3\ldots n \tag{1}$$

and having the special property

$$(n+1)!=(n+1)n!. \tag{2}$$

It can equally well be defined by the *Euler integral*,

$$n! = \int_0^\infty dt\, e^{-t} t^n \tag{3}$$

which supplements the value $0! = 1$ which would otherwise be meaningless.

The generalizations of (2) and (3) for any complex number $z = x + iy$ are

$$\Gamma(z+1) = z\,\Gamma(z); \qquad \Gamma(n+1) = n! \tag{4}$$

and

$$\Gamma(z) = \int_0^\infty dt\, e^{-t} t^{z-1} \quad (\text{if } \operatorname{Re} z > 0). \tag{5}$$

This function is meromorphic and has poles along the negative real axis at $z = -n$ $(n = 0, 1, 2, \ldots)$ with residues $(-1)^n/n!$.

Special values.

$$\Gamma(1) = 0! = 1; \qquad \Gamma(n) = (n-1)!; \tag{6}$$

$$\Gamma(\tfrac{1}{2}) = \sqrt{\pi}; \qquad \Gamma(\tfrac{3}{2}) = (\tfrac{1}{2})! = \tfrac{1}{2}\sqrt{\pi}; \tag{7}$$

$$\Gamma(n + \tfrac{3}{2}) = (n + \tfrac{1}{2})! = \frac{(2n+1)!}{2^{2n+1}\, n!}\sqrt{\pi}. \tag{8}$$

Relations between functions of different arguments.

$$\Gamma(z)\,\Gamma(1-z) = \frac{\pi}{\sin \pi z}; \tag{9}$$

$$\Gamma(2z) = \frac{1}{\sqrt{\pi}}\, 2^{2z-1}\, \Gamma(z)\,\Gamma(z + \tfrac{1}{2}). \tag{10}$$

Infinite series or product expansions. To compute the complex number

$$\Gamma(x + iy) = \xi e^{i\eta} \tag{11a}$$

we may use the expansions

$$\xi = \Gamma(x) \prod_{n=0}^\infty \left\{1 + \frac{y^2}{(x+n)^2}\right\}^{-\frac{1}{2}} \tag{11b}$$

and

$$\eta = y\left\{-C + \sum_{n=1}^\infty \left(\frac{1}{n} - \frac{1}{y}\tan^{-1}\frac{y}{x+n-1}\right)\right\} \tag{11c}$$

with

$$C = \int_0^\infty dt\, e^{-t} \log \frac{1}{t} = 0.577215\ldots \tag{11d}$$

the Euler constant. For $x=1$ we get

$$\xi^2 = |\Gamma(1+iy)|^2 = \frac{\pi y}{\sinh \pi y}. \tag{12}$$

Asymptotic behaviour. For $|z| \gg 1$ and $|\arg z| < \pi$ (i.e. excluding the negative real axis with its poles) we may use *Stirling's formula*

$$\log \Gamma(z) = \left(z - \frac{1}{2}\right) \log z - z + \frac{1}{2} \log 2\pi + O\left(\frac{1}{z}\right) \tag{13}$$

or

$$\Gamma(z) \simeq \sqrt{\frac{2\pi}{z}}\, e^{z(\log z - 1)}. \tag{14}$$

The formula is often used for

$$z! = \Gamma(z+1) = z\,\Gamma(z) \simeq \sqrt{2\pi z}\, e^{z(\log z - 1)} \left\{1 + \frac{1}{12z} + \frac{1}{288 z^2} + \cdots\right\}. \tag{15}$$

The last series is semiconvergent. Putting it $=1$, we get the following comparison:

n	$n!$	$\sqrt{2\pi n}\, e^{n(\log n - 1)}$
0	1	0
1	1	0.925
2	2	1.920
3	6	5.836
4	24	23.506
5	120	118.01

Bessel functions

The differential equation

$$u'' + \frac{1}{z} u' + \left(1 - \frac{\nu^2}{z^2}\right) u = 0 \tag{1}$$

is solved by

$$u = A\, J_\nu(z) + B\, N_\nu(z) \tag{2}$$

or by

$$u = C_1\, H_\nu^{(1)}(z) + C_2\, H_\nu^{(2)}(z). \tag{3}$$

The function J_ν is called *Bessel function* (in the proper sense), the function N_ν *Neumann function.* If ν is not an integer, the definition

$$N_\nu(z) = \frac{1}{\sin \pi \nu} (\cos \pi \nu \, J_\nu(z) - J_{-\nu}(z)) \tag{4}$$

may be used, otherwise (i.e. for $\nu = n$ with $n = 0, \pm 1, \pm 2, \ldots$),

$$J_{-n}(z) = (-1)^n J_n(z) \tag{5}$$

is no longer linearly independent of J_n. The function $N_n(z)$ still may then be defined by its asymptotics (see below).

The Bessel function J_ν can be defined by its power series

$$J_\nu(z) = \left(\frac{z}{2}\right)^\nu \sum_{n=0}^{\infty} \frac{(-1)^n}{n! \, \Gamma(\nu+n+1)} \left(\frac{z}{2}\right)^{2n}, \tag{6}$$

converging in the whole z plane cut along the negative real axis, since $z=0$ is a branch point.

The functions $H_\nu^{(1)}$ and $H_\nu^{(2)}$ are called *Hankel functions* of the first and second kind. They are defined by

$$H_\nu^{(1)}(z) = J_\nu(z) + i N_\nu(z); \qquad H_\nu^{(2)}(z) = J_\nu(z) - i N_\nu(z). \tag{7}$$

The fundamental system (2) of solutions has real values for real z; its Wronskian is $2/(\pi z)$. The Wronskian of (3) is $-4i/(\pi z)$. If ν is not integer, J_ν and $J_{-\nu}$ form a third fundamental system of solutions with the Wronskian $-2 \sin \pi \nu/(\pi z)$.

Recurrence relations. For each one of the four types of functions defined by (2) and (3) there hold the relations,

$$u_{\nu-1} + u_{\nu+1} = \frac{2\nu}{z} u_\nu; \qquad u_{\nu-1} - u_{\nu+1} = 2 u'_\nu \tag{8a}$$

or

$$u_{\nu+1} = \frac{\nu}{z} u_\nu - u'_\nu; \qquad u_{\nu+1} = \frac{2\nu}{z} u_\nu - u_{\nu-1}. \tag{8b}$$

Asymptotic behaviour. With the abbreviation

$$\zeta = z - \frac{\pi}{2}(\nu + \tfrac{1}{2}) \tag{9a}$$

for $|z| \gg 1 + |\nu|$ and $|\arg z| < \pi$, i.e. for large values of $|z|$ in the z plane cut along the negative real axis, there hold the asymptotic formulae

$$\left.\begin{aligned} J_\nu(z) &\to \sqrt{\frac{2}{\pi z}} \cos \zeta; & N_\nu(z) &\to \sqrt{\frac{2}{\pi z}} \sin \zeta; \\ H_\nu^{(1)}(z) &\to \sqrt{\frac{2}{\pi z}} e^{i\zeta}; & H_\nu^{(2)}(z) &\to \sqrt{\frac{2}{\pi z}} e^{-i\zeta}. \end{aligned}\right\} \tag{9b}$$

Functions of the type $z^{-\frac{1}{2}} H_\nu^{(1,2)}(z)$ for real z describe outgoing and incoming spherical waves, respectively.

Modified Bessel functions. The functions

$$I_\nu(x) = i^{-\nu} J_\nu(i x) \tag{10}$$

and

$$K_\nu(x) = \frac{i\pi}{2} i^\nu H_\nu^{(1)}(i x) \tag{11a}$$

or, if ν is not an integer,

$$K_\nu(x) = \frac{\pi}{2\sin\nu\pi}\left[I_{-\nu}(x) - I_\nu(x)\right] \tag{11b}$$

have real values for real positive x. The last function is of special interest because of its asymptotic behaviour for large x:

$$K_\nu(x) \to \sqrt{\frac{\pi}{2x}}\, \mathrm{e}^{-x}. \tag{12}$$

More formulae for K_0 and K_1 are given in Problem 185. — In Problem 99 the differential equation

$$u'' - g^2 x^{-n} u = 0 \tag{13a}$$

has been solved by

$$u = \sqrt{x}\, K_{\frac{1}{2\lambda}}\left(\frac{g}{\lambda} x^{-\lambda}\right) \quad \text{with } \lambda = \frac{n-2}{2}. \tag{13b}$$

In some optical diffraction problems the function

$$\mathrm{Ai}(x) = \frac{1}{\pi}\sqrt{\frac{x}{3}}\, K_{\frac{1}{3}}\left(\tfrac{2}{3} x^{\frac{3}{2}}\right) \tag{14a}$$

plays a role. It is called the *Airy function.* Its analytic continuation to negative x values yields

$$\mathrm{Ai}(-x) = \tfrac{1}{3}\sqrt{x}\left\{J_{\frac{1}{3}}\left(\tfrac{2}{3} x^{\frac{3}{2}}\right) + J_{-\frac{1}{3}}\left(\tfrac{2}{3} x^{\frac{3}{2}}\right)\right\}. \tag{14b}$$

In this book, the Airy function has been used in Problem 40 where it is shown in Fig. 28. It might also be used with advantage in Problem 117 where, however, regress to the functions J_ν and I_ν with $\nu = \pm\frac{1}{3}$ has been preferred.

Spherical Bessel functions. The index values $\nu = l + \frac{1}{2}$ with integer $l = 0, 1, 2, \ldots$ play an important role, because they occur in factorizing the solutions of the wave equation in spherical polar coordinates. It is usual to introduce the four standard types[1]

[1] In the literature the functions here denoted by j_l etc are often denoted by $\hat{j}_l$ etc with $j_l = \frac{1}{z}\hat{j}_l$. This has the advantage that $h_l^{(1,2)}(z) = \frac{1}{z}\hat{h}_l^{(1,2)}(z)$ then become outgoing and incoming spherical waves.

$$j_l(z) = \sqrt{\frac{\pi z}{2}}\, J_{l+\frac{1}{2}}(z); \tag{15}$$

$$n_l(z) = \sqrt{\frac{\pi z}{2}}\, N_{l+\frac{1}{2}}(z) = (-1)^{l+1}\sqrt{\frac{\pi z}{2}}\, J_{-(l+\frac{1}{2})}(z); \tag{16}$$

$$h_l^{(1)}(z) = j_l(z) + i\, n_l(z); \qquad h_l^{(2)}(z) = j_l(z) - i\, n_l(z). \tag{17}$$

They are solutions of the differential equation

$$u'' + \left(1 - \frac{(l+\frac{1}{2})^2}{z^2}\right) u = 0 \tag{18}$$

and have very simple asymptotics:

$$j_l(z) \to \sin\left(z - \frac{l\pi}{2}\right); \qquad n_l(z) \to -\cos\left(z - \frac{l\pi}{2}\right);$$
$$h_l^{(1)} \to i^{-(l+1)} e^{iz}; \qquad h_l^{(2)}(z) \to i^{l+1} e^{-iz}. \tag{19}$$

The most important Wronskians are

$$j_l n_l' - n_l j_l' = 1; \qquad h_l^{(1)} h_l^{(2)\prime} - h_l^{(2)} h_l^{(1)\prime} = -2i. \tag{20}$$

For $|z| \ll l+\frac{1}{2}$ we have

$$j_l(z) \simeq \frac{2^l l!}{(2l+1)!} z^{l+1}; \qquad n_l(z) = -\frac{(2l)!}{2^l l!} z^{-l}. \tag{21}$$

Recurrence relations. For each one of the four types of functions defined by (15)–(17) there hold the relations

$$u_{l+1} = \frac{2l+1}{z} u_l - u_{l-1}; \qquad u_{l+1} = \frac{l+1}{z} u_l - u_l'. \tag{22}$$

The first of these relations may also be used for constructing functions with negative l.

The spherical Bessel functions are elementary functions. The simplest of them are assembled in the following survey.

$$j_0 = \sin z; \qquad n_0 = -\cos z;$$

$$j_1 = \frac{\sin z}{z} - \cos z; \qquad n_1 = -\frac{\cos z}{z} - \sin z;$$

$$j_2 = \left(\frac{3}{z^2} - 1\right)\sin z - \frac{3}{z}\cos z; \qquad n_2 = -\left(\frac{3}{z^2} - 1\right)\cos z - \frac{3}{z}\sin z$$

and

$$h_0^{(1)} = -i\,e^{iz}; \qquad h_0^{(2)} = i\,e^{-iz};$$

$$h_1^{(1)} = \left(-\frac{i}{z} - 1\right)e^{iz}; \qquad h_1^{(2)} = \left(\frac{i}{z} - 1\right)e^{-iz};$$

$$h_2^{(1)} = \left(-\frac{3i}{z^2} - \frac{3}{z} + i\right)e^{iz}; \qquad h_2^{(2)} = \left(\frac{3i}{z^2} - \frac{3}{z} - i\right)e^{-iz}.$$

Legendre functions

The differential equation

$$(1-z^2)u'' - 2zu' + \nu(\nu+1)u = 0 \tag{1}$$

is of the hypergeometric type with the singularities at $z = \pm 1$ and ∞. The complete solution can be written

$$u_\nu = A\,{}_2F_1\left(-\nu, \nu+1, 1; \frac{1-z}{2}\right) + B z^{-\nu-1}\,{}_2F_1\left(\frac{\nu}{2}+1, \frac{\nu}{2}+\frac{1}{2}, \nu+\frac{3}{2}; \frac{1}{z^2}\right), \tag{2}$$

where the factor of A is called a Legendre function of the first kind, $P_\nu(z)$, and the factor of B is (save of a normalization factor) a Legendre function of the second kind, $Q_\nu(z)$.

If ν is an integer, $l = 0, 1, 2, \ldots$, the function of the first kind becomes a polynomial. For $z = x$ with real x and $|x| \leq 1$ or $x = \cos\vartheta$, Legendre polynomials have a simple geometrical meaning being connected with spherical harmonics according to

$$P_l(\cos\vartheta) = \sqrt{\frac{4\pi}{2l+1}}\,Y_{l,0}. \tag{3}$$

Properties of Legendre polynomials. The polynomials form an orthogonal set:

$$\int_{-1}^{+1} dx\,P_l(x)\,P_{l'}(x) = \frac{1}{l+\frac{1}{2}}\,\delta_{ll'}. \tag{4}$$

The first polynomials are

$$P_0(x) = 1; \quad P_1(x) = x; \quad P_2(x) = \tfrac{3}{2}x^2 - \tfrac{1}{2}; \quad P_3(x) = \tfrac{5}{2}x^3 - \tfrac{3}{2}x; \quad P_4(x) = \tfrac{35}{8}x^4 - \tfrac{15}{4}x^2 + \tfrac{3}{8}. \tag{5}$$

They are either even or odd functions of x, according to whether the index l is even or odd, so that

$$P_l(-x)=(-1)^l P_l(x). \tag{6}$$

Polynomials of higher order than $l=1$ may be derived from the recurrence relation

$$(l+1)P_{l+1}(x)+l P_{l-1}(x)=(2l+1)x P_l(x). \tag{7}$$

The derivatives are connected with the polynomials by

$$(1-x^2)P_l'=l(P_{l-1}-x P_l)=(l+1)(x P_l-P_{l+1}) \tag{8a}$$

whence there follows

$$(2l+1)P_l=P_{l+1}'-P_{l-1}'. \tag{8b}$$

At $x=\pm 1$ we have

$$P_l(\pm 1)=(-1)^l; \tag{9a}$$

the n'th derivative at $x=\pm 1$ becomes

$$\frac{d^n P_l(\pm 1)}{dx^n}=(\mp 1)^{l+n}\frac{(l+n)!}{2^n n!(l-n)!}. \tag{9b}$$

The definition of the polynomials by (2) may be written, with $x=\cos\vartheta$,

$$P_l(\cos\vartheta)={}_2F_1\left(l+1,\ -l,\ 1;\ \sin^2\frac{\vartheta}{2}\right)=\sum_{n=0}^{l}(-1)^n\frac{(l+n)!}{n!^2(l-n)!}\sin^{2n}\frac{\vartheta}{2}$$

$$=1-\frac{l(l+1)}{1!^2}\sin^2\frac{\vartheta}{2}+\frac{(l-1)l(l+1)(l+2)}{2!^2}\sin^4\frac{\vartheta}{2}\ \ldots \tag{10}$$

If $l\gg 1$ and $\left|\sin\frac{\vartheta}{2}\right|$ of the order of $1/l$, this series simplifies to the Bessel series so that we get

$$P_l(\cos\vartheta)\simeq J_0\left((2l+1)\sin\frac{\vartheta}{2}\right), \tag{11}$$

with an error of the order of $1/l^2$. The zeros of P_l are still given rather accurately even up to large angles ϑ. For the example $l=10$ the approximation has been represented in Fig. 55 (p. 275 of vol. I).

For geometrical relations, cf. spherical harmonics.

Legendre functions of the first kind. Expansions. If ν is not an integer, the expansion (10) is no longer finite so that $P_\nu(x)$ becomes a transcendental with singularities at $x = \pm 1$. It may then be expanded into a series of Legendre polynomials,

$$P_\nu(x) = \frac{\sin \pi \nu}{\pi} \sum_{n=0}^{\infty} \frac{(-1)^n (2n+1)}{(\nu - n)(\nu + n + 1)} P_n(x). \tag{12}$$

This is a special case of the general type of expansion

$$f(x) = \sum_{n=0}^{\infty} (2n+1) f_n P_n(x) \tag{13a}$$

with coefficients

$$f_n = \tfrac{1}{2} \int_{-1}^{+1} dx\, f(x) P_n(x) \tag{13b}$$

which is always possible since the Legendre polynomials form a *complete* orthogonal set.

The following examples of such expansions (with $|x| < 1$) are often useful:

$$e^{ixy} = \frac{1}{y} \sum_{l=0}^{\infty} (2l+1) i^l j_l(y) P_l(x); \tag{14}$$

$$\frac{\sin(y\sqrt{2(1-x)})}{y\sqrt{2(1-x)}} = \sum_{l=0}^{\infty} (2l+1) j_l^2(y) P_l(x). \tag{15}$$

The series (14) and (15) are generally used with $y = kr$ and $x = \cos\vartheta$ so that $y\sqrt{2(1-x)} = 2kr \sin\frac{\vartheta}{2}$. They hold for all real values of y. Another important example may even be used to define the Legendre polynomials:

$$\frac{1}{\sqrt{1 - 2xy + y^2}} = \sum_{n=0}^{\infty} y^n P_n(x) \quad \text{if } |y| < 1. \tag{16}$$

Legendre functions of the second kind. The expansion

$$\frac{1}{z - x} = \sum_{n=0}^{\infty} (2n+1) Q_n(z) P_n(x) \tag{16a}$$

where z is any complex number, except real values between -1 and $+1$, leads to

$$Q_n(z) = \frac{1}{2} \int_{-1}^{+1} \frac{dx\, P_n(x)}{z - x}. \tag{16b}$$

The functions $Q_n(z)$ are called Legendre functions of the second kind. From the symmetry of (16a) in z and x, it follows at once that they satisfy the differential equation (1) of the polynomials $P_n(x)$. They have, however, branch points at $z=\pm 1$ with logarithmic singularities. The simplest of these functions are

$$Q_0(z) = \frac{1}{2}\log\frac{z+1}{z-1}; \qquad Q_1(z) = P_1(z)Q_0(z) - 1;$$
$$Q_2(z) = P_2(z)Q_0(z) - \frac{3}{2}z. \tag{17}$$

The higher Q_n's may be determined from the recurrence relation (7) which holds for the Q's as well as for the P's. Their general form is

$$Q_n(z) = P_n(z)Q_0(z) - W_{n-1}(z) \tag{18}$$

with W_{n-1} a polynomial of degree $n-1$, either even or odd, according to whether the index n is even or odd.

Spherical harmonics

In factorizing the solutions of the wave equation in spherical polar coordinates defined by

$$x = r\sin\vartheta\cos\varphi; \qquad y = r\sin\vartheta\sin\varphi; \qquad z = r\cos\vartheta \tag{1}$$

with z the polar axis, the angular part of the solution satisfies the differential equation

$$\frac{1}{\sin\vartheta}\frac{\partial}{\partial\vartheta}\left(\sin\vartheta\frac{\partial u}{\partial\vartheta}\right) + \frac{1}{\sin^2\vartheta}\frac{\partial^2 u}{\partial\varphi^2} + l(l+1)u = 0, \tag{2}$$

where the separation parameter l is integer, $l=0,1,2,\ldots$ By further factorization,

$$u = \Theta(\vartheta)\,\mathrm{e}^{im\varphi} \tag{3a}$$

with $m=0,\pm 1,\pm 2,\ldots$ we get

$$\frac{1}{\sin\vartheta}\frac{d}{d\vartheta}\left(\sin\vartheta\frac{d\Theta}{d\vartheta}\right) + \left[l(l+1) - \frac{m^2}{\sin^2\vartheta}\right]\Theta = 0 \tag{3b}$$

or, using

$$t = \frac{z}{r} = \cos\vartheta$$

as variable,

$$(1-t^2)\frac{d^2\Theta}{dt^2} - 2t\frac{d\Theta}{dt} + \left[l(l+1) - \frac{m^2}{1-t^2}\right]\Theta = 0. \tag{3c}$$

This differential equation is a slight generalization of the one of the Legendre polynomial $P_l(t)$ into which it passes over for $m=0$, but it remains of hypergeometric type for $m \neq 0$. Its only regular solution, in arbitrary normalization, may be written

$$\Theta_{l,m} = (1-t^2)^{\frac{m}{2}} \frac{d^m P_l(t)}{dt^m} \tag{3d}$$

for $m \geq 0$, and $\Theta_{l,-m}$ turns out to be the same function. Since the polynomial P_l cannot be differentiated more than l times without vanishing, there exist $2l+1$ regular solutions with integer $|m| \leq l$ for every value of l.

We still have a free hand concerning normalization. We adopt its most customary form, which best reflects the geometrical meaning of these solutions regular on the entire surface of the unit sphere,

$$Y_{l,m}(\vartheta, \varphi) = \frac{1}{\sqrt{2\pi}} \mathscr{P}_l^m(\vartheta)\, e^{im\varphi} \tag{4}$$

with

$$\oint d\Omega\, |Y_{l,m}|^2 = 1 \tag{5}$$

or

$$\int_0^\pi d\vartheta \sin\vartheta\, |\mathscr{P}_l^m(\vartheta)|^2 = 1. \tag{6}$$

Eq. (3d) then has to be normalized according to

$$\mathscr{P}_l^m = \sqrt{\frac{2l+1}{2}\frac{(l-m)!}{(l+m)!}}\,(1-t^2)^{\frac{m}{2}} \frac{d^m P_l(t)}{dt^m} \tag{7}$$

for $m \geq 0$. To include negative values of m, we define

$$\mathscr{P}_l^{-m}(\vartheta) = (-1)^m \mathscr{P}_l^m(\vartheta). \tag{8}$$

In this standard form the spherical harmonics $Y_{l,m}$ up to $l=3$ are tabulated on p. 174 of vol. I. For $m=0$ we have the useful relations

$$\mathscr{P}_l^0 = \sqrt{l+\tfrac{1}{2}}\, P_l \quad \text{and} \quad Y_{l,0} = \sqrt{\frac{2l+1}{4\pi}}\, P_l. \tag{9}$$

Recurrence relations. There exist several important relations, connecting spherical harmonics with such of neighbouring values of l and m, viz.

$$\sin\vartheta\, e^{i\varphi}\, Y_{l,m} = a_{l,m}\, Y_{l+1,m+1} - a_{l-1,-m-1}\, Y_{l-1,m+1}; \tag{10a}$$

$$\sin\vartheta\, e^{-i\varphi}\, Y_{l,m} = -a_{l,-m}\, Y_{l+1,m-1} + a_{l-1,m-1}\, Y_{l-1,m-1}; \tag{10b}$$

$$\cos\vartheta\, Y_{l,m} = b_{l,m}\, Y_{l+1,m} + b_{l-1,m}\, Y_{l-1,m} \tag{11}$$

with the abbreviations

$$a_{l,m} = \sqrt{\frac{(l+m+1)(l+m+2)}{(2l+1)(2l+3)}}; \quad b_{l,m} = \sqrt{\frac{(l+m+1)(l-m+1)}{(2l+1)(2l+3)}}. \tag{12}$$

By repeated application of the relations (10) and (11) higher powers of $\sin\vartheta$ and $\cos\vartheta$ may be multiplied by $Y_{l,m}$ so that indices differing by more than ± 1 from l and m may appear on the right-hand side.

Derivatives. If the operators

$$r\left(\frac{\partial}{\partial x} \pm i\frac{\partial}{\partial y}\right) = e^{\pm i\varphi}\left\{\sin\vartheta\, r\frac{\partial}{\partial r} + \cos\vartheta\frac{\partial}{\partial\vartheta} \pm \frac{i}{\sin\vartheta}\frac{\partial}{\partial\varphi}\right\}$$

and

$$r\frac{\partial}{\partial z} = \cos\vartheta\, r\frac{\partial}{\partial r} - \sin\vartheta\frac{\partial}{\partial\vartheta} \tag{13}$$

are applied to a spherical harmonic, the following results are obtained:

$$r\left(\frac{\partial}{\partial x} \pm i\frac{\partial}{\partial y}\right) Y_{l,m} = \mp l\, a_{l,\pm m}\, Y_{l+1,m+1} \mp (l+1)\, a_{l-1,\mp m-1}\, Y_{l-1,m\pm 1};$$

$$r\frac{\partial}{\partial z} Y_{l,m} = -b_{l,m}\, Y_{l+1,m} + (l+1)\, b_{l-1,m}\, Y_{l-1,m} \tag{14}$$

where $a_{l,m}$ and $b_{l,m}$ are defined by (12).

In the theory of angular momentum, the three hermitian operators L_x, L_y, L_z with

$$L_z = -i\left(x\frac{\partial}{\partial y} - y\frac{\partial}{\partial x}\right) \tag{15}$$

etc. cyclic, play a large role. In spherical polar coordinates we may write

$$L_\pm = L_x \pm i L_y = i e^{\pm i\varphi}\left(\mp i\frac{\partial}{\partial\vartheta} + \cot\vartheta\frac{\partial}{\partial\varphi}\right) \tag{16a}$$

and

$$L_z = -i\frac{\partial}{\partial\varphi}. \tag{16b}$$

Applied to a spherical harmonic, L_+ and L_- turn out to be shift operators which raise or lower the index m by 1:

$$L_+ Y_{l,m} = -\sqrt{l(l+1)-m(m+1)}\, Y_{l,m+1}; \tag{17a}$$

$$L_- Y_{l,m} = -\sqrt{l(l+1)-m(m-1)}\, Y_{l,m-1}. \tag{17b}$$

On the other hand, $Y_{l,m}$ is an eigenfunction of L_z:

$$L_z Y_{l,m} = m\, Y_{l,m}. \tag{18}$$

The same holds for the second-order operator

$$L^2 = L_x^2 + L_y^2 + L_z^2 = \tfrac{1}{2}(L_+ L_- + L_- L_+) + L_z^2, \tag{19}$$

viz.

$$L^2 Y_{l,m} = l(l+1)\, Y_{l,m}. \tag{20}$$

There further hold the relations

$$L_- L_+ Y_{l,m} = [l(l+1)-m(m+1)]\, Y_{l,m}; \tag{21a}$$

$$L_+ L_- Y_{l,m} = [l(l+1)-m(m-1)]\, Y_{l,m}. \tag{21b}$$

The operators L_i here defined are identical with the angular momentum component operators used in the text of this book, except for a factor $\hbar$.

Orthogonality and expansion. The spherical harmonics satisfy the orthonormality relations

$$\oint d\Omega\, Y^*_{l',m'}\, Y_{l,m} = \delta_{ll'}\, \delta_{mm'}. \tag{22}$$

They form a complete set of functions in the sense that any regular function on the surface of the unit sphere may be expanded into a series

$$f(\vartheta,\varphi) = \sum_{l=0}^{\infty} \sum_{m=-l}^{+l} f_{l,m}\, Y_{l,m}(\vartheta,\varphi) \tag{23a}$$

with coefficients

$$f_{l,m} = \oint d\Omega\, Y^*_{l,m}(\vartheta,\varphi)\, f(\vartheta,\varphi). \tag{23b}$$

A few important expansions are the following ones.

1. Expansion of the Legendre polynomial $P_l(\cos\gamma)$ where γ is the angle between the directions to the points (ϑ,φ) and (ϑ',φ') on the unit sphere:

$$P_l(\cos\gamma) = \frac{4\pi}{2l+1} \sum_{m=-l}^{+l} Y^*_{l,m}(\vartheta',\varphi')\, Y_{l,m}(\vartheta,\varphi). \tag{24}$$

This formula is equivalent to a transformation by rotating the polar axis through an angle γ. It mixes only spherical harmonics of the same order l. It is also called the addition theorem.

2. Plane wave. If the plane wave propagates along the polar axis, only Legendre polynomials occur:

$$e^{ikz} = e^{ikr\cos\vartheta} = \frac{1}{kr} \sum_{l=0}^{\infty} \sqrt{4\pi(2l+1)}\, i^l j_l(kr)\, Y_{l,0}(\vartheta). \tag{25}$$

This formula has been proved in extenso in Problem 81. Incidentally, it leads by inversion to an integral representation of the spherical Bessel functions:

$$j_l(z) = i^{-l} \frac{z}{2} \int_{-1}^{+1} dt\, e^{izt} P_l(t). \tag{26}$$

If the plane wave runs in the direction of a vector $\boldsymbol{k}$ with polar angles Θ, Φ, the expansion may be generalized to

$$e^{i\boldsymbol{k}\cdot\boldsymbol{r}} = \frac{4\pi}{kr} \sum_{l=0}^{\infty} i^l j_l(kr)\, Y^*_{l,m}(\Theta,\Phi)\, Y_{l,m}(\vartheta,\varphi). \tag{27}$$

3. Spherical wave, to be used as Green's function of the wave equation. If $\boldsymbol{r}$ and $\boldsymbol{r}'$ are vectors the directions of which are defined by polar angles ϑ, φ and ϑ', φ', and γ is the angle between them, then

$$\frac{e^{ik|\boldsymbol{r}-\boldsymbol{r}'|}}{4\pi|\boldsymbol{r}-\boldsymbol{r}'|} = \frac{1}{rr'} \sum_{l=0}^{\infty} \sqrt{\frac{2l+1}{4\pi}}\, \Gamma_l(r,r')\, Y_{l,0}(\cos\gamma) \tag{28a}$$

with

$$\Gamma_l(r,r') = \begin{cases} \dfrac{i}{k} j_l(kr)\, h_l^{(1)}(kr') & \text{for } r<r', \\[2ex] \dfrac{i}{k} j_l(kr')\, h_l^{(1)}(kr) & \text{for } r>r'. \end{cases} \tag{28b}$$

In the limit $k \to 0$ this yields the well-known formula

$$\frac{1}{|\boldsymbol{r}-\boldsymbol{r}'|} = \begin{cases} \dfrac{1}{r'} \displaystyle\sum_{l=0}^{\infty} \left(\frac{r}{r'}\right)^l P_l(\cos\gamma) & \text{for } r<r', \\[2ex] \dfrac{1}{r} \displaystyle\sum_{l=0}^{\infty} \left(\frac{r'}{r}\right)^l P_l(\cos\gamma) & \text{for } r>r'. \end{cases} \tag{29}$$

The hypergeometric series

The differential equation

$$z(1-z)v'' + \{(c-2\lambda) - (a+b+1-2\lambda-2\mu)z\}\, v'$$
$$+ \left\{\frac{\lambda(\lambda-c+1)}{z} + \frac{\mu(\mu-a-b-c)}{1-z} - [(\lambda+\mu)(\lambda+\mu-a-b)+ab]\right\} v = 0 \tag{1}$$

has regular singularities only at $z=0,1,\infty$. Putting

$$v(z)=z^{\lambda}(1-z)^{\mu}u(z) \tag{2}$$

it can be brought into the standard form

$$z(1-z)u''+[c-(a+b+1)z]u'-ab\,u=0 \tag{3}$$

which is called the hypergeometric or Gaussian differential equation. For all values of its three parameters a, b, c, except $c=-n$ with $n=0,1,2,\ldots$, it has a solution which is regular and does not vanish at $z=0$ which, when normalized according to $u(0)=1$, is called the hypergeometric series ${}_2F_1(a,b,c;z)$. Solving the differential equation (3) by series expansion at the origin we find[2]

$$\begin{aligned}{}_2F_1(a,b,c;z) = 1 &+ \frac{ab}{c}\frac{z}{1!} + \frac{a(a+1)b(b+1)}{c(c+1)}\frac{z^2}{2!}\\ &+ \frac{a(a+1)(a+2)b(b+1)(b+2)}{c(c+1)(c+2)}\frac{z^3}{3!}+\cdots\end{aligned} \tag{4a}$$

or

$${}_2F_1(a,b,c;z) = \frac{\Gamma(c)}{\Gamma(a)\Gamma(b)}\sum_{n=0}^{\infty}\frac{\Gamma(a+n)\Gamma(b+n)}{\Gamma(c+n)\,n!}z^n. \tag{4b}$$

This function is invariant with respect to exchanging the parameters a and b. If $a=-n$ or $b=-n$ $(n=0,1,2,\ldots)$ it becomes a polynomial of degree n, called a Jacobi polynomial according to the definition

$$J_n(p,q;z) = {}_2F_1(-n,p+n,q;z).$$

These polynomials form an orthogonal set according to

$$\int_0^1 dz\, z^{q-1}(1-z)^{p-q}J_m J_n=0 \quad \text{for } m\neq n.$$

The hypergeometric series does not exist for $c=-n$; in that case however the limiting process

$$\lim_{c\to -n}\frac{{}_2F_1(a,b,c;z)}{\Gamma(c)} \tag{5}$$

$$=\frac{\Gamma(a+n+1)\Gamma(b+n+1)}{\Gamma(a)\Gamma(b)}\frac{z^{n+1}}{(n+1)!}\,{}_2F_1(a+n+1,b+n+1,n+2;z)$$

leads to a solution of the differential equation (3).

---

[2] The notation ${}_2F_1$ was introduced by Pochhammer who generalized the hypergeometric series to ${}_nF_m$ with $n+m$ parameters of which products of n appear in the numerator and of m in the denominator of the series in the same way as 2, resp. 1 in Eq. (4a). In our context we need only ${}_2F_1$ and the confluent series ${}_1F_1$.

If none of the three parameters a, b, c are either zero or negative integers, the series (4a, b) converges absolutely for $|z|<1$. Its analytic continuation beyond this circle can be made unique by a branch cut extending from $z=1$ to $z=\infty$.

The following formulae may serve for its continuation beyond $|z|<1$:

$$
{}_2F_1(a,b,c;z) = \frac{\Gamma(c)\,\Gamma(c-a-b)}{\Gamma(c-a)\,\Gamma(c-b)}\,{}_2F_1(a,b,a+b-c+1;1-z) \tag{6}
$$

$$
+\frac{\Gamma(c)\,\Gamma(a+b-c)}{\Gamma(a)\,\Gamma(b)}(1-z)^{c-a-b}\,{}_2F_1(c-a,c-b,c-a-b+1;1-z)
$$

and

$$
{}_2F_1(a,b,c;z) = \frac{\Gamma(c)\,\Gamma(b-a)}{\Gamma(b)\,\Gamma(c-a)}(-z)^{-a}\,{}_2F_1\left(a,a-c+1,a-b+1;\frac{1}{z}\right) \tag{7}
$$

$$
+\frac{\Gamma(c)\,\Gamma(a-b)}{\Gamma(a)\,\Gamma(c-b)}(-z)^{-b}\,{}_2F_1\left(b,b-c+1,b-a+1;\frac{1}{z}\right).
$$

The last formula determines the asymptotic behaviour of ${}_2F_1$ for $z\to\infty$:

$$
{}_2F_1(a,b,c;z) \to \frac{\Gamma(c)\,\Gamma(b-a)}{\Gamma(b)\,\Gamma(c-a)}(-z)^{-a} + \frac{\Gamma(c)\,\Gamma(a-b)}{\Gamma(a)\,\Gamma(c-b)}(-z)^{-b}. \tag{8}
$$

The general solution of the hypergeometric differential equation for $|z|<1$ is

$$
u = C_1\,{}_2F_1(a,b,c;z) + C_2\,z^{1-c}\,{}_2F_1(a+1-c,b+1-c,2-c;z). \tag{9}
$$

Only for integer $c=0, \pm 1, \pm 2, \ldots$ the two special solutions used in (9) become identical, as is easily seen by applying Eq. (5). The second solution then has a logarithmic singularity at $z=0$.

In the following, the most important formulae for the practical use of the hypergeometric series have been collected.

$$
{}_2F_1(a,b,c;z) = \frac{a(c-b)}{c(a-b)}\,{}_2F_1(a+1,b,c+1;z) + \frac{b(c-a)}{c(b-a)}\,{}_2F_1(a,b+1,c+1;z)
$$

$$
= \frac{c-a}{c}\,{}_2F_1(a,b,c+1;z) + \frac{a}{c}\,{}_2F_1(a+1,b,c+1;z),
$$

$$z\,{}_2F_1(a,b,c;z)=\frac{c-1}{a-b}[{}_2F_1(a-1,b,c-1;z)-{}_2F_1(a,b-1,c-1;z)]$$

$$=\frac{c-1}{a-1}[{}_2F_1(a-1,b,c-1;z)-{}_2F_1(a-1,b-1,c-1;z)]$$

$$=\frac{c-a}{a-b}\,{}_2F_1(a-1,b,c;z)+\frac{c-b}{b-a}\,{}_2F_1(a,b-1,c;z)+{}_2F_1(a,b,c;z),$$

$$(1-z)\,{}_2F_1(a,b,c;z)=\frac{c-1}{a-1}\,{}_2F_1(a-1,b-1,c-1;z)+\frac{a-c}{a-1}\,{}_2F_1(a-1,b,c;z).$$

Derivatives:

$$\frac{d}{dz}\,{}_2F_1(a,b,c;z)=\frac{ab}{c}\,{}_2F_1(a+1,b+1,c+1;z),$$

$$z(1-z)\frac{d}{dz}\,{}_2F_1(a,b,c;z)=\frac{a(c-a)}{a-b}\,{}_2F_1(a-1,b,c;z)+\frac{b(c-b)}{b-a}\,{}_2F_1(a,b-1,c;z)$$

$$+\frac{b(c-b)-a(c-a)}{a-b}\,{}_2F_1(a,b,c;z).$$

The confluent series

If, in the hypergeometric differential equation, we perform the limiting process $b\to\infty$, $z=x/b$, we get Kummer's differential equation,

$$\frac{d^2u}{dx^2}+(c-x)\frac{du}{dx}-au=0. \tag{1}$$

The singularity at $z=1$ has been shifted to $x=\infty$, so that in the complex x plane there is a regular singularity still at $x=0$ but an irregular singularity at $x=\infty$ caused by the confluence of the two singularities at $z=1$ and $z=\infty$. Hence the name of the solution.

The general solution of (1) is

$$u=C_1\,{}_1F_1(a,c;x)+C_2\,x^{1-c}\,{}_1F_1(a-c+1,2-c;x) \tag{2}$$

with the so-called confluent series ${}_1F_1$ being defined by

$${}_1F_1(a,c;z)=1+\frac{a}{c}\frac{z}{1!}+\frac{a(a+1)}{c(c+1)}\frac{z^2}{2!}+\frac{a(a+1)(a+2)}{c(c+1)(c+2)}\frac{z^3}{3!}+\cdots \tag{3a}$$

or

$${}_1F_1(a,c;z)=\frac{\Gamma(c)}{\Gamma(a)}\sum_{n=0}^{\infty}\frac{\Gamma(a+n)}{\Gamma(c+n)}\frac{z^n}{n!}. \tag{3b}$$

This series converges absolutely in the whole z plane. To make it unique, a branch cut has to be made from $z=0$ to $z=\infty$ which, in the standard notation used in this book, runs along the positive imaginary axis.

Only for $c=-n$, $n=0,1,2,\ldots$ is the series (3a, b) not defined, but in this case

$$\lim_{c\to -n}\frac{{}_1F_1(a,c;z)}{\Gamma(c)}=\frac{\Gamma(a+n+1)}{\Gamma(a)}\frac{z^{n+1}}{(n+1)!}\,{}_1F_1(a+n+1,n+2;z) \tag{4}$$

is a solution of (1) (with z instead of x).

The asymptotic behaviour of the confluent series for $|z|\to\infty$ is given by

$${}_1F_1(a,c;z)\to \mathrm{e}^{-i\pi a}\frac{\Gamma(c)}{\Gamma(c-a)}z^{-a}+\frac{\Gamma(c)}{\Gamma(a)}\mathrm{e}^z z^{a-c}. \tag{5}$$

This formula does not hold for $a=-n$, $n=0,1,2,\ldots$ where, according to (3), the function ${}_1F_1(-n,c;z)$ becomes a polynomial of degree n. Of special interest among these are the Laguerre polynomials

$$L_n^{(m)}(z)=\frac{(n+m)!}{n!m!}\,{}_1F_1(-n,m+1;z) \tag{6}$$

and the Hermite polynomials (cf. Problem 30)

$$\left.\begin{aligned} H_{2n}(z)&=(-1)^n\frac{(2n)!}{n!}\,{}_1F_1(-n,\tfrac{1}{2};z^2),\\ H_{2n+1}(z)&=(-1)^n\frac{(2n+1)!}{n!}\,2z\,{}_1F_1(-n,\tfrac{3}{2};z^2).\end{aligned}\right\} \tag{7}$$

Finally, we again list some important formulae for the confluent series:

$$\begin{aligned} {}_1F_1(a,c;z)&=\frac{a}{c}\,{}_1F_1(a+1,c+1;z)+\frac{c-a}{c}\,{}_1F_1(a,c+1;z)\\ &=\mathrm{e}^z\,{}_1F_1(c-a,c;-z);\\ z\frac{d}{dz}\,{}_1F_1(a,c;z)&=a\{{}_1F_1(a+1,c;z)-{}_1F_1(a,c;z)\};\\ \frac{d}{dz}\,{}_1F_1(a,c;z)&=\frac{a}{c}\,{}_1F_1(a+1,c+1;z).\end{aligned}$$

Some functions defined by integrals

Error integral and related functions. The error integral is defined by

$$\operatorname{erfc} z = \int_z^\infty dt\, e^{-t^2};$$

as an alternative also the definition

$$\operatorname{erf} z = \int_0^z dt\, e^{-t^2} = z\, {}_1F_1(\tfrac{1}{2}, \tfrac{3}{2}; -z^2)$$

is often used. Both integrals are connected by the complete integral

$$\operatorname{erfc} z + \operatorname{erf} z = \int_0^\infty dt\, e^{-t^2} = \tfrac{1}{2}\sqrt{\pi}\,.$$

There holds the power expansion

$$\operatorname{erf} z = z - \frac{z^3}{1!\,3} + \frac{z^5}{2!\,5} - \cdots;$$

the function $\operatorname{erfc} z$ can for real positive argument $z \gg 1$ be represented by the semiconvergent series

$$\operatorname{erfc} z \to \frac{e^{-z^2}}{2z} \sum_{n=0}^\infty (-1)^n \frac{(2n)!}{2^{2n} n!} z^{-2n} = \frac{e^{-z^2}}{2z}\left(1 - \frac{1}{2z^2} + \frac{3}{4z^4} \cdots\right).$$

From the identity

$$F(z,\beta) = \int_z^\infty dt\, e^{-\beta t^2} = \beta^{-\frac{1}{2}} \operatorname{erfc}(\sqrt{\beta}\, z)$$

there follows by differentiation,

$$\frac{\partial F}{\partial \beta} = -\int_z^\infty dt\, t^2 e^{-\beta t^2} = -\frac{1}{2}\beta^{-\frac{3}{2}} \operatorname{erfc}(\sqrt{\beta}\, z) - \frac{z}{2\beta} e^{-\beta z^2}$$

so that, with $\beta = 1$, we get the reduction formula

$$\int_z^\infty dt\, t^2 e^{-t^2} = \frac{1}{2} \operatorname{erfc} z + \frac{z}{2} e^{-z^2}.$$

In this manner, by repeated differentiations, all integrals of the form

$$\int_z^\infty dt\, t^{2n} e^{-t^2}$$

may be finally reduced to the error function. For $z=0$, this leads to the special formula for the complete integrals,

$$\int_0^\infty dt\, t^{2n} e^{-t^2} = \sqrt{\pi}\,\frac{(2n-1)!}{2^{2n}(n-1)!}.$$

With $t^2=x$, this may be written as an Euler integral so that

$$\int_0^\infty dt\, t^{2n} e^{-t^2} = \tfrac{1}{2}\Gamma(n+\tfrac{1}{2}).$$

Exponential integral. This function is defined by

$$\mathrm{Ei}\, z = \int_{-\infty}^{z} \frac{dt}{t} e^{t}$$

and is of special interest for negative real values $z=-x$ where we write

$$\mathrm{E}_1(x) = -\mathrm{Ei}(-x) = \int_x^\infty \frac{dt}{t} e^{-t}.$$

There holds the power series

$$\mathrm{E}_1(x) = -C + \log\frac{1}{x} + x - \frac{x^2}{2!2} + \frac{x^3}{3!3} - \cdots$$

where

$$C = \int_0^\infty dt\, e^{-t} \log\frac{1}{t} = 0.577215\ldots$$

is the Euler constant. For $x \gg 1$, there holds the semiconvergent series

$$\mathrm{E}_1(x) = \frac{e^{-x}}{x}\left(1 - \frac{1!}{x} + \frac{2!}{x^2} - \cdots\right).$$

The exponential integral can be generalized to

$$\mathrm{E}_n(x) = \int_x^\infty \frac{dt}{t^n} e^{-t}.$$

These integrals may, by partial integration, be reduced to $\mathrm{E}_1(x)$ according to

$$\mathrm{E}_n(x) = \frac{1}{n-1}\left[x^{n-1} e^{-x} - \mathrm{E}_{n-1}(x)\right].$$

Index

for both volumes

Numbers refer to problems, not to pages. Volume I comprises the Problems 1—128, Volume II the Problems 129—219. The letter A refers to the Mathematical Appendix at the end of Volume II.

Airy function 40, 41, 117, A
Alkali atom, dielectric susceptibility 159
— —, inelastic scattering 166
Amaldi correction for Thomas-Fermi atoms 173
Amplitude structure 26, 27
Angular distribution of dipole radiation emitted 214
— — of photoelectrons 186
Angular momentum see also orbital momentum 4
— —, commutation relations 50, 51
— —, commutators with tensor 53, 54
— — components for a spinning top 46
— — expansion of plane waves 81, 136, 205
— — — of scattering amplitude 82
— — and Laplacian 49
— — operator 42
— — operator components in spherical polar coordinates 48
— — operators determined by infinitesimal rotation 47
— — originating magnetic moment 127
— — in relativistic theory 201
— — replaced by complex variable 113
— — for two particles on a circle 148
Anharmonic oscillator 35, 69, 70
Anomalous scattering 85, 112
— — of protons by protons 165
Anticommutation properties of Dirac matrices 189
— — of Pauli matrices 131
Antisymmetrized product 152
Atomic radius 173
Axial vector see Pseudovector

Background integral 113
Backward scattering amplitude 21, 22
Band structure of energy spectrum 28, 29
Barrier 19, 21, 22, 23
Bessel functions, formulae A
Bethe-Peierls formula 90
Binding energy 90
Bloch's theorem 28
Bohr magneton 127
Born approximation 94, 96, 97, 98, 102, 105, 106, 107, 183, 184, 211
Born integral, divergence 105, 108
Born-Oppenheimer approximation 44, 161, 163
Bose quantization 210, 213
Bound state determined by low-energy scattering 90, 147
Breit-Wigner formula 114
Bremsstrahlung 219
Brillouin zones 29

Calogero's equation 97, 100
— —, linearized approximation 98, 99, 101, 102
Canonical equations 10
Capacity of a potential hole 25, 63, 68, 106
Central forces in momentum space 76, 77, 91
Central-force field, relativistic electron 201
— —, spin electron 133

Centre-of-mass motion, separation from internal motion 150
— —, three-atomic molecule 149
Charge conjugation 194
Charge density 1
Circle with two particles 148
Circular cylindrical coordinates A
— oscillator 42
Classical dynamics for space averages 3, 4, 5
— interaction integral 44, 155, 156, 163
— turning point 40, 117—124
Clifford algebra 192, 194, 197
— numbers 191, 192
Closed-shell configuration 61
Collision parameter 185
Collision parameter integral for scattering amplitude 104
Commutation relations 7, 8
— — of spin components 129
— — of wave operators 210
Commutator with hamiltonian 10
Compound state 113, 114
Compton effect, unrelativistic cross section formula 218
Conduction current in relativistic theory 199
Conduction electrons in a metal 167, 168
Confluent series 30, 42, 65, 67, 69, 70, 110, 111, 202, 203, 204
— —, formulae A
Conservation of charge 1
— of energy 5
— of probability 1
Continuity equation 1, 21
Continuous spectrum 26, 219
Convergence of spherical harmonics series 83, 103
Coordinate formulae A
Coulomb excitation 185
— scattering, anomalous scattering 112
— —, extended charge 108, 112
— —, partial-wave expansion 111
— —, phases in WKB approximation 123
— —, point charge 110, 111
Coupling parameter, power expansion 102, 105
Cross section see also scattering cross section
Cross section for bremsstrahlung 219
— — for Compton scattering 218
— — and transition probability 183
Crystal lattice see periodic potential
Current see electric current
Current density of probability 1, 16, 17, 80, 126
Curvilinear coordinates 13, 46

Degeneracy of eigenvalues 42, 66
— of gases 167
Delta function, Fourier integral 14
Density of final states 183, 186, 211, 213
— of mass 1
— of momentum 1
— of probability 1, 16, 17, 126
— of states in a Fermi gas 167
Depolarization of plane Dirac wave by a potential jump 208
Derivatives of an operator 8, 10, 11
Deuterium, spectroscopic discovery 150
Deuteron, bound state and scattering length 147
—, central-force models 72, 75
—, hard-core potential 91, 92
—, tensor interaction 144, 145
Diamagnetism 128, 160
Dielectric susceptibility 159
Differential cross section of scattering 80
Diffusion equation 16
Dipole-dipole interaction 161, 162
Dipole, magnetic 127
Dipole radiation 213, 214, 215
— —, selection rules 216
Dipole transitions 43, 79, 213—216
Dirac equation, charge conjugation
— —, iteration 189
— —, Lorentz invariance 191
— —, one-dimensional problems 197, 207
— —, parity transformation 193
— — split-up in two 200
— —, standard form 189
— hamiltonian 189, 200
— perturbation method 181, 182, 183
Dispersion law of relativistic material waves 189
Dispersion of light 187
Dissociation energy 44, 69, 70, 163
Doublet, spin functions 146, 147, 194

Effective range 88, 89
Eigenfunctions of harmonic oscillator (table) 30
— of hydrogen atom (table) 67
Eigenspinors of total angular momentum 133, 137, 142, 201
Eigenvalue condition 18, 25
Eigenvalues of energy see energy levels
— of Mathieu equation 148
Eigenvectors of spin operators 130
Eikonal 115
Elastic scattering see scattering
Electric current density 1, 213
— — — in Dirac theory 198, 199, 207, 209
Electrical quadrupole see quadrupole
Electron gas of atomic electrons see Thomas-Fermi atom
— — of conduction electrons in a metal 167, 168
— — in a white dwarf 171
Electron spin resonance 138
Ellipsoidal coordinates 44, A
Elliptic integrals 170
Emission of a photon 213, 214
Energy bands 28, 29
Energy conservation 5
— flux vector 5
Energy levels of anharmonic oscillator 35
— — of Coulomb potential 67, 203
— — of gravitation field over earth's surface 40, 119
— — of harmonic oscillator 30
— — of Hulthén potential 68
— — of hydrogen atom, relativistic theory 203
— — of Pöschl-Teller holes 38, 39
— — of rectangular hole 18
— — of rectangular hole with division wall 19
— — of spherical well 62
— — of symmetrical top 46
— — of two-atomic molecules 69, 70, 71
Energy, total, of an atom in Thomas-Fermi approximation 174
— of vacuum 212
Equation of continuity 1, 21
Equilibrium distance in neutral hydrogen molecule 163
Error integral A
Eulerian angles 46, 55
Exchange integral 44
Exchange integral in excited helium 155, 156
— — in lithium ground state 158
— — in neutral hydrogen molecule 163
— — in many-body problems 153
Excitation degeneracy 162
Expectation value 3, 4, 7, 9, 12
— — of spin in plane Dirac waves 196
— —, time derivative 9
— — of angular momentum 4, 58
Exponential integral (formulae) A
— potential 75
— —, scattering 107

Fermi energy 167, 168
— gas 167
— quantization 210, 213
Field emission 169, 170
Final state density 183, 186
Fine structure of hydrogen atom, relativistic theory 203
— —, unrelativistic theory 136
Floquet's theorem 28
Form factor 108
Forward scattering amplitude 21, 22
Four-current see electric current
Fourier integral 14, 15, 17
— transform 14, 34, 76, 184
— — of potential 77
Free fall in quantum theory 40, 119
Fresnel's reflection formulae 45

Gamma function, formulae A
Gauge transformation 125, 126
Generators of the rotational group 52
g-factor of electron 136
Golden Rule 182, 183, 211, 213
Good quantum number 133
Green's function in three dimensions, partial wave expansion 94
— — for partial waves 94, 96
Group velocity 16, 17

Hamiltonian depending upon time 11
— of spin-orbit coupling 136
Hankel functions 63, 82, 83, 117, 185
— —, formulae A
Hard-core potential for deuteron 91, 92

Hard sphere, scattering 84, 109
Harmonic oscillator in Hilbert-space 31, 33
— — in matrix notation 33
— — in momentum space 34
— —, Schrödinger theory 30
— —, table of lowest eigenfunctions 30
— —, WKB approximation 118
— vibrations of a linear molecule 149
Heisenberg representation 10
Heisenberg's uncertainty rule 17, 40
Heitler-London approximation 163
Helicity 137
—, expectation value 196, 208
— of neutrino 200
— operator in Dirac theory 190
— in one-dimensional Dirac problems 197
— of a plane Dirac wave 190, 195
— of a plane Dirac wave, states of mixed helicity 195, 196
Helium, excited states 155, 156
—, ground state 154
Hermite polynomials 30, 32, A
Hermitian conjugate 6, 31
— operator 6, 7, 59
High-energy scattering 104
Hilbert space 6, 10, 12, 31, 33, 50
— —, its construction for angular momentum 56
— —, its construction for an harmonic oscillator 31
— — of spin operators 129, 139
Homogeneous electrical field, motion of electrons 41
Hulthén potential 68
Hydrogen atom 67
— —, lifetimes of excited states 215
— —, relativistic theory 202, 203
— —, spectral line intensities 217
— — as a two-body problem 150
Hydrogen eigenfunctions (table) 67
— — in momentum space 78
Hydrogen molecule, ionized 44
— —, neutral, ground state 163
— —, scattering of slow neutrons 147
Hydrogen star 171
Hypergeometric series 37, 38, 39, 46, 64, 68, 207
— —, formulae A

Image force, effect upon field emission 170
Index of refraction for light waves 187
— — — for particle waves 45, 115
Induced dipole moment 187
Inelastic scattering 166
Infinitesimal rotation 47
Inhomogeneous differential equation 94
Integral equation for momentum-space wave function 14, 77
— — for radial part of wave function 94
Intensities of spectral lines 213, 217
Intermolecular potentials 104
Intrinsic magnetic moment of electron 136, 138
Ionization energies of helium and two-electron ions 154, 155
Ionization in stellar matter 171
Irreducible representation of a matrix system 189
Isotope shift in electron binding 73

Kepler problem in momentum space 78
— —, relativistic radial solutions at positive energies 204
— —, — theory for bound states 202
— —, unrelativistic radial solutions at positive energies 111
— —, — theory for bound states 67
— —, WKB approximation 120
Kernel, symmetrization 94
Kinetic energy density 5
— — operator 13, 46, 49
Klein-Gordon equation 189
Klein-Nishina formula 218
Klein's paradox 202, 207
Kratzer's molecular potential 69
K shell binding energies 154
— — screening constants 178
Kummer's differential equation see confluent series

Laguerre polynomials A
Landé factor 135
Larmor frequency 138
Legendre functions and polynomials A
Lennard-Jones potential 104
Level density 26

Lifetime of an excited state 215
Line intensity 213, 217
Line shape 182
Lithium ground state 157, 158
Logarithmic derivative 20, 22, 23, 82—86, 89, 92, 101
— phase 69, 110, 111, 112
Lorentz covariants 192
— invariance of Dirac equation 191
— transformation, infinitesimal 192
low-energy scattering 83, 88, 89
— — and bound state 90, 147
— — by Pöschl-Teller potential 93
Lyman lines of hydrogen 217

Magnetic dipole see also magnetic moment 127
— field in Schrödlinger equation 125
— fields originating spin resonance 138
— moment of deuteron 145
— — of electron 136, 138
— — originated by angular momentum 127
— — of spin state, expectation value 135
— properties of a Fermi gas 168
— — of neon 160
— quantum number 127
— resonance, spin flip 188
Magnetization 128
Magneton of Bohr 127
Mass density 1
Mathieu equation 148
Matrix of an operator 6, 33
Measurement of an observable 12
Metal, paramagnetic susceptibility 168
Metric used in four-space 189
Mixture of S and D state 143, 144, 145
Modified Bessel functions, formulae A
Molecular potentials 44, 69, 70
Molecule of hydrogen see hydrogen molecule
— as a symmetrical top 46
—, three-atomic, modes of vibration 149
—, two-atomic 69, 70, 71
Momentum density 1
— operator 3, 7, 8, 10
— space 14, 15, 34, 76, 77
Momentum density, Born scattering 184
Momentum, total, of Schrödinger field 1, 3, 210
Morse potential 70, 71
Multiplication table of spin algebra 131
Muonic atom 74

Negative-power potential, scattering length 99, 100
Neon, diamagnetic susceptibility 160
Neumann functions, formulae A
— series 94, 105
Neutrino theory 200
Normal Zeeman effect 127, 216
Normalization 1, 14, 15
— volume 183
Nuclear radius, effect on electron binding energy 73

Observable, repeated measurement 12
Opacity 19, 21, 22, 86
Opaque wall 19, 20, 21, 27, 86
Operator of magnetic moment 135
Operators acting on particle numbers 210
Optical theorem 104
Orbital momentum, expectation value in relativistic theory 203
— — in spin states 133
Orthogonal system 2
Orthogonality of spherical harmonics 57
Ortho-helium 155, 156
Ortho-hydrogen, spin functions 147
Overlap integral 44, 156, 158, 163
Oscillator see harmonic, anharmonic, circular, spherical oscillator
Oscillator strength 187

Parabolic coordinates 110, A
Para-hydrogen, spin functions 147
Paramagnetic resonance 180
Paramagnetism 128, 168
Par-helium 155, 156
Parity 18, 19, 20, 22, 25, 26, 143
Parity mixing operators 200
Parity transformation of Dirac equation 193
Partial cross section 84, 87
Partial wave 81, 82, 205
— — expansion for Coulomb scattering 111

Partial wave expansion of plane wave 81
— — — of scattering amplitude 82, 83
— — — of three-dimensional Green's function 94
Particle resonance 113, 114
Particle number operator 210
Pauli algebra 131
— matrices 52, 129, 130, 189
— principle 152, 167
Periodic perturbation, photoeffect 186
— — of a two-level system 180
— potential 28, 29
Periodicity cube 15, 210
Permutation of particles 152, 153
Perturbation see also periodic perturbation
— by a light wave 186
— method of Dirac 181—183
Perturbation theory for anharmonic oscillators 35
— — of atom-atom interaction 161, 162
— — of dielectric susceptibility 159
— — for isotope shift 73
— — of magnetic susceptibilities 128
— — for muonic atom 74
— — for three-dimensional Stark effect 79
— — for two-dimensional Stark effect 43
— — of a two-level system 179, 180
Phase angle, behaviour at resonance 27
Phase average 58
Phase function 104, 124
Phase shift 82, 84, 85, 86, 87, 93
— —, determination by successive steps 96, 97
— — determined from integral equation 94
— — in exponential potential 107
— — in WKB approximation 121, 122
— — in Yukawa potential 106
Phase velocity 16
Photoeffect 186
Photon emission probability 213
Photon number in quantized radiation field 212
Plane Dirac wave 190
— — —, angular momentum expansion 205
— — —, incident on potential jump 208
Plane wave, expansion into partial waves 81, 136, 205
— —, oblique incidence 45
— —, one-dimensional 16
— — of spin particles, relativistic 190
— — of spin particles, unrelativistic 137
Poisson equation 156, 172, 173, 174
Polarizability of an atom 159
Polarization current 199
Polarization of dipole radiation emitted 214, 216
— of plane Dirac wave by a potential jump 208
Polytrope 171
Pöschl-Teller potentials 38, 39, 93
Potential energy density 5
— step 37
— — in Dirac theory 207
— wall see barrier
— well 18, 25, 62
Pressure of electron gas 167, 171
Principal quantum number 67
Probability conservation 1
— of photon emission 213
Proton-proton scattering 165
Pseudoscalar in Dirac theory 192, 193
Pseudovector in Dirac theory 192, 193

Quadrupole moment 61
— — of deuteron 145
— — of a spin electron state in a central field 134
— tensor 54, 61
Quantization of radiation field 212
— of Schrödinger field 210
Quartet, spin functions 146, 147
Quasipotential 104, 124
Quaternions 131

Radial momentum operator 59
— WKB functions 116
Radiation condition of Sommerfeld 80
—, dipole emission 213, 214
— field, quantization 212

Radiative transition, probability 215
Real state 26
Rectangular barrier 23
— hole 25, 63
— —, scattering 89, 101
Reduced mass 72, 75, 150
Reflected intensity 21, 22, 23, 37, 39, 45, 207, 209
Reflection of Dirac particles at a potential step 207, 209
— law 45, 208
Refraction index see index of refraction
— law 45, 208
Regge pole and trajectory 113, 114
Resonance absorption 182
— denominator 182, 186, 187
— field strength 138
— level 27
— in periodic perturbation 180, 182
— scattering 84, 86
Riccati equation 115
Rigid body in quantum mechanics 46
— rotator 43, 79
Ritz approximation for helium states 154, 156
— — for neutral hydrogen molecule 163
— — for solution of Thomas-Fermi differential equation 177
— variational method 44, 72, 74, 75
Rotational group 46, 52
Rotations of two-atomic molecules 69, 70, 71
Rotator 43, 79
Running wave 16
Rutherford scattering 108, 110
— — of equal particles 164

Sabatier transform 124
Scalar in Dirac theory 192
Scattered wave, interference with incoming plane wave 80
Scattering amplitude 80, 82, 105
— —, collision parameter integral 104
— —, convergence of partial wave expansion 83, 103
— —, partial wave expansion 82, 83
Scattering, application of the Golden Rule 183, 211
— cross section, definition 80
— of Dirac electrons by a central-force potential 206
— of equal particles 164
Scattering at a hard sphere 84, 109
—, inelastic 166
Scattering length 84, 92, 95
— — , different signs 88, 147
— —, negative-power potential 99, 100
— —, proton-neutron system 147
— —, square-well potential 89, 101
— —, Yukawa potential 102
Scattering at low and high energies see low- and high-energy scattering
—, one-dimensional model 21, 22, 23
— of neutrons by molecular hydrogen 147
— of protons by protons 165
— in quantized-wave picture 211
— at a spherical cavity 86, 88
Schrödinger field, quantization 210
— representation 10
Schwinger's variational principle 95
Screened hydrogen functions 154, 155, 156, 157, 160
— nucleus, effective potential 156
Screening 73
— constants 154, 157, 160
— of K electrons in heavy atoms 178
S-D-mixture 144, 145
Selection rules for dipole radiation 216
Self-adjoint see hermitian
Shadow effect 109
Shape-independent approximation 88
Short-range attraction between two protons 165
Short-range force 112
Singlet, spin function 139
Slater determinant 152, 153, 158
Sommerfeld's radiation condition 80
Sommerfeld-Watson transformation 113
Space average 3, 4, 5
Spectrum of bremsstrahlung 219
— of wave numbers 17
Spherical Bessel functions 62, 63, 81, 83, 87, 94, 108, 109
— — —, formulae A
— — —, integral representation 81
Spherical charge distribution, scattering 108
Spherical harmonics expansion of plane waves 81, 136, 205
— —, formulae A
— — of order 2, tensorial quality 54

Spherical harmonics, orthogonality 57
— — of the second kind 106
— —, table 67
— —, transformation under rotation 55
Spherical oscillator 65
— polar coordinates A
Spin algebra see also Clifford algebra 131, 141
Spin-dependent central force 140, 147
— tensor force 143
Spin electron in central field 133, 201
— exchange operator 140
— expectation value of plane Dirac wave 196
— flip in magnetic fields 188
— one particles 52
— and orbital momentum, coupling for one electron 133
— orbit coupling 136
— resonance 138
— of three-electron state 146
— of two-electron state 139
Spin vector under space rotation 132
— — operator 129
Spinor, Lorentz transformation 191
— in one-dimensional Dirac problem, algebraic properties 197
— in unrelativistic theory 132
Square well see also potential well and rectangular hole
— — of finite depth 25, 63, 89
Square-well potential, scattering 89, 101
Standard representation of Dirac matrices 189
Standing wave 18
Stark effect of three-dimensional rotator 79
— — of two-dimensional rotator 43
Stationary state 16
Statistical methods 167—178
Step see potential step
Stirling's formula A
Successive approximations to solution of integral equation 94, 96, 105
Susceptibilities, diamagnetic and paramagnetic 128, 160, 168
Susceptibility, dielectric 159
Symmetrical top 46
Symmetrization for helium states 155, 156
— for many-body system 152
Symmetrization of Rutherford scattering of equal particles 164, 165
— of three-particle spin functions 146
— of two-particle spin functions 139

Tensor, commutators with angular momentum 53, 54
— in Dirac theory 192
— force 143
Thomas factor 136
Thomas-Fermi atom 172—178
Thomson cross section 218
Three-atomic linear molecule 149
Three electrons see lithium
Tietz approximation of Thomas-Fermi atom 176—178
Time derivative of an expection value 9
— — of an operator 10
Time reversal 16
Torque 4
Total angular momentum 133, 142, 201
Total reflection 45
Transition probability 182, 183, 186, 211, 213
Transmitted intensity 21, 22, 23, 39, 45, 207, 209
Transmission of Dirac particles through a potential step 207
Transverse wave 212
Triplet, eigenspinors of total angular momentum 142
—, spin functions 139
—, tensor force properties 143, 144
Tunnel effect 23
— — in field emission 169, 170
Turning point, classical 40, 117—124
Two electrons see also helium
— — in atomic ground state 154
Two-level system 179, 180
Two particles on a circle 148

Uncertainty rule 17
Unitary matrix in two dimensions 130
— transformation 10, 50

Valence vibrations of three-atomic linear molecule 149
Van der Waals force 161
Variational approach 44, 72, 74, 75
— method equivalent to Thomas-Fermi differential equation 177

Variational method for helium ground state 154
— — for neutral hydrogen molecule 163
Variational principle of Schrödinger 2
— — of Schwinger 95
Vector in Dirac theory 192
Vector particles 52
Vector potential 125
— —, expansion into plane waves 212
Vertex 219
Vibrations of two-atomic molecules 69, 70
Virial theorem applied to excited helium states 155
— — for Coulomb forces 151, 175
Virtual level 26, 27

Wave group 17
— packet 17
Well size see also capacity 63
Wentzel-Kramers-Brillouin see WKB method
White dwarf 171
Wigner-Eckart theorem 133
WKB method, boundary condition of Langer 117, 118
— —, determination of phase shifts 121—123
— — — of energy levels 118—120
— —, radial wave function 116
— —, transmission coefficient of potential barrier 169
Wood-Saxon potential 64
Work function 169
Wronskian 24, 28

X rays, continuous spectrum 219
— —, isotope shift 73
— —, K shell binding energies 178
— —, screening constants 178

Yukawa attractive potential, binding energies 72
— potential, scattering 102, 106, 108

Zeeman effect, normal 127, 216
Zero-point energy of radiation field 212

Die Grundlehren der mathematischen Wissenschaften in Einzeldarstellungen mit besonderer Berücksichtigung der Anwendungsgebiete

Neu seit 1967

14. Klein: Elementarmathematik vom höheren Standpunkte aus. 1. Band: Arithmetik, Algebra, Analysis. DM 24,—; US $ 6.60
15. Klein: Elementarmathematik vom höheren Standpunkte aus. 2. Band: Geometrie. DM 24,—; US $ 6.60
16. Klein: Elementarmathematik vom höheren Standpunkte aus. 3. Band: Präzisions- und Approximationsmathematik. DM 19,80; US $ 5.50
22. Klein: Vorlesungen über höhere Geometrie. DM 28,—; US $ 7.70
26. Klein: Vorlesungen über nicht-euklidische Geometrie. DM 24,—; US $ 6.60
27. Hilbert/Ackermann: Grundzüge der theoretischen Logik. DM 38,—; US $ 10.50
30. Lichtenstein: Grundlagen der Hydromechanik. DM 38,—; US $ 10.50
32. Reidemeister: Vorlesungen über Grundlagen der Geometrie. DM 18,—; US $ 5.00
38. Neumann: Mathematische Grundlagen der Quantenmechanik. DM 28,—; US $ 7.70
40. Hilbert/Bernays: Grundlagen der Mathematik I. DM 68,—; US $ 18.70
43. Neugebauer: Vorlesungen über Geschichte der antiken mathematischen Wissenschaften. 1. Band: Vorgriechische Mathematik. DM 48,—; US $ 13.20
50. Hilbert/Bernays: Grundlagen der Mathematik II. DM 84,—; US $ 23.10
61. Maak: Fastperiodische Funktionen. DM 38,—; US $ 10.50
67. Byrd/Friedman: Handbook of Elliptic Integrals for Engineers and Scientists. 2nd edition, revised. DM 64,—; US $ 17.60
68. Aumann: Reelle Funktionen. DM 68,—; US $ 18.70
76. Tricomi: Vorlesungen über Orthogonalreihen. DM 68,—; US $ 18.70
78. Lorenzen: Einführung in die operative Logik und Mathematik. DM 54,—; US $ 14.90
87. van der Waerden: Mathematische Statistik. DM 68,—; US $ 18.70
94. Funk: Variationsrechnung und ihre Anwendung in Physik und Technik. DM 120,—; US $ 33.00
114. MacLane: Homology. DM 62,—; US $ 15.50
116. Hörmander: Linear Partial Differential Operators. DM 42,—; US $ 10.50
117. O'Meara: Introduction to Quadratic Forms. DM 68,—; US $ 18.70
120. Collatz: Funktionalanalysis und numerische Mathematik. DM 58,—; US $ 16.00
123. Yosida: Functional Analysis. DM 66,—; US $ 16.50
124. Morgenstern: Einführung in die Wahrscheinlichkeitsrechnung und mathematische Statistik. DM 38,—; US $ 10.50
127. Hermes: Enumerability. Decidability. Computability. DM 39,—; US $ 10.80
133. Haupt/Künneth: Geometrische Ordnungen. DM 68,—; US $ 18.70
134. Huppert: Endliche Gruppen I. DM 156,—; US $ 42.90
135. Handbook for Automatic Computation. Vol 1/Part a: Rutishauser: Description of ALGOL 60. DM 58,—; US $ 14.50
136. Greub: Multilinear Algebra. DM 32,—; US $ 8.00
137. Handbook for Automatic Computation. Vol. 1/Part b: Grau/Hill/Langmaack: Translation of ALGOL 60. DM 64,—; US $ 16.00
138. Hahn: Stability of Motion. DM 72,—; US $ 19.80
139. Mathematische Hilfsmittel des Ingenieurs. Herausgeber: Sauer/Szabó. 1. Teil. DM 88,—; US $ 24.20
140. Mathematische Hilfsmittel des Ingenieurs. Herausgeber: Sauer/Szabó. 2. Teil. DM 136,—; US $ 37.40

141. Mathematische Hilfsmittel des Ingenieurs. Herausgeber: Sauer/Szabó. 3. Teil. DM 98,—; US $ 27.00
142. Mathematische Hilfsmittel des Ingenieurs. Herausgeber: Sauer/Szabó. 4. Teil. DM 124,—; US $ 34.10
143. Schur/Grunsky: Vorlesungen über Invariantentheorie. DM 32,—; US $ 8.80
144. Weil: Basic Number Theory. DM 48,—; US $ 12.00
145. Butzer/Berens: Semi-Groups of Operators and Approximation. DM 56,—; US $ 14.00
146. Treves: Locally Convex Spaces and Linear Differential Equations. DM 36,—; US $ 9.90
147. Lamotke: Semisimpliziale algebraische Topologie. DM 48,—; US $ 13.20
148. Chandrasekharan: Introduction to Analytic Number Theory. DM 28,—; US $ 7.00
149. Sario/Oikawa: Capacity Functions. DM 96,—; US $ 24.00
150. Iosifescu/Theodorescu: Random Processes and Learning. DM 68,—; US $ 18.70
151. Mandl: Analytical Treatment of One-dimensional Markov Processes. DM 36,—; US $ 9.80
152. Hewitt/Ross: Abstract Harmonic Analysis. Voll. II. Structure and Analysis for Compact Groups. Analysis on Locally Compact Abelian Groups. DM 140,—; US $ 38.50
153. Federer: Geometric Measure Theory. DM 118,—; US $ 29.50
154. Singer: Bases in Banach Spaces I. DM 112,—; US $ 30.80
155. Müller: Foundations of the Mathematical Theory of Electromagnetic Waves. DM 58,—; US $ 16.00
156. van der Waerden: Mathematical Statistics. DM 68,—; US $ 18.70
157. Prohorov/Rozanov: Probability Theory. DM 68,—; US $ 18.70
158. Constantinescu/Cornea: Potential Theory on Harmonic Spaces. In preparation
159. Köthe: Topological Vector Spaces I. DM 78,—; US $ 21.50
160. Agrest/Maksimov: Theory of Incomplete Cylindrical Functions and their Applications. In preparation
161. Bhatia/Szegö: Stability Theory of Dynamical Systems. DM 58,—; US $ 16.00
162. Nevanlinna: Analytic Functions. DM 76,—; US $ 20.90
163. Stoer/Witzgall: Convexity and Optimization in Finite Dimensions I. DM 54,—; US $ 14.90
164. Sario/Nakai: Classification Theory of Riemann Surfaces. DM 98,—; US $ 27.00
165. Mitrinović: Analytic Inequalities. DM 88,—; US $ 26.00
166. Grothendieck/Dieudonné: Eléments de Géométrie Algébrique I. DM 84,—; US $ 23.10
167. Chandrasekharan: Arithmetical Functions. DM 58,—; US $ 16.00
168. Palamodov: Linear Differential Operators with Constant Coefficients. DM 98,—; US $ 27.00
169. Rademacher: Topics in Analytic Number Theory. In preparation
170. Lions: Optimal Control of Systems Governed by Partial Differential Equations. DM 78,—; US $ 21.50
171. Singer: Best Approximation in Normed Linear Spaces by Elements of Linear Subspaces. DM 60,—; US $ 16.50
172. Bühlmann: Mathematical Methods in Risk Theory. DM 52,—; US $ 14.30
173. F. Maeda/S. Maeda: Theory of Symmetric Lattices. DM 48,—; US $ 13.20
174. Stiefel/Scheifele: Linear and Regular Celestial Mechanics. DM 68,—; US $ 18.70
175. Larsen: An Introduction to the Theory of Multipliers. DM 84,—; US $ 23.10
176. Grauert/Remmert: Analytische Stellenalgebren. DM 64,—; US $ 17.60
177. Flügge: Practical Quantum Mechanics I. DM 70,—; US $ 19.30
178. Flügge: Practical Quantum Mechanics II. DM 60,—; US $ 16.50
179. Giraud: Cohomologie non abélienne. En préparation